Composites Science and Technology

Series Editor

Mohammad Jawaid, Laboratory of Biocomposite Technology, Universiti Putra Malaysia, INTROP, Serdang, Malaysia

This book series publishes cutting edge research monographs comprehensively covering topics in the field of composite science and technology. The books in this series are edited or authored by top researchers and professional across the globe. The series aims at publishing state-of-the-art research and development in areas including, but not limited to:

- Conventional Composites from natural and synthetic fibers
- Advanced Composites from natural and synthetic fibers
- Chemistry and biology of Composites and Biocomposites
- Fatigue damage modelling of Composites and Biocomposites
- Failure Analysis of Composites and Biocomposites
- Structural Health Monitoring of Composites and Biocomposites
- Durability of Composites and Biocomposites
- Biodegradability of Composites and Biocomposites
- Thermal properties of Composites and Biocomposites
- Flammability of Composites and Biocomposites
- Tribology of Composites and Biocomposites
- Applications of Composites and Biocomposites

Review Process

The proposal for each volume is reviewed by the main editor and/or the advisory board. The chapters in each volume are individually reviewed single blind by expert reviewers (at least two reviews per chapter) and the main editor.

Ethics Statement for this series can be found in the Springer standard guidelines here - https://www.springer.com/us/authors-editors/journal-author/journal-author-helpdesk/before-you-start/before-you-start/1330#c14214

Shamsher Bahadur Singh · C. V. R. Murty
Editors

RC Structures Strengthened with FRP for Earthquake Resistance

Springer

Editors
Shamsher Bahadur Singh
Department of Civil Engineering
Birla Institute of Technology and Science (BITS) Pilani
Pilani, Rajasthan, India

C. V. R. Murty
Department of Civil Engineering
Indian Institute of Technology Madras
Chennai, Tamil Nadu, India

ISSN 2662-1819 ISSN 2662-1827 (electronic)
Composites Science and Technology
ISBN 978-981-97-0101-8 ISBN 978-981-97-0102-5 (eBook)
https://doi.org/10.1007/978-981-97-0102-5

This Springer imprint is published by the registered company Springer Nature Singapore Pte Ltd.
The registered company address is: 152 Beach Road, #21-01/04 Gateway East, Singapore 189721, Singapore

Paper in this product is recyclable.

Contents

Design of Concrete Beams Reinforced with FRP Bars

T. V. V. S. S. Padmanabha Rao, K. K. Bajpai, S. B. Singh, and C. V. R. Murty

1 Introduction

The built environment is said to be acceptable if it complies with five tenets, namely *safety*, *functionality*, *sustainability*, *aesthetics* and *economy*. The first three are non-negotiable, even in the order of their compliance; the last two can be swapped based on the financial strength of the owner of the facility. On the matter of concrete structures made with FRP bars used as reinforcement, two aspects dominate, namely *safety* and *sustainability*. In particular, the concerns are with regard to *earthquake structural safety* and *durability* of such structures.

(a) **Earthquake Structural Safety**

The *earthquake structural safety* of concrete structures reinforced with FRP bars built in areas likely to sustain earthquake ground shaking of moderate to severe intensities. As on date, the responses of concrete beams reinforced with FRP bars have not been studied *comprehensively and adequately* under different cyclic loading histories, especially shear behaviour of beams with transverse reinforcement also made of FRP bars, distortional shear behaviour of beam-column joints, plastic hinge rotational behaviour of beams, and axial compression-bending moment behaviour of columns. Detailed R&D programs are yet to be undertaken by neutral research

T. V. V. S. S. Padmanabha Rao
Primero, Montreal, QC, USA

K. K. Bajpai
Department of Civil Engineering, IIT Kanpur, Kanpur, India

S. B. Singh
Department of Civil Engineering, BITS Pilani, Pilani, India

C. V. R. Murty (✉)
Department of Civil Engineering, P. S. Rao Institute Chair Professor, IIT Madras, Chennai, India
e-mail: cvrm@iitm.ac.in

S. B. Singh and C. V. R. Murty (eds.), *RC Structures Strengthened with FRP for Earthquake Resistance*, Composites Science and Technology,
https://doi.org/10.1007/978-981-97-0102-5_1

laboratories, which demonstrate quantitatively that good earthquake behaviour can be guaranteed of concrete structures reinforced with FRP bars. Often, the *high* cost and the *low* availability of the FRP bars and their brittleness are quoted as reasons for not undertaking the much needed studies. Hence, a clear statement cannot be made yet, which recommends the use of concrete structures reinforced with FRP bars in areas prone to severe to moderate earthquake shaking. During earthquakes, concrete structures reinforced with the conventional steel bars are still collapsing, indicating earthquake structural safety remains unresolved.

(b) **Durability Aspect**

Concrete structures (such as buildings and bridges), which are subjected to aggressive environment, are facing corrosion problems—over 30% of steel produced is used to replace older corroded steel [1]. With over 20 varieties of cements, 80 varieties of admixtures, and uncounted types of coarse and fine aggregates and of water, the contribution to corrosion has not been *quantified* under these multitudes of combinations of the ingredients of concrete. In addition, there is another dimension imposed by the levels of detrimental environment. Corrosion is leading to deterioration in the capacity of the concrete structure (especially its *stiffness*, *strength* and *deformability*), and, in turn, this is leading to loss of its *serviceability* (especially its *service life*).

Engineers have been attempting many measures to control this problem—*epoxy-coated steel bars*, *cathodic protection*, *increased concrete cover thickness*, and *polymer concrete*. But, each of these "remedies" could only impede, but not eliminate corrosion problems. Also, they brought additional shortcomings. For instance, (a) *epoxy-coating on steel bars* reduce bond between steel and concrete, (b) *cathodic protection* increases maintenance effort, (c) *increased concrete cover thickness* leads to early spalling of concrete results in large sudden loss of strength of member, and (d) *polymer concrete* requires high precision engineering.

Clearly, the target is to have *non-corrosive* concrete structures. This points to two needs, namely:

(1) Replace bars made of *corrosive materials* (like steel) with bars made of non-corrosive materials; and
(2) Examine the roles of ingredients of concrete (like admixtures, water quality and quantity, impurities in sand, and chemical composition of aggregates) to understand how corrosion can be impeded.

(c) **This Chapter**

This chapter discusses flexural design of *concrete beams reinforced with FRP bars* under monotonic loading. Its purpose is to understand the feasibility of replacing conventional longitudinal steel bars with FRP bars. It presents:

(1) Procedure for structural design for bending moment, and
(2) Behaviour of concrete beams reinforced with FRP bars [2].

2 Critical Issues in Design for Bending Moment

The implications should be understood of material behaviour, especially of FRP bars on the *behaviour*, and hence on the *design* of concrete beams reinforced with FRP bars.

2.1 *Material*

Commonly used FRP bars in civil engineering infrastructures are of three types, namely *glass fiber reinforced polymer* (GFRP), *aramid fiber reinforced polymer* (AFRP) and *carbon fiber reinforced polymer* (CFRP). The range of values of salient properties of steel bars and the said three types of FRP bars are presented in Table 1. The following are the salient observations from these properties:

(1) The steel bars are *ductile* and hence have *yield strength.* Such ductile behaviour is absent in FRP bars; they are *brittle* with *no post-yield behaviour*.
(2) The upper bound *tensile strengths* of FRP bars are larger than those of steel bars.
(3) The *modulus of elasticity* of FRP bars varies over a large range; GFRP and AFRP have values smaller than those of the steel bars.
(4) The *rupture strain* of FRP bars is much smaller than those of the steel bars.
(5) The *coefficient of thermal expansion* of FRP bars is direction dependent unlike steel bars. Especially, the *coefficient of thermal expansion* of FRP bars in the longitudinal direction is smaller than those of steel bars (and even negative for CFRP and AFRP bars), while that in the transverse direction it is much higher than that of steel bars.

The major implications of the above observations are:

Table 1 Salient properties of steel and FRP reinforcing bars [3]

Material property	Steel	GFRP	CFRP	AFRP
Modulus of elasticity (GPa)	195–205	35–51	120–580	41–125
Nominal yield strength (MPa)	276–517	Not applicable		
Tensile strength (MPa)	483–690	483–1,600	600–3,690	1,720–2,540
Yield strain (%)	1.4–2.5	Not applicable		
Rupture strain (%)	6.0–12.0	1.2–3.1	0.5–1.7	1.9–4.4
Coefficient of thermal expansion ($\times 10^{-6}$/°C)				
(a) Longitudinal direction	11.7	6–10	−2 to 0	−6 to −2
(b) Transverse direction	11.7	21–23	23–32	60–80

(a) The fact (high strength and low ductility) together with brittle nature of FRP bars suggests that the ductile *under-reinforced design* of concrete beams with FRP bars is not practicable; only *over-reinforced design* of concrete beams with FRP bars should be admitted; and
(b) The higher *coefficient of thermal expansion* of FRP bars in the transverse direction can have adverse impact on the bond between the FRP bar and concrete.

2.2 *Design Philosophy*

Traditionally, concrete beams with steel bars were designed to behave in a ductile manner. This was possible because steel bars have reasonable *post-yield strain range* with *large elongation* at rupture. But, once the FRP bars replace the conventional steel bars, the fact that FRP bars are brittle needs to be recognised; stress–strain curve of FRP bars is linear till failure and this needs to be considered in the design method. FRP bars are made of *fibers* and *binder matrix*. The actual *stress–strain curve* of FRP bars and its two constituents are depicted in Fig. 1. The brittle nature of FRP bars confirms that *under-reinforced* design philosophy (which depends on the brittle failure of FRP bars in tension) is inadmissible in concrete beams reinforced with FRP bars. Therefore, concrete beams with FRP bars should be designed based on *over-reinforced* design philosophy. Thus, a major change is needed in the philosophy of design of concrete beams—*over-reinforced* design should be accepted in concrete beams reinforced with FRP bars (against the conventional *under-reinforced* design admitted in concrete beams reinforced with steel bars).

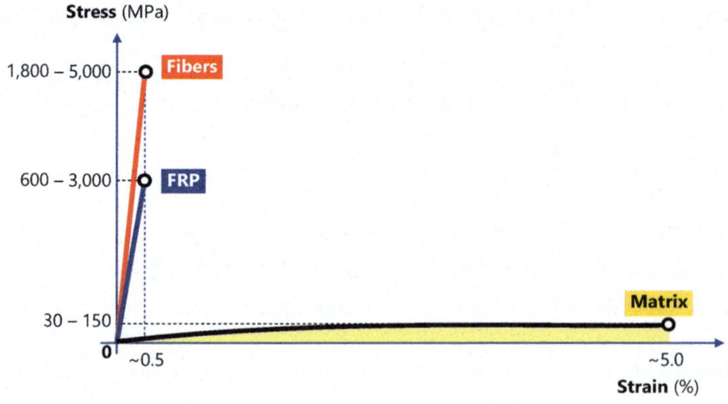

Fig. 1 Stress–strain curves of FRP bars and its constituents (fibers and binder matrix) [4]

3 Structural Design

The response of concrete beams reinforced with FRP bars has not been studied *adequately* under different cyclic loading histories, especially shear behaviour of beams with transverse reinforcement also made of FRP bars, distortional shear behaviour of beam-column joint, plastic hinge rotational behaviour of beams, and axial compression-bending moment behaviour of columns. Hence, a clear statement cannot be made yet, which recommends the use of RC structures reinforced with FRP bars in areas prone to severe to moderate earthquake shaking.

3.1 Design Properties of Materials

The main properties needed in the design of concrete beams reinforced with FRP bars are:

(a) **Stress–Strain Curves**

Unconfined Concrete:

The *actual* stress–strain curves of concrete in compression and FRP bars in tension are required to design concrete beams reinforced with FRP bars. Typically, the stress–strain curve of *unconfined* concrete in compression is used in the estimation of the *design* strength capacity and that of *confined* concrete in the estimation of the *overstrength* capacity. In this chapter, the discussion is restricted to *design* bending moment capacity of beams, and hence stress–strain curve of *unconfined* concrete alone is presented. The expression for *design stress–strain curve* (f_c, ε_c) of unconfined concrete in compression (Fig. 2) is:

$$f_c(\varepsilon_c) = \begin{cases} \frac{0.67 f_{ck}}{\gamma_M}\left[2\left(\frac{\varepsilon_c}{\varepsilon_{cp}}\right) - \left(\frac{\varepsilon_c}{\varepsilon_{cp}}\right)^2\right] & 0 \le \varepsilon_c \le \varepsilon_{cp} \\ \frac{0.67 f_{ck}}{\gamma_M} & \varepsilon_{cp} < \varepsilon_c \le \varepsilon_{cu} \end{cases} \quad (1)$$

where

ε_{cp} Strain at *design peak strength* of *unconfined concrete* (Table 2) $0.00190 + 0.000005 f_{ck}$,

ε_{cu} *Ultimate strain* in *unconfined concrete* $= 0.003775 - 0.00001375 f_{ck}$ (Table 2), and

γ_M *Partial safety factor for* concrete, taken as 1.5.

The shape of the monotonic stress–strain ($f_c - \varepsilon_c$) curve of *unconfined concrete* in compression is a combination of a *parabola* and a *rectangle*; this is derived from the experimental stress–strain curves obtained from tests. The parabolic portion (in the range $0 \le \varepsilon_c \le \varepsilon_{cp}$) increases with increase in the grade of concrete, because

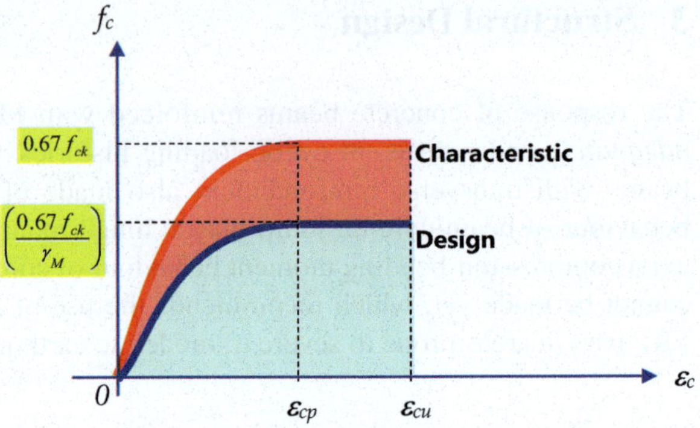

Fig. 2 *Design stress–strain curve* of *unconfined concrete* in compression

Table 2 *Strains at design peak strength,* ε_{cp} and *ultimate strain* ε_{cu} of *unconfined concrete* in compression

Characteristic compressive strength f_{ck} (MPa) of concrete	ε_{cp}	ε_{cu}
20	0.002 00	0.003 50
25	0.002 03	0.003 43
30	0.002 05	0.003 36
35	0.002 08	0.003 29
40	0.002 10	0.003 23
45	0.002 13	0.003 16
50	0.002 15	0.003 09
55	0.002 18	0.003 02
60	0.002 20	0.002 95
65	0.002 23	0.002 88
70	0.002 25	0.002 81
80	0.002 30	0.002 68
90	0.002 35	0.002 54
100	0.002 40	0.002 40

ε_{cp} increases with increase in f_{ck}. And, the rectangular portion (in the range $\varepsilon_{cp} < \varepsilon_c \leq \varepsilon_{cu}$) reduces with increase in the grade of concrete, because ε_{cp} increases with f_{ck} and ε_{cu} decreases with increase in f_{ck}.

When extreme fiber of concrete reaches the strain ε_{cu}, the full stress-block is developed in concrete (Fig. 3), wherein the parameter α of average stress and the parameter β of the location of the centroid of the compressive force are listed in Table 3 for different grades of concrete.

FRP Bars:

There is no *overstrength* capacity in FRP bars, because it is a brittle material. Hence, the *actual* stress–strain curve of FRP bar is used as such for estimating the *overstrength* capacity; it is used with a partial safety factor for strength (of FRP material) for estimating the design bending moment capacity. The expression for *design*

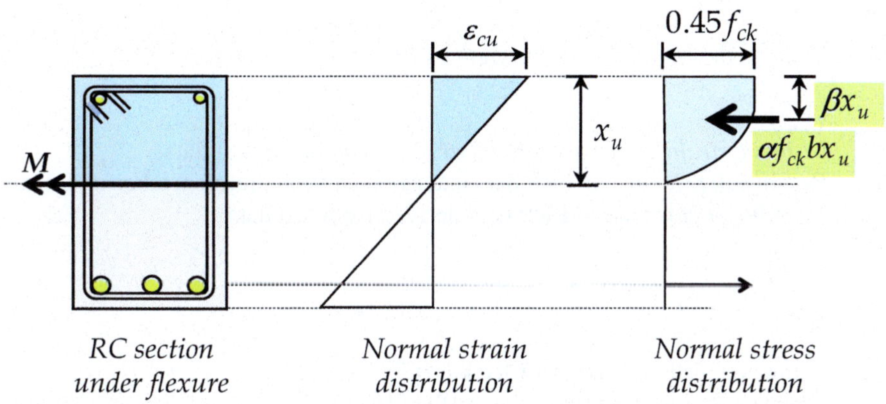

Fig. 3 Parameters of full compression stress block of *unconfined concrete*

Table 3 Parameters of full compression stress block of *unconfined concrete*

Characteristic compressive strength f_{ck} (MPa) of unconfined concrete	α	β
20	0.364	0.416
25	0.361	0.413
30	0.358	0.411
35	0.355	0.409
40	0.352	0.407
45	0.349	0.404
50	0.346	0.402
55	0.342	0.399
60	0.338	0.396
65	0.334	0.393
70	0.330	0.391
75	0.325	0.388
80	0.321	0.386
85	0.316	0.383
90	0.311	0.380
95	0.305	0.377
100	0.300	0.375

stress–strain curve (f_f, ε_f) of FRP bars in tension and compression (Fig. 4) is:

$$f_f(\varepsilon_f) = \begin{cases} 0 & \varepsilon_{fd} \le \varepsilon_{fu} \\ E_f \varepsilon_f & -\varepsilon_{fd} \le \varepsilon_f \le +\varepsilon_{fd} \\ 0 & \varepsilon_f \le -\varepsilon_{fd} \end{cases}, \tag{2}$$

where

ε_{fd} Strain at *design maximum strength* of *FRP bar in tension* $\left(\frac{\varepsilon_{fu}}{\gamma_M}\right)$,
ε_{fu} *Ultimate strain* capacity of *FRP bars in tension* from test data,
γ_M *Partial safety factor for* FRP bars in tension and compression, taken as 1.5, and
E_f *Modulus of elasticity* of FRP bars derived from test data $\left(\frac{f_{fu}}{\varepsilon_{fu}}\right)$,

wherein

f_{fu} Ultimate tensile strength of FRP bar (Fig. 4), and
ε_{fu} Ultimate tensile strain of FRP bar (Fig. 4).

(a) **Modulus of Elasticity**

The *modulus of elasticity* E_s of steel bars in tension is taken as 200 GPa. But, the *modulus of elasticity* E_f of FRP bars in tension is not the same for all types and grades of FRP bars. Hence, unlike in the steel bars, E_f is not taken as a predetermined value, but is dependent critically on the test data of bars to be used in the beams.

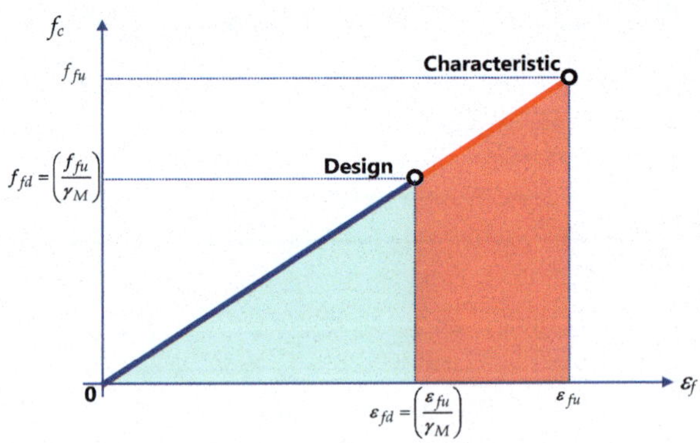

Fig. 4 *Design stress–strain curve* of FRP bars in tension and in compression

3.2 *Design Bending Moment Capacity M_L*

The *Criteria for Strength* of the structure or part thereof shall be assessed from material reaching a *design limit state* (either *strain* or *stress*) at any critical cross-section of a member. The *design bending moment capacity* M_L of a member at any section (including the effect of *Partial Safety Factors* γ_M for *Material Strengths*) shall not be less than the corresponding demand at that section produced by the probable most unfavourable combination of loads on the structure (arrived at by using the appropriate *partial safety factors* γ_L for *loads*).

The aim when specifying the design for *strength* of a cross-section is to ensure sufficient margin between the *capacity* and *demand* at each section. The *Partial Safety Factors* γ_M for materials strength and γ_L for loads are applied to the respective *characteristics values.* The values of γ_M and γ_L are so chosen that the members and hence the structure possess a *known* level of margin of *strength.* The *load combinations* and the *partial safety factors* to be considered in design are chosen as per the design standard. For ensuring the specified objective, the design is based on characteristic values for material strengths and applied loads, which take into account the variations in the material strengths and in the loads to be supported. The characteristic values should be based on statistical data if available; where such data are not available, they should be based on experience. The *design values* are derived from the characteristic value through the use of *partial safety factors*, one for *material strengths* and the other for *loads.* In the absence of data, these factors should be specified in the standard (considering *material*, *loading* and *limit state* being considered).

3.2.1 Limiting Strains

The method of estimating design strengths discussed hereunder is applicable to *shallow* concrete members (whose l/d ratios are greater than 5, and not to *deep concrete members*). The design bending moment capacity M_L of sections of concrete members should be that corresponding to when either of the following limit states of *strain* are reached:

(1) Limiting strain $\varepsilon_{c,\text{lim}}$ of concrete in compression:

$$\varepsilon_{c,\text{lim}} = \varepsilon_{cu}, \tag{3}$$

and

(2) Limiting strain $\varepsilon_{f,\text{lim}}$ of FRP bars in tension and compression:

$$\varepsilon_{f,\text{lim}} = \frac{\varepsilon_{fu}}{\gamma_M} = 0.67\varepsilon_{fu}, \tag{4}$$

where ε_{cu} is as per Table 2 depending on the grade of concrete, and ε_{fu} as per test on type and grade of the FRP bar being used.

The limiting strain in concrete is the maximum strain that unconfined concrete can resist without significant loss in strength. But, the limiting strain in FRP bars is restricted to within two-thirds of its rupture strain in tension. This is necessary because the failure of FRP bars is *brittle*, with *no yield plateau*. The design of members cannot be such that the bar reaches its rupture strain. Hence, sufficient margin is kept by considering the *design limiting strain* of FRP bars *in tension* to be two-thirds of its rupture strain in tension. There is insufficient study of FRP bars in compression in concrete beams. Hence, as an interim measure, the *design limiting strain* of FRP bars *in compression* is considered to be the same as that in tension; this has the underlying assumption that the FRP bars are held from buckling by *closely-spaced closed-loop* transverse reinforcement hoops.

When estimating the *design bending moment capacity* M_L of *shallow concrete* members that are reinforced with *FRP bars*, each section of the member should comply with the basic *principles of mechanics*, namely:

(a) *Equilibrium* of forces at a section,
(b) *Compatibility* of strains at a section, and
(c) *Constitutive Law* of constituent materials,

at the limit states of strain specified above.

3.2.2 Assumptions

The following assumptions shall be made when estimating the *design bending moment capacity* M_L of *shallow concrete* members reinforced with *FRP bars*:

(1) Plane sections normal to the axis of bending before bending, remain plane and normal to the axis of bending even after bending.
(2) The tensile strength of the concrete is ignored.
(3) The monotonic *design* stress–strain curve of:
    (a) *Concrete* in compression, i.e., $(f_c - \varepsilon_c)$, is as per Eq. (1), and
    (b) FRP bars in tension and compression, *i.e.*, $(f_f - \varepsilon_f)$, as per Eq. (2).
(4) The *design* strength of:
    (a) *Concrete* in *bending compression* is $f_{cbc,d} = (0.67 f_{ck}) / \gamma_M = 0.45 f_{ck}$, where the *Partial Safety Factor* γ_M is 1.5 for concrete in compression, and
    (b) *FRP Bars* in *direct tension and direct compression* are $f_{fc,d} = f_{ft,d} = (f_{fu}) / \gamma_M = 0.67 f_{fu}$, where *Partial Safety Factor* γ_M is 1.5 for FRP bars in axial tension and compression.

The bases behind these assumptions are:

(1) *Bernoulli's Hypothesis*: The distribution of strain across the depth of the section is *linear*, which has been observed in numerous tests on slender reinforced concrete members.
(2) By ignoring the tensile strength of concrete, a conservative estimate is obtained of bending moment capacity. Estimates can be made taking the tensile strength of concrete into account, but such estimates are justified rarely.
(3) The monotonic stress–strain ($f_c - \varepsilon_c$) curve of concrete in compression has been idealised in the current form to simplify the design process. The *assumed* stress–strain curve of concrete is a combination of a *parabola* and a *rectangle*; it is referred to as the *stress block*. Classically, bending compressive stress has been related to the compressive strength of a cylinder. The apparent strength of concrete in the compression zone of beam or column at failure is observed to be approximately 0.85 times the cylinder compressive strength of the same concrete. Since cube strength is used in India to reflect the *characteristic compressive strength* f_{ck} of concrete, *cube characteristic strength* f_{ck} is converted to *cylinder characteristic strength* by a factor of 0.80. Then, *direct* characteristic compressive stress from a cylinder is converted to *bending* characteristic compressive stress by a factor 0.85. Hence, the *characteristic* bending compressive stress in a flexure member is obtained as $0.80 \times 0.85 f_{ck} = 0.67 f_{ck}$. And, the *design* bending compressive strength is obtained by applying the *Partial Safety Factor* γ_M to the *characteristic* value. γ_M of 1.5 is justified for concrete, because it is a brittle material and is prepared at site.

    And, the monotonic stress–strain curve ($f_f - \varepsilon_f$) of FRP bars in tension and compression are considered to be linear. In compression, it is assumed that the FRP bar is held against buckling by the transverse reinforcement spaced closely. These transverse reinforcements are closed loops made of FRP too. They are fabricated specially at the factory, with fibers woven into the desired rectangular shapes with desired diameters, but with no joints.
(4) The *design* axial tensile and compressive strengths are obtained by applying the *Partial Safety Factor* γ_M to the *characteristic* value.

    (a) For concrete in compression, γ_M is taken as 1.5, as in conventional design of concrete beams with steel bars, and
    (b) For FRP bars in tension and compression, γ_M is suggested as 1.5, as against the value of 1.15 adopted in conventional design of concrete beams with steel bars. This higher value is tied with two facts, namely linear *stress–strain behaviour* of FRP bars and limiting strain needs to be significantly less than the rupture strain.

3.2.3 Design Limiting Bending Moment Capacity M_L

The Limit State Method of design of RC members as enumerated in the Indian Concrete Code IS 456 is extended to estimate the design limiting bending moment capacity M_L. The method of estimating M_L of a concrete member reinforced with FRP bars subjected to bending moment is described hereunder for a member with a rectangular cross-section.

(a) **Limiting Strains**

A concrete member shall be said to have reached its *limit state* at a section under the action of *bending moment*, when at a section either concrete or FRP bar reaches the corresponding *limiting strain* $\varepsilon_{c,\text{lim}}$ or $\varepsilon_{f,\text{lim}}$, respectively. And, the bending moment resisted by the concrete member when the member reaches the said *limit state* at a section is called the *limiting moment capacity*. The same is taken as M_L of that section.

The bending moment capacity of a section is based on the reaching of limiting strain states. These limiting values need not be reached simultaneously. If at any section:

(1) Concrete reaches the limiting strain $\varepsilon_{c,\text{lim}}$ before FRP bar reaches $\varepsilon_{f,\text{lim}}$, it is called a *compression-controlled over-reinforced* section;
(2) Both concrete and steel reach simultaneously the corresponding limiting strains $\varepsilon_{c,\text{lim}}$ and $\varepsilon_{s,\text{lim}}$, respectively, it is called a *balanced over-reinforced* section; and
(3) FRP bar reaches the limiting strain $\varepsilon_{f,\text{lim}}$ before concrete reaches $\varepsilon_{c,\text{lim}}$, it is called an *tension-controlled under-reinforced* section.

The third option is preferable in the design of concrete members reinforced FRP bars, because it gives the best behaviour.

(b) **The Method**

RC slender beams, slabs, footings and rafts shall be designed to be under-reinforced with FRP reaching the limiting strain in tension before concrete reaches the limiting strain in compression. The *design bending moment capacity* M_L of RC beams shall be obtained by solving the following set of governing nonlinear equations:

(a) *Equilibrium* of forces and moment at a section,
(b) *Compatibility* of strains at a section, and
(c) *Constitutive law* of constituent materials.

The *limiting moment capacity* M_L of a *doubly-reinforced rectangular* concrete section designed to be *over-reinforced* is estimated as given hereunder (Fig. 5).

(1) *Equilibrium of forces and moment at a section*

Force Equilibrium:

$$f_{c,avg} b x_L + \left(f_{fc} - f_{cfc}\right) A_{fc} = f_{ft} A_{ft}, \tag{5}$$

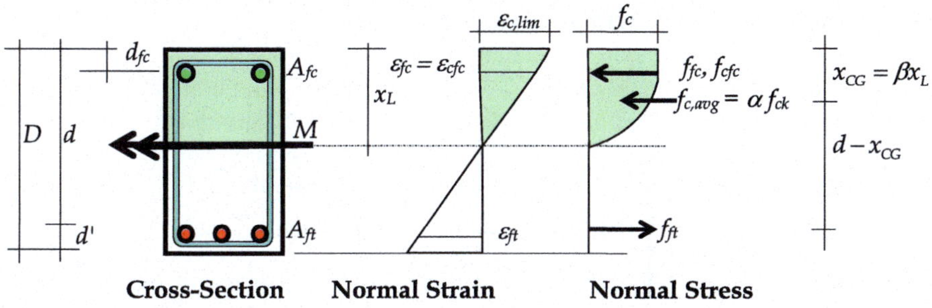

Fig. 5 Distributions of strain and stress across a *rectangular* concrete section

and

Moment Equilibrium:

$$M_L = f_{c,ave} b x_L (x_L - x_{CG}) + (f_{fc} - f_{cfc}) A_{fc} (x_L - d_{fc}) + f_{ft} A_{ft} (d - \beta x_L), \tag{6}$$

where

$$\alpha = 0.45\left[1 - \frac{1}{3}\left(\frac{\varepsilon_{cp}}{\varepsilon_{cu}}\right)\right], \tag{7}$$

and

$$\beta = \frac{\left[\frac{1}{2} - \frac{1}{3}\left(\frac{\varepsilon_{cp}}{\varepsilon_{cu}}\right) + \frac{1}{12}\left(\frac{\varepsilon_{cp}}{\varepsilon_{cu}}\right)^2\right]}{\left[1 - \frac{1}{3}\left(\frac{\varepsilon_{cp}}{\varepsilon_{cu}}\right)\right]}. \tag{8}$$

(2) *Compatibility of strains at a section*

$$\varepsilon_c = \varepsilon_{c,\lim}, \tag{9}$$

$$\varepsilon_{fc} = \left(\frac{x_L - d_{fc}}{x_L}\right)\varepsilon_{c,\lim}, \tag{10}$$

$$\varepsilon_{cfc} = \left(\frac{x_L - d_{fc}}{x_L}\right)\varepsilon_{c,\lim}, \tag{11}$$

and

$$\varepsilon_{ft} = \left(\frac{d - x_L}{x_L}\right)\varepsilon_{c,\lim}, \tag{12}$$

(3) *Constitutive law of constituent materials*

$$f_{c,avg}(\varepsilon_c) = \alpha_{\max} f_{ck}, \tag{13}$$

$$f_{cfc}(\varepsilon_{cfc}) = \begin{cases} 0.45 f_{ck}\left[2\left(\frac{\varepsilon_{csc}}{\varepsilon_{cp}}\right) - \left(\frac{\varepsilon_{csc}}{\varepsilon_{cp}}\right)^2\right] & 0 \le \varepsilon_{cfc} \le \varepsilon_{cp} \\ 0.45 f_{ck} & \varepsilon_{cp} < \varepsilon_{cfc} \le \varepsilon_{cu} \end{cases}, \tag{14}$$

$$f_{fc}(\varepsilon_{fc}) = E_f \varepsilon_{fc}, \tag{15}$$

and

$$f_{ft}(\varepsilon_{ft}) = E_f \varepsilon_{ft}. \tag{16}$$

When the section is *over-reinforced*, strain ε_c in extreme fiber of concrete section on the compression side should be the limiting strain $\varepsilon_{c,\lim}$ specified, and that in the extreme fiber of FRP bar on the tension side less than the limiting strain $\varepsilon_{f,\lim}$, where

$$\varepsilon_{ft,\lim} = \frac{\varepsilon_{fu}}{\gamma_M}, \tag{17}$$

and

$$\varepsilon_{c,\lim} = \varepsilon_{cu}. \tag{18}$$

Thus, the *limiting depth* x_L of neutral axis shall be equal to the *depth* x_{bal} of neutral axis of the balanced section, given by:

$$x_{bal} = \frac{\varepsilon_{c,\lim}}{\varepsilon_{c,\lim} + \varepsilon_{f,\lim}} d. \tag{19}$$

Therefore, x_L shall be estimated as per the following step-wise procedure:

(a) Estimate x_{bal} using Eq. (19).
(b) Use x_{bal} as an approximation of x_L, and estimate ε_{fc}, ε_{cfc} and ε_{ft} using Eqs.(10)–(12).
(c) Use ε_c, ε_{fc} and ε_{cfc}, and estimate f_{cfc}, f_{fc} and f_{ft} using Eqs. (14)–(16).
(d) Use f_{cfc}, f_{fc} and f_{ft}, and estimate x_L using Eq. (5).
(e) Check if x_L is converged. If yes, go to Step (f), else go back to Step (b), and use the value of x_L from Step (d).
(f) Estimate M_L using Eq. (6).

The above *step-wise iterative procedure* (i.e., Step (a) to Step (e)) of an *over-reinforced* concrete beam reinforced with FPR bars, to estimate x_L is depicted in Table 4.

Table 4 Step-wise procedure to estimate limiting depth x_L of neutral axis of a concrete beam reinforced with FPR bars

x_L assumed	ε_{fc}	ε_{cfc}	ε_{ft}	f_{cfc}	f_{fc}	f_{ft}	x_L estimated
mm				MPa	MPa	MPa	mm
	Equation (10)	Equation (11)	Equation (12)	Equation (14)	Equation (15)	Equation (16)	Equation (5)
x_{bal}							
…							
…							
…	ε_{fc}						x_L

When a beam needs to be *designed*, the only known quantity is external *design bending moment demand* M_D on the beam. So, to begin with, M_L may be taken as M_D. All geometric and material properties and areas of steel in the section are assumed, and hence considered to be known; the state of stress in the beam needs to be determined. Thus, the quantities assumed are: (a) *Geometric properties* b, D, d and d_{fc}, (b) *Material properties* f_{ck} and f_{fu}, and (c) *Areas of FRP* bars A_{ft} and A_{fc}. There are 8 equations, namely: 2 *equilibrium equations*, 3 *compatibility conditions*, and 3 *constitutive relations*. And, there are 8 *unknown* quantities, namely the pairs of strains and stresses (ε_{cfc},f_{cfc}), (ε_{fc},f_{fc}) and (ε_{ft},f_{ft}), the depth of neutral axis x_L, and bending moment demand M_D. Again, there are 8 unknowns and 8 equations, and is solvable, even though the solution is iterative.

When the strength of a beam needs to be *assessed*, the *design bending moment capacity* M_L needs to be ascertained with all details (geometric and material properties) of the section given. The quantities given are: (a) *Geometric properties* b, D, d and d_{fc}, (b) *Material properties* f_{ck} and f_{fu}, and (c) *Areas of FRP bars* A_{ft} and A_{fc}. (ε_c,f_c) are known as concrete is required to reach its limiting strain $\varepsilon_{c,\mathrm{lim}}$. Further, 8 equations are available, namely 2 *equilibrium equations*, 3 *compatibility conditions*, and 3 *constitutive relations*. And, there are 8 *unknown* quantities, namely the 3 pairs of strains and stresses (ε_{cfc},f_{cfc}), (ε_{fc},f_{fc}) and (ε_{ft},f_{ft}), the depth of neutral axis x_L, and bending moment capacity M_L. There are 8 unknowns and 8 equations, and is solvable, even though the solution is iterative.

3.3 Design Shear Force Capacity V_L

Again, the *Limit State Method* of design of RC members as enumerated in the *Indian Concrete Code IS 456* is extended to estimate the *limiting design shear force capacity* V_L. At the critical section, V_L is estimated by adding contributions to resistance of both concrete and transverse FRP bars. Unlike in the estimation of the *design bending moment capacity* M_L, where limiting strains formed the basis of estimating M_L, here V_L is estimated by a procedure that considers limiting stresses as the basis of estimating V_L.

(a) **Limiting Stresses**

At the critical section of a concrete beam, contributions of concrete in shear and of transverse FRP bars in tension are used to estimate the *limiting design shear force capacity* V_L. In particular:

(i) Concrete shall be assumed to have reached its *limiting stress* of the *Design Shear Strength* τ_c of concrete, and
(ii) Transverse reinforcing FRP bars shall be assumed to have reached its *limiting stress* of $0.67\,f_{fu} \leq 0.004\,E_f$, the *design tensile strength* of FRP bars in tension, wherein upper bound is based on strain of 0.004 to prevent the loss of aggregate interlock.

And, the shear force resisted by the concrete member when it reaches these said *limiting stresses* simultaneously at a section are taken as the *design shear force capacity* V_L of that section.

(b) **The Method**

Normally, in rectangular sections, only the web is considered in the estimation of V_L. Also, transverse reinforcement required to resist shear force is provided *only* in the form of vertical closed-loop stirrups.

The *Design Shear Force Capacity* V_L of a beam of rectangular cross-section is:

$$V_L = V_{LC} + V_{LF}, \tag{20}$$

where

V_{LC} Contribution of concrete to V_L.
$\tau_C bd$, and
V_{LF} Contribution of transverse reinforcing FRP bars to V_L.
$0.67 f_{fu} A_{SV} \left(\frac{d}{s_V}\right)_{INT} \leq 0.004\, E_f\, A_{SV} \left(\frac{d}{s_V}\right)_{INT}$,

in which

A_{SV} Cross-sectional area of all transverse reinforcement legs of FRP in one plane
$N_{SV}\left[\pi\left(\frac{\varphi_{SV}}{2}\right)^2\right]$,
d Effective depth of the member,
s_V Spacing of the stirrups along the length of the member,
f_{fu} Characteristic tensile strength of FRP used in transverse reinforcement,
N_{SV} Number of legs in transverse reinforcement at each section of the member,
φ_{SV} Diameter of the bar used as transverse reinforcement,
$(\ldots)_{INT}$ Integral value of the number,
τ_C *Design shear strength* of concrete as specified in conventional concrete design codes, and
$\tau_{C,\max}$ *Maximum design shear stress* of concrete as specified in the standard.

The *design shear force capacity* V_L of concrete member is considered as the sum of the shear strengths V_{LF} offered by concrete and the shear strength V_{DF} offered by FRP transverse reinforcement. V_{LC} arises from τ_C. Also, in arriving at the *Design Shear Strength* V_{LF} offered by shear reinforcement, a constant-angle truss model is considered. The angle of shear crack (or compression strut) is taken as 45° to get a conservative estimate of V_{LF}. Although different theories have been proposed to explain the action of shear reinforcement, the truss analogy theory is accepted generally. The truss analogy assumes that in the imaginary pin jointed truss the *longitudinal tension* is carried by longitudinal reinforcing bars, *transverse tension* by transverse reinforcing bars, and the *longitudinal and diagonal compression* by concrete.

Conventionally, in the design of concrete beams reinforced with steel bars, the design of shear reinforcements allows the yielding of shear reinforcements at the factored loads. Thus, the member was expected to undergo ductile failure characteristics. But, the shear strengths of a beam cannot be increased indefinitely by the addition of unlimited amounts of transverse reinforcement. This is because where large shears are carried, the diagonal compressive struts are formed within the core concrete, which may crush the web concrete. Such failures are brittle. Hence, and upper limit of $\tau_{c,\max}$ is imposed on τ_c. The partial safety factor γ_M in concrete under shear is taken as 1.5.

But, when FRP bars are used, the transverse reinforcement cannot be permitted to yield. Thus, a larger γ_M of 1.5 is recommended over the ultimate tensile strength of FRP bars, unlike 1.15 of steel bars. Further, the design method when steel bars are used requires that shear reinforcement is estimated with f_y limited to 415 MPa. The reasons for are:

(1) The grade of steel influences the width of cracks; larger the grade of steel, the larger is the crack width.
(2) Higher strength steel bars imply smaller area, and when these small area bars are used as stirrups, the bars may rupture at the corners during bar bending.

(c) **The Gap Area**

When the complete concrete structure is to be reinforced with FRP bars, transverse reinforcement too is required to be made of FRP. But, straight FRP bars made at the factory cannot be bent at site; they will rupture. Hence, it is necessary to prefabricate the stirrups of different configurations in the factory (Fig. 6). Detailed studies are required to develop suitable technology to prefabricate FRP stirrups and put them through a path of qualification, certification and development as a technology.

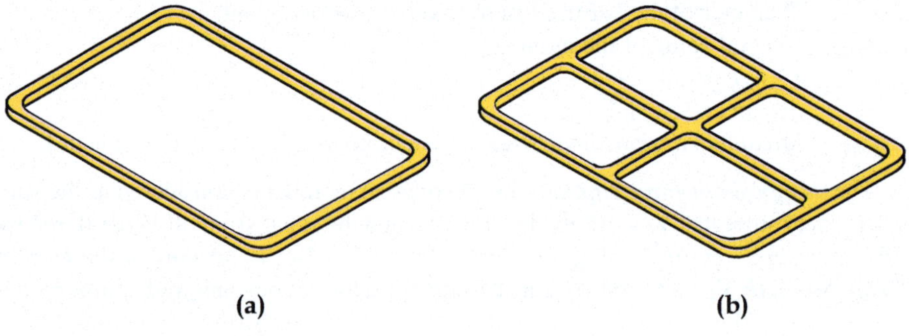

Fig. 6 Prefabricated stirrups for use in concrete beams of *rectangular* cross-section: **a** 2-legged stirrups, and **b** 3-legged stirrups

4 Experimental Study

An experimental study was conducted of the beams reinforced with GFRP bars and conventional steel bars [2]. The details of the test specimens and its fabrications with typical calculations, test setup, load application system, instrumentation for response measurement, data acquisition and control system and the loading history used in the experiment are presented hereunder.

4.1 Test System

Two-point loading is employed to subject simply-supported concrete beams to pure bending (Fig. 7). 2 m long beams with clear span of 1.8 m and shear span of 0.6 m are subjected to monotonic vertical load using actuators.

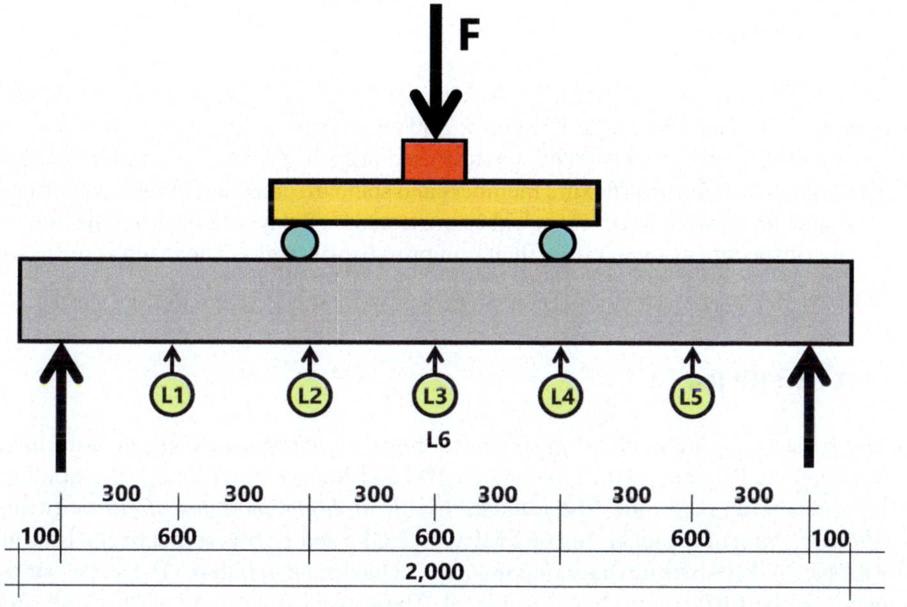

Fig. 7 Set-up to undertake experimental study of concrete beams of *rectangular* cross-section reinforced with FRP bars

4.2 Specimen Details

12 beams were tested, of which 8 were provided with longitudinal reinforcement made of GFRP bars and remaining 4 of conventional steel bars. 4 beams with GFRP bars and all 4 beams with steel bars were provided with shear reinforcement of 2-legged 6 mm diameter mild steel stirrups spaced at 110 mm centers, and remaining 4 beams with GFRP bars with 2-legged 6 mm diameter mild steel stirrups spaced at 90 centers. All beams have cross-section of 150 mm × 200 mm (Fig. 8). The details of reinforcement in the 12 test specimens are given in Table 5.

4.3 Properties of Material Used

The properties of different materials used in the specimen are presented hereunder.

4.3.1 Concrete

Six 150 mm × 150 mm standard concrete cubes, four 150 mm × 300 mm standard cylinders and four 100 mm × 100 mm × 500 mm flexure beams were cast with each specimen. The mix proportion of concrete used in each specimen is given in Table 6. The specimen, cylinders, flexure members and standard cubes are immersed in water tank and cured for 28 days. These cubes are tested on 7 days, 28 days and the day of testing of specimens to ascertain their compressive strengths. The results are shown in Table 7.

4.3.2 GFRP Bars

The rebars used are made of *glass fiber reinforced polymer* (GFRP; manufacturer M/s Hughes Brothers, with Fiber Aslan 100 and Matrix Vinyl Ester); the nominal bar diameter is 6.35 mm. The *ultimate tensile strength*, *modulus of elasticity* and *ultimate strain* of the rebars are 662 MPa, 33.7 GPa and 1.96%, respectively. Like all FRP bars, GFRP bars too have *linear elastic* behavior up to failure. The stress–strain curve of the FRP bar is shown in Fig. 9. These were used in an earlier study for strengthening masonry walls [5].

4.3.3 Steel Reinforcement

The grade of steel used in longitudinal reinforcement is Fe415 and in transverse reinforcement Fe250. The *yield* and *ultimate* strengths of the 8 mm diameter bars are 428.6 and 510.0 MPa, respectively. The *modulus of elasticity* and *ultimate strain* of

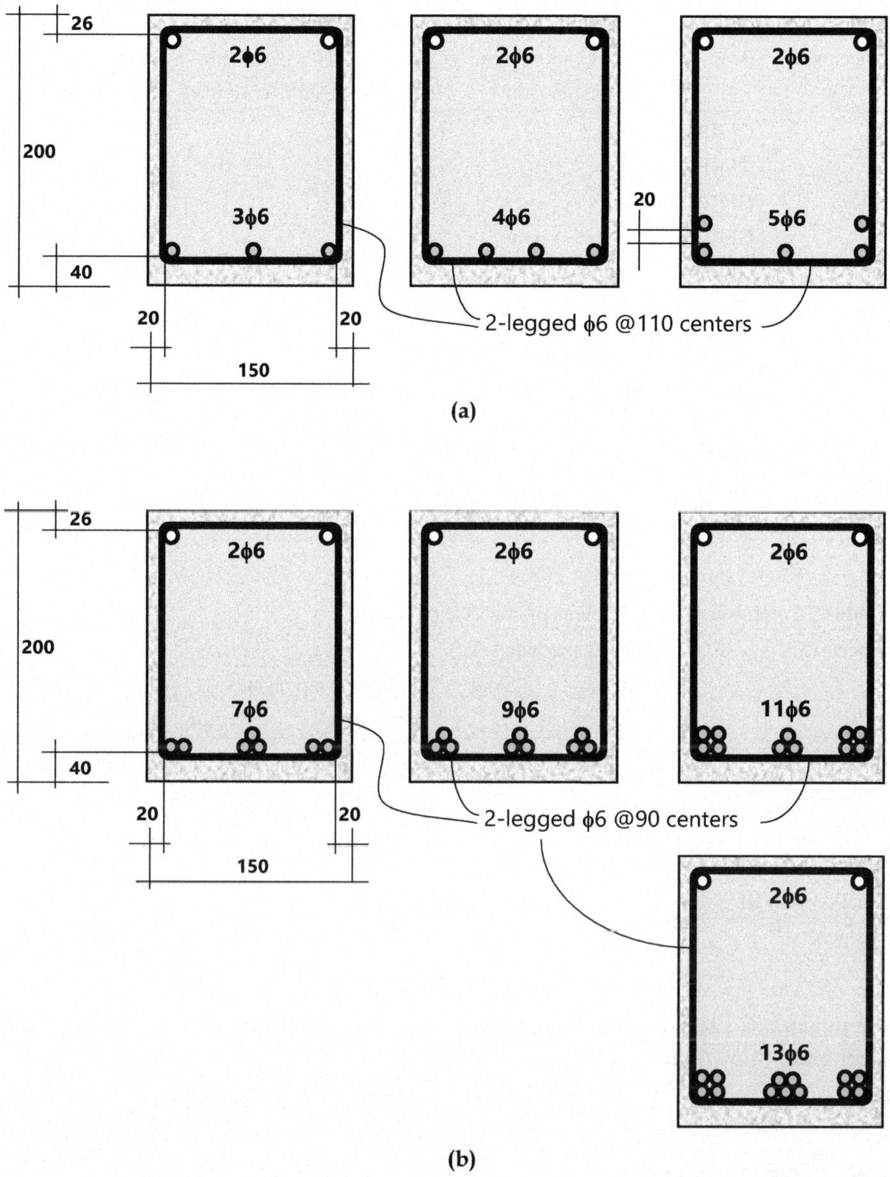

Fig. 8 Cross-section of concrete beams with longitudinal bars: **a** GFRP3a, 3b, 4 and 5, and Steel3a, 3b, 4 and 5, and **b** GFRP7, 9, 11 and 13

Table 5 Details of reinforcement in concrete beam specimen studied

Specimen	Beam	Longitudinal reinforcement		Transverse reinforcement	
		Diameter (mm)	Number	Diameter (mm)	Spacing (mm)
I	GFRP3a	6	3	6	110
II	GFRP3b		3		
III	GFRP4		4		
IV	GFRP5		5		
V	GFRP7	6	7	6	90
VI	GFRP9		9		
VII	GFRP11		11		
VIII	GFRP13		13		
I	Steel3a	8	3	6	110
II	Steel3b		3		
III	Steel4		4		
IV	Steel5		5		

Table 6 Composition of ingredients of concrete per batch

Specimen	Quantity per batch (kg)				
	Cement	Sand	Course aggregates		Water
			10 mm	20 mm	
I, II, III	35	76.3	69.1	69.1	22.4
Proportion	1	2.18	1.975	1.975	0.64
IV	36	79.2	68.8	68.8	23.76
Proportion	1.00	2.20	1.91	1.91	0.66
V, VI, VII, VIII	35	72.8	67.2	67.2	21.7
Proportion	1.00	2.08	1.92	1.92	0.62

the rebars are 171 GPa and 31%, respectively. The stress–strain curve of the 8 mm steel bar is shown in Fig. 10.

4.4 Sensors Used

Responses of the beams are measured using load cell, displacement meters (LVDT) and strain gauges.

Table 7 Strengths of concrete in the 12 specimen

Specimen		Cube compressive strength (MPa)			Modulus of rupture (MPa)
		7 days	28 days	Day of testing	
I	GFRP 3a	12.5	23.3	27.7	3.85
	Steel 3a	11.7	22.0	27.0	3.9
II	GFRP 3b	16.8	24.4	29.8	4.3
	Steel 3b	18.2	25.2	33.2	4.0
III	GFRP 4	14.9	21.5	26.9	3.5
	Steel 4	14.6	25.0	27.8	3.75
IV	GFRP 5	12.9	20.8	23.7	3.1
	Steel 5	12.9	22.0	22.7	3.3
V	GFRP 7	20.0	–	22.5	
VI	GFRP 9	21.3	–	24.0	
VII	GFRP 11	19.0	–	21.0	
VIII	GFRP 13	16.3	–	20.75	

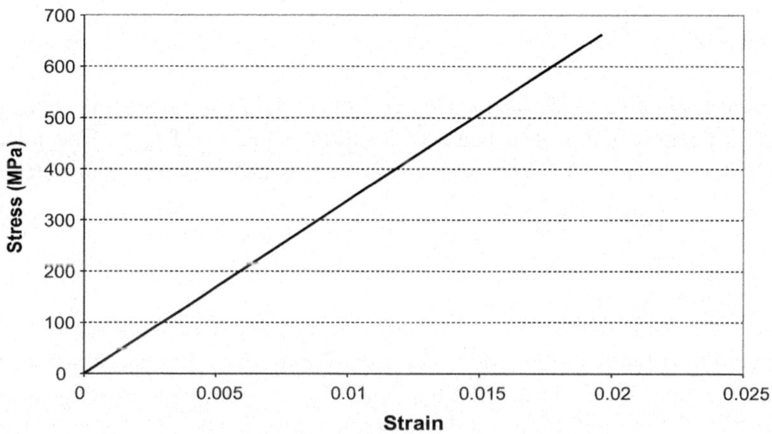

Fig. 9 *Stress–strain curve* of GFRP bar of 6 mm diameter

4.4.1 Load Cell

A compression load cell (Make: Sensotec; Model: TH/1590–05) of capacity 500 kN was used to record load applied during flexural loading of beams in a compression testing machine.

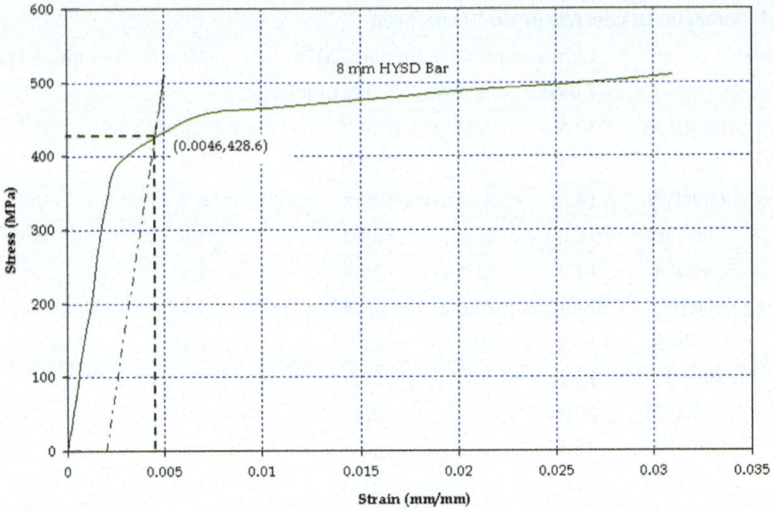

Fig. 10 *Stress-strain curve* of Steel bar of 6 mm diameter

4.4.2 LVDTs

Six *linear variable displacement transducers* (LVDTs) were used to measure the vertical deflection under each beam at locations shown in Fig. 7. The maximum displacement range of LVDTs used was ±50 mm. All LVDTs were calibrated before use.

4.4.3 Strain Gauges

Strain gauges (Manufacturer: Measurements Group, Inc., Raleigh, North Carolina; Model CEA-06-125UW-120; and range 20,000 micro-strains) were used to measure strains on each longitudinal bar at the mid-span of the beam. The 5 mm long gauges were pasted with a fully encapsulated grid and exposed copper-coated integral solder tabs. The strain limits were approximately 5% for gauge lengths 3.2 mm and larger. The resistance of gauges at 24 °C is 120.0 ± 0.3% Ohms and Gauge Factor is 2.07 ± 0.5%.

4.5 Results and Observations

4.5.1 Experimental Study

The normalised *load–deflection* and *load-strain* curves of the beams *GFRP3a* and *Steel3a* are presented in Fig. 11a and b, respectively. At the normalised moment of ~0.12, the beam with GFRP bars shows a drastic drop in stiffness; this is attributed to the initial cracking of the beam, when tensile bending stress in concrete at the bottom face of the beam reaches the modulus of rupture. Thereafter, the beam with GFRP bars has only linear behaviour till failure; the maximum normalised bending moment is ~0.38. On the other hand, the beam with steel bars undergoes yielding of longitudinal steel bars at a normalised moment of ~0.50, reaching a maximum normalised bending moment of ~0.60, and finally failing at a normalised bending moment of ~0.56. In general, the behaviour of all other beams is similar (Figs. 12, 13, 14 and 15). All these curves are shown together in Fig. 16. In Figs. 11, 12, 13, 14, 15 and 16, *F* represents the applied load and *a* the shear span in consistent units.

The salient *observations* and *inferences* from the tests are:

(1) Cracks appear at the mid-span in steel reinforced beams, and within the constant moment region in GFRP reinforced beams. These cracks are flexural cracks; they start from bottom of the beam and progress vertically towards the top. As the applied load increases, new cracks appear within the shear span. They start vertically and bend along a curved path towards the top owing to combined shear and flexural stresses.

(2) No cracks developed within a distance of d from the support. This may be due to the compressive stresses induced due to friction between beam and support.

(3) Beams with less FRP reinforcement have lower stiffness. So, the amount of reinforcement is important, not only from the point of view of *flexural capacity*, but also from the point of view of *serviceability*.

(4) There is no significant difference in the initial cracking moments of GFRP reinforced beams.

(5) At a load where the steel beam yields, the deflections of GFRP reinforced beams are much larger than that of steel reinforced beams. So, serviceability considerations may govern design of GFRP reinforced beams. Also, GFRP reinforced beams show brittle behaviour with sudden rupture. So, the design philosophy for GFRP reinforced beams needs to consider *higher factor of safety* against ultimate load.

(6) With increase in percentage of longitudinal FRP reinforcements, the ultimate bending moment capacity of GFRP reinforced beams increases. But, this increase is limited by the strain capacity of concrete in compressive of *over-reinforced* beams, and not proportional to the increase in reinforcement percentage.

(7) GFRP reinforced beams with large percentage of GFRP bars fail in compression of concrete (GFRP5, 7, 9 and 11), and demonstrate a flat region in the load-deformation behaviour (which may be interpreted as *ductility*, even

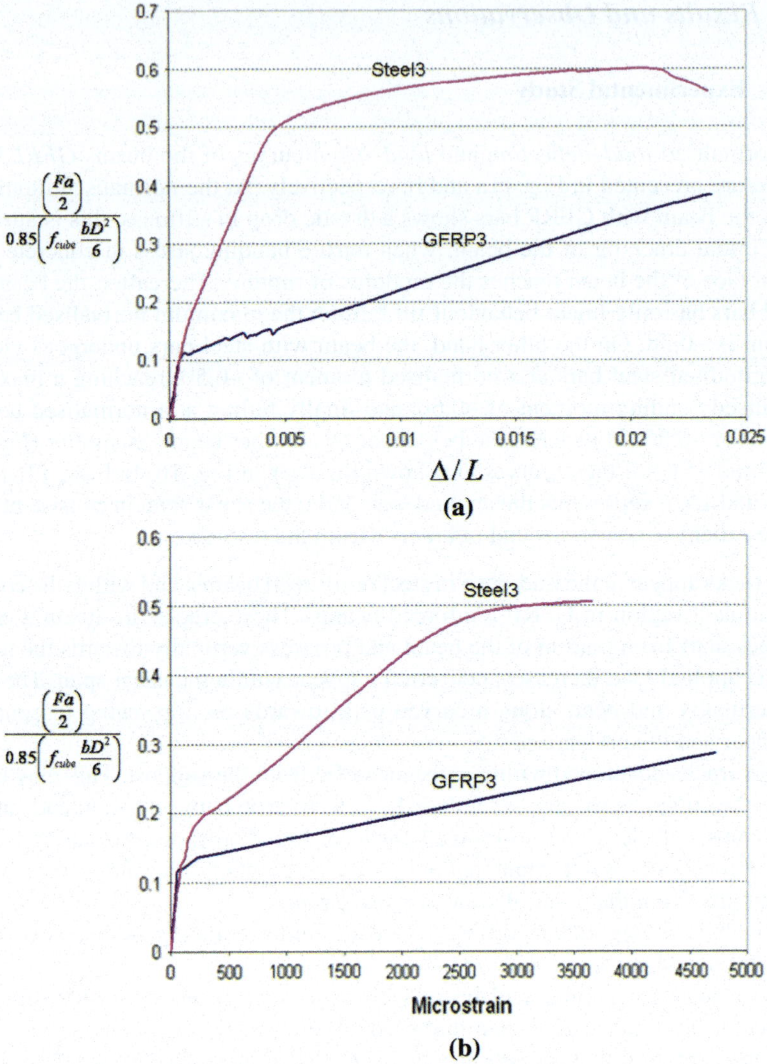

Fig. 11 Monotonic behaviour of GFRP3a and Steel3a beams: **a** Load–deflection curve, and **b** Load-strain curve

through GFRP does not yield), unlike beams with low percentage of GFRP bars (GFRP3a, 3b, 4 and 5), which fail in tension of GFRP bars (Table 8). Beams GFRP5, 7, 9 and 11 sustain lesser deflection at service loads than beams GFRP3a, 3b, 4 and 5, owing to higher initial stiffness of the beam and lower stress in FRP bar.

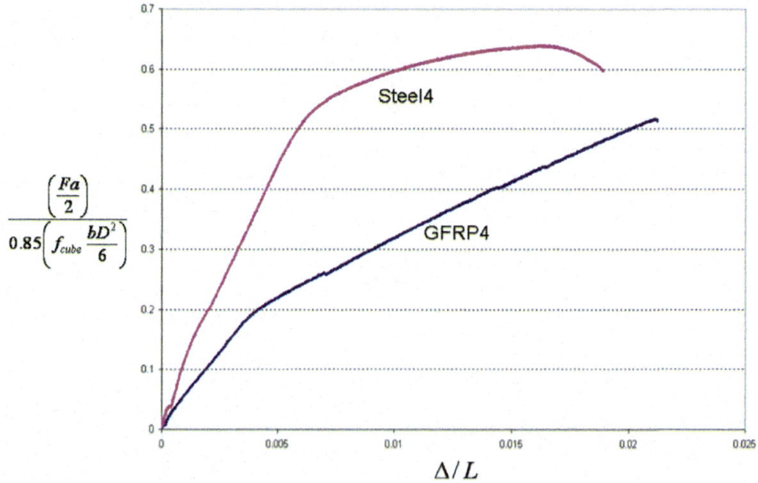

(a)

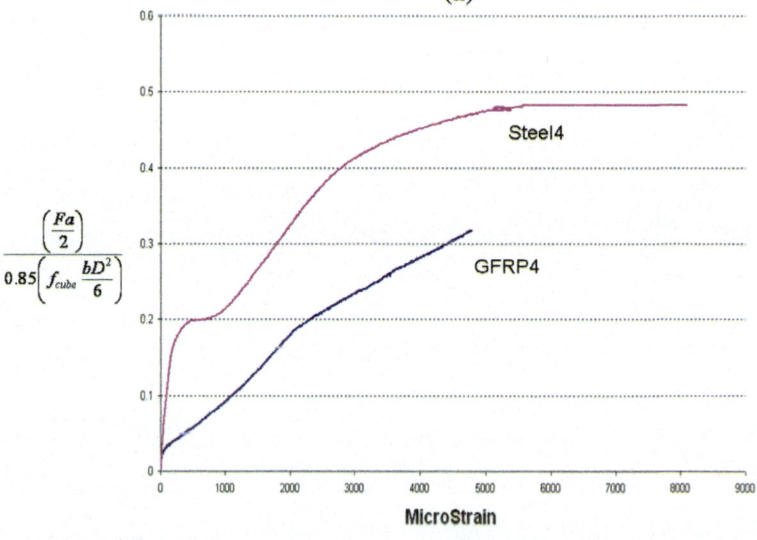

(b)

Fig. 12 Monotonic behaviour of GFRP4 and Steel4 beams: **a** Load–deflection curve, and **b** Load-strain curve

4.5.2 Analytical Study

The *ultimate bending moment capacity* is estimated; the method mentioned in Sect. 3 of this chapter is employed without using the partial safety factor γ_M of 1.5 for concrete and GFRP. These estimates are shown in Table 9.

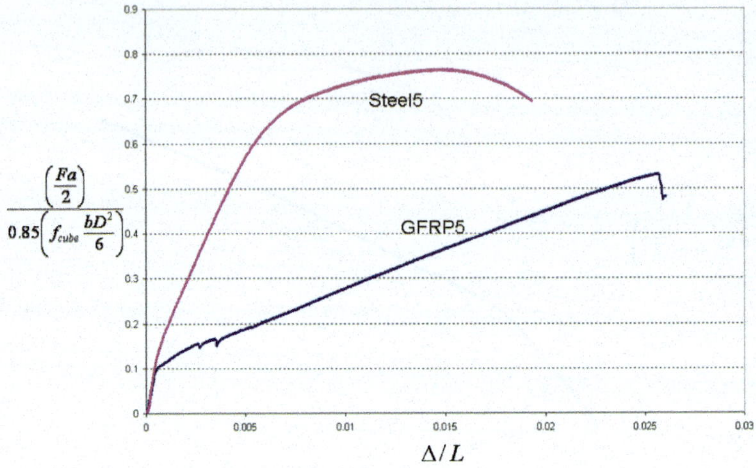

(a)

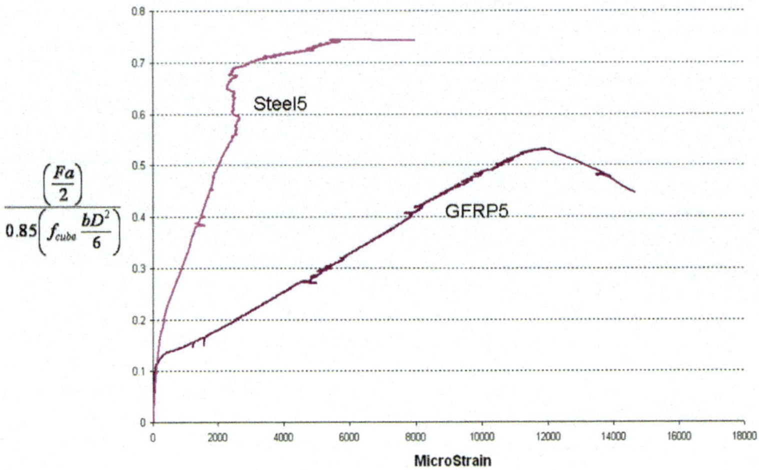

(b)

Fig. 13 Monotonic behaviour of GFRP5 and Steel5 beams: **a** Load–deflection curve, and **b** Load-strain curve

The design philosophy of concrete beams reinforced with GFRP bars suggests that concrete will crush at failure. But, the tests show that as the percentage of area of longitudinal bars increases, the failure mode transits from *rupture of GFRP bars* to *crushing of concrete* (Table 10). Clearly, when the percentage of FRP (A_{ft}/bd) exceeds the balanced percentage $(A_{ft,b}/bd)$ of steel given by:

$$\left(\frac{A_{ft,b}}{bd}\right) = \left(\frac{f_{c,avg}}{f_{fu}}\right)\left(\frac{x_b}{d}\right), \tag{21}$$

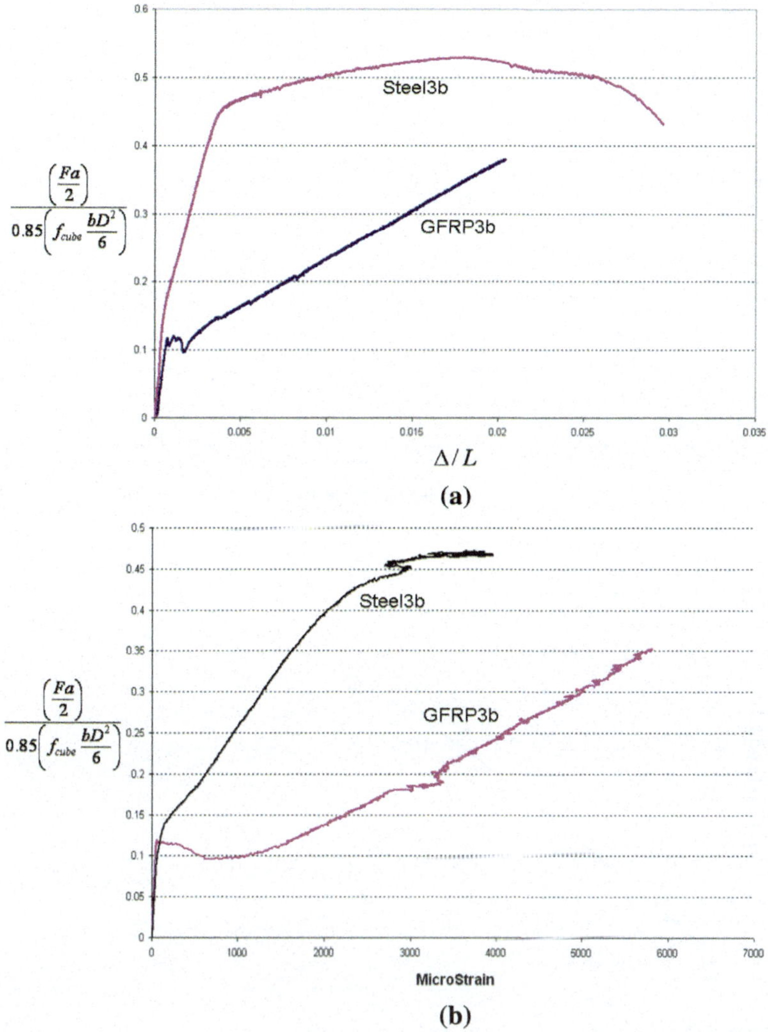

Fig. 14 Monotonic behaviour of GFRP3b and Steel3b beams: **a** Load–deflection curve, and **b** Load-strain curve

the beam will fail by crushing of concrete, else by rupture of longitudinal steel bars. In Eq. (21), x_b is estimated using Eq. (19). Also, Eq. (21) itself is derived from Eq. (5) neglecting the term corresponding to the compressive force in FRP bars on the compression side.

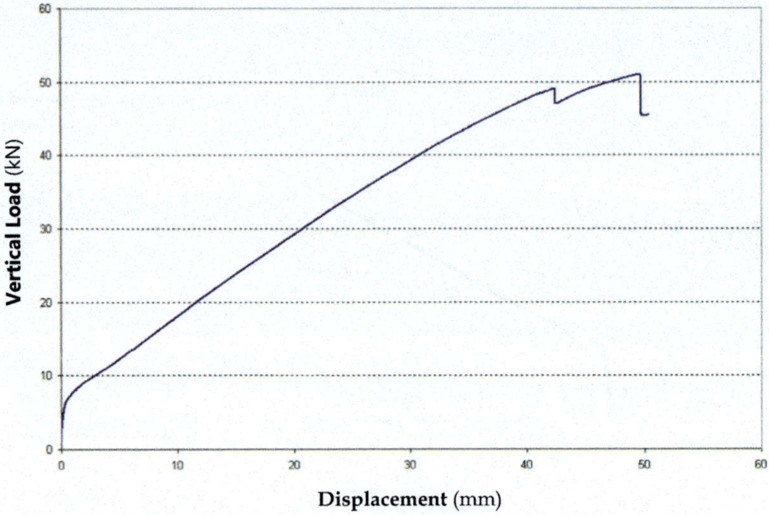

(a)

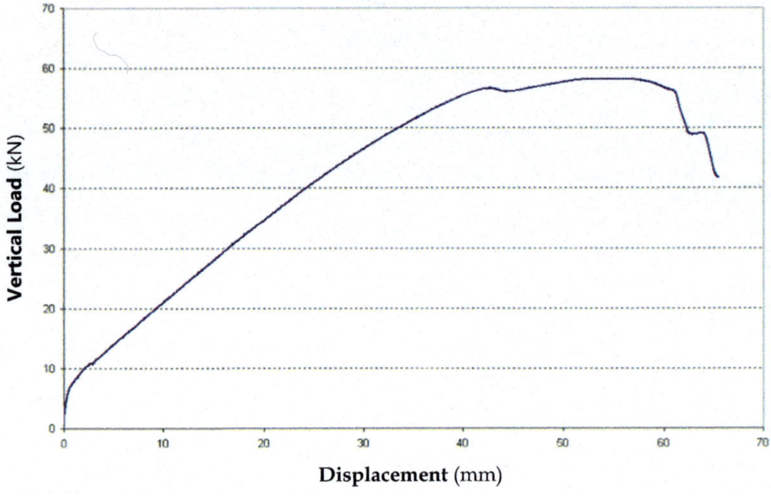

(b)

Fig. 15 Monotonic load–deflection behaviour of GFRP5, GFRP7, GFRP9, 3b and Steel3b beams

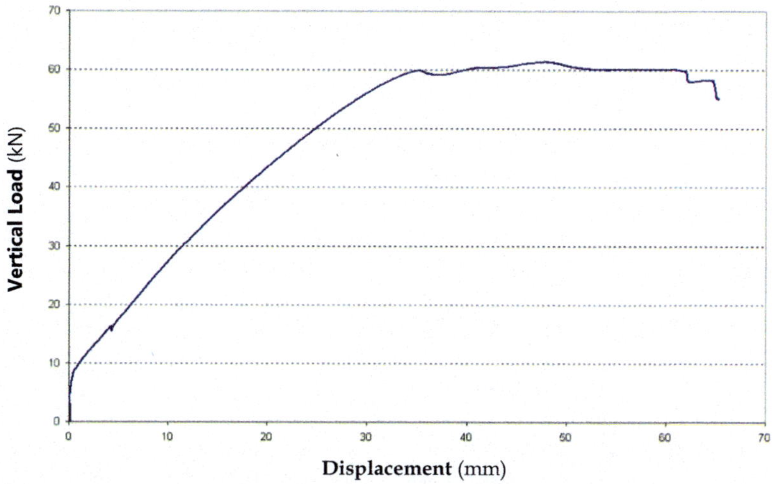

(c)

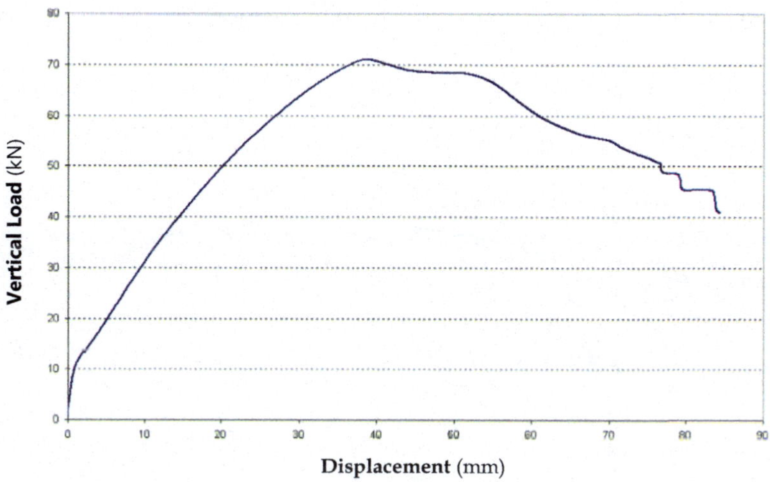

(d)

Fig. 15 (continued)

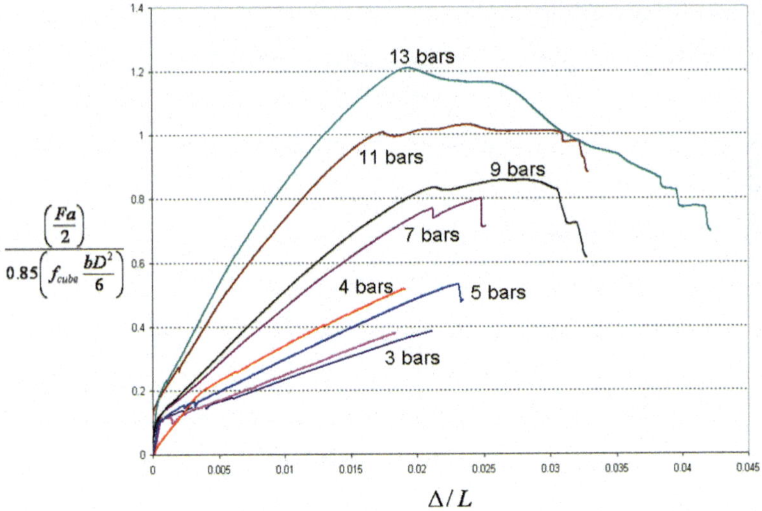

(a)

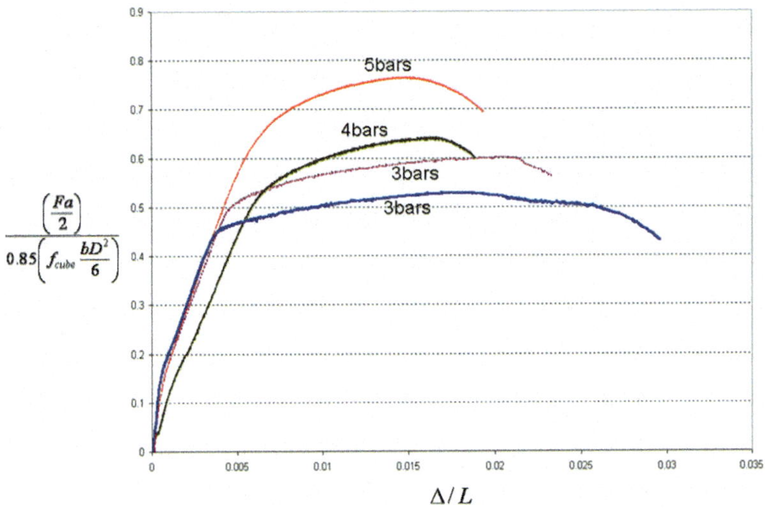

(b)

Fig. 16 Monotonic load–deflection behaviour: **a** GFRP reinforced beams, and **b** Steel reinforced beams

Table 8 Stiffness (kN/mm) and ductility of beams

Specimen	Stiffness (kN/mm)	Ductility factor
GFRP3a	0.56	1
GFRP 3b	0.46	1
GFRP 4	0.69	1
GFRP 5	0.61	1
GFRP 7	0.99	1.45
GFRP 9	1.23	1.87
GFRP 11	1.47	2.46
GFRP 13	2.17	2.16
Steel 3a	3.77	4.60
Steel 3b	3.42	7.06
Steel 4	4.15	2.78
Steel 5	4.10	3.01

Table 9 Summary of test results for GFRP reinforced beams–load capacity

Specimen	Ultimate moment capacity			ρ_{actual}	Sub-type of over-reinforced failure mode
	Experimental estimate (kN)	Analytical estimate (kN)	Experimental/ Analytical	$100\left(\frac{A_f}{bd}\right)$	
GFRP 3a	9.0	9.4	0.96	0.43	Rupture of longitudinal bars
GFRP 3b	13.1	11.2	1.17	0.58	
GFRP 4	12.2	10.5	1.16	0.77	
GFRP 5	7.71	8.7	0.89	0.43	
GFRP 7	15.3	12.6	1.21	1.01	Crushing of concrete
GFRP 9	17.4	14.5	1.20	1.30	
GFRP 11	18.6	14.7	1.26	1.59	
GFRP 13	21.3	15.5	1.37	1.87	

Table 10 Summary of test results for GFRP reinforced beams–strain capacity

Specimen	Ultimate strain		$\rho_{balanced}$	ρ_{actual}	Sub-type of over-reinforced failure mode
	Concrete	FRP bar		$100\left(\frac{A_f}{bd}\right)$	
GFRP 3a	0.0035	0.0200	1.46	0.43	Rupture of longitudinal bars
GFRP 3b	0.0035	0.0200	1.31	0.58	
GFRP 4	0.0035	0.0200	1.16	0.77	
GFRP 5	0.0035	0.0200	0.59	0.43	
GFRP 7	0.0044	0.0200	0.43	1.01	Crushing of concrete
GFRP 9	0.0052	0.0200	0.46	1.30	
GFRP 11	0.0038	0.0200	0.40	1.59	
GFRP 13	0.0047	0.0200	0.40	1.87	

5 Closing Comments

The *preliminary* experimental study reported in this chapter on concrete beams reinforced with GFRP bars is encouraging. It is possible to design concrete beams with FRP bars and seek the desired behaviour. In particular, the following aspects should be considered in the said design:

(1) Concrete beams reinforced with FRP bars should be designed as *over-reinforced* beams, and not as *under-reinforced* beams.
(2) The stiffness of concrete beams reinforced with FRP bars increases with increase in area of longitudinal bars.
(3) The failure mode depends critically on the ultimate strain of concrete in bending compression. Hence, detailed experimental program is required to estimate ultimate strain of concrete, for use in design criteria. Also, this result will guide in more accurately determining the balanced percentage of FRP reinforcements of concrete beams.
(4) Concrete beams reinforced with FRP bars are less stiff than conventional concrete beams reinforced with steel bars. Hence, design of concrete beams reinforced with FRP bars, *deformation* is likely to govern the design rather than *strength.*
(5) If concrete beams are required to be used in environments involving *nuclear radiation*, a technology development program is necessary to produce prefabricated transverse reinforcement made of FRP too. In such a case, all reinforcement is made of FRP material.

In the domain of concrete beams reinforced with FRP bars, the residual agenda is significant for future work in this area. More experimental and analytical studies should be conducted to investigate the flexural behaviour of concrete beams reinforced with FRP bars, involving a wider range of:

(a) FRP reinforcements, namely GFRP, CFRP and AFRP bars.
(b) Concrete beam cross-section sizes, grades of concrete (from low to high), percentage of longitudinal FRP reinforcement, and percentage of transverse FRP reinforcement.

Also, the effect of confinement of concrete should be quantified arising from the use of prefabricated transverse FRP reinforcement on ductility of concrete beams reinforced with FRP bars.

References

1. Iannuzzi M, Frankel GS (2022) The carbon footprint of steel corrosion. NPJ Mater Degradradation 6(101). https://doi.org/10.1038/s41529-022-00318-1
2. Padmanabha Rao TVVSS (2006) Flexural behaviour of GFRP reinforced concrete beams. M.Tech. Thesis, Department of Civil Engineering, Indian Institute of Technology Kanpur

3. ACI 440 (2000) Guide for the design and construction of concrete reinforced with FRP bars. ACI Committee 440.01R-05 Recommendations. American Concrete Institute, Farmington Hills, MI, USA
4. ISIS (2001) Reinforcing concrete structures with fiber reinforced polymers. In: Design Manual No.3, Intelligent Sensing for Innovative Structures (ISIS) Research Network of the Canadian Network of Centers of Excellence, Canada
5. Bajpai K, Duthinh D (2003) Bending performance of masonry walls strengthened with near-surface mounted FRP bars. In: 9th North American masonry conference. Clemson(USA), pp 1052–1063

FRP Based Earthquake Retrofitting of RC Columns

S. B. Singh and C. V. R. Murty

Abstract Fiber reinforced polymer (FRP) materials are being used for the last two decades as reinforcement in new concrete structures and prestressing tendons in pretressed concrete structures. In addition, they are being used at large scale in external strengthening and/or retrofitting of deficient structures to restore the strength and stiffness of the structure to the design values or to meet the current code provisions. But, to make such structures capable of resisting the effects of strong earthquake shaking, at least a minimum stiffness, sufficient strength and adequate ductility are to be ensured. FRP laminates and FRP fabric sheets are used in external strengthening to provide the needed stiffness, strength, and ductility by suitably orientating the fibers and creating confinement. These FRP materials possess high tensile strength with moderate stiffness and are non-corrodible; they can be used efficiently for improving performance of structural elements, without increasing the earthquake forces. In this chapter, a unified design approach has been presented for earthquake retrofitting of RC columns; a design example is provided to demonstrate the effectiveness of FRP for seismic strength and ductility enhancements.

Keywords Design approach · Ductility · Fibre Reinforced Polymer (FRP) · Seismic forces · Strength · Stiffness · RC columns

S. B. Singh (✉)
Department of Civil Engineering, Birla Institute of Technology and Science, Pilani 333031, India
e-mail: sbsingh@pilani.bits-pilani.ac.in; sbsinghbits@gmail.com

C. V. R. Murty
Department of Civil Engineering, P. S. Rao Institute Chair Professor, Indian Institute of Technology Madras, Chennai, India
e-mail: cvrm@iitm.ac.in

S. B. Singh and C. V. R. Murty (eds.), *RC Structures Strengthened with FRP for Earthquake Resistance*, Composites Science and Technology,
https://doi.org/10.1007/978-981-97-0102-5_2

1 Introduction

The fiber reinforced polymer (FRP) materials are advanced composite materials having high specific tensile strength and high specific stiffness as reflected by their corresponding *strength-to-weight ratio* and *stiffness-to-weight ratio*, respectively. Most importantly, these advance composite materials have high corrosion resistance and are considered practically to be non-corrodible. These materials could be used in different forms, such as reinforcement bars, laminates, strips, tendons and fabric sheets. Also, mechanical characteristics of these materials depend on the kinds of fibers and bonding resins used. Currently, carbon fiber reinforced polymer (CFRP), glass fiber reinforced polymer (GFRP), aramid fiber reinforced polymer (AFRP) and basalt fiber reinforced polymer (BFRP) are being used in civil engineering applications. In addition, natural fiber based FRP also are being used at large scale for developing green FRP composites with significantly reduced cost and high sustainability of its use.

Since these FRP materials have high tensile strength, they are being used for external strengthening of structures to meet the strength, stiffness and ductility demands on structures that have deteriorated over time due to aging effects or have been damaged during when natural actions (such as earthquake and tsunami) appear on them [1, 2]. Recently, the background was provided [3] to European seismic design provisions and stated that a large part of research effort is directed towards the confinement models mostly for static loading situations, such as for shear and anchorage. Performance-based criteria were presented for global and local retrofit of structures permitting acceptable and repairable damage during *severe intensity* earthquake ground shaking and limited damage during *low intensity* earthquake ground shaking. A review on fiber reinforced polymers for structural retrofitting [4] presented scope and uses of FRP materials in earthquake strengthening of RC, masonry and steel structures. Concepts of displacement-based earthquake design were introduced [5] for a direct and rational retrofit. Primary strategies and principles were suggested [6] for earthquake retrofit of existing structures with main focus on local strengthening. Results of experimental tests were presented [7] on effects of strengthening of 2 m high columns with and without internal steel reinforcements and strengthened with textile-reinforced concrete; an increase of 85% in axial compressive strength was reported with respect to that of non-strengthened concrete columns.

Results of a full-scale tests were presented [8] on under-designed RC structure in the as-built and FRP retrofitted configurations, along with retrofit criteria and calculation procedures to design the amount and layout of FRP for improving the earthquake performance of the deficient structures. The results highlight the effectiveness of FRP wrapping in improving the global performance of under-designed RC columns with regard to ductility and energy dissipation capacity. In another study [9], the principles were presented of the design of GFRP retrofit and verified with experimental tests on both the as-built and GFRP retrofitted structure. In an analytical study [10], a uniaxial model was used for concrete confined with FRP wrapping, steel jackets or conventional transverse reinforcement. A relation was provided between axial and lateral strains to estimate the ultimate strength and ultimate strain of concrete confined with FRP wrapping.

2 Earthquake Retrofitting of RC Structures Using FRP

RC structures need 7 virtues to be fully earthquake resistant, namely: (1) Good structural configuration, (2) At least a minimum overall lateral stiffness, (3) Sufficient overall lateral strength, (4) Large overall lateral deformability, (5) Good overall lateral ductility, (6) Acceptable collapse mechanism, and (7) Good energy dissipation. The order in which these are to be ensured is as given above. These are to be ensured at the global and local levels. In the context of retrofitting of RC structures, the global virtues should be ensured first and the local virtues next.

In general, it is observed that FRP wrapping to RC members is effective only to enhance some of the local aspects, like axial strength capacity. Hence, the EuroCode [6] permits the use of FRP wrapping to increase the shear strength of individual members, enhance ductility of the concrete members with effective confinement, and eliminate the chances of lap splice failure. For effective earthquake retrofitting using external strengthening (wrapping) with FRP, the following are critical points:

(1) Structural configuration cannot be corrected by using FRP wrapping method;
(2) Stiffness irregularities in the structure cannot be solved by using FRP wrapping method. Overall stiffness cannot be altered significantly by wrapping columns with FRP sheets;
(3) Strength irregularities in the structure cannot be solved by using FRP wrapping method. Overall strength can be modified marginally by strengthening select members,
(4) Overall deformability of the structure is addressed only marginally, and hence no appreciable change in global ductility.
(5) FRP strengthening is regarded as selective intervention technique to assist in the following:

    (a) Increasing the flexural strength of deficient RC beams at mid-span sections (though not at the faces of the columns) and of deficient RC columns at mid-height (though not at the faces of the beams and slabs) using FRP composites with fibers oriented along axis of members;
    (b) Increasing the shear capacity of RC members using FRP composites with fibers placed in direction transverse to the axis of members;
    (c) Increasing the section ductility of RC beams and columns by confinement in critical zones through wrapping of FRP composites;
    (d) Improving the efficiency of lap splice zones of RC beams and columns by providing confinement using FRP wrapping;
    (c) Preventing buckling of longitudinal bars under compression in RC beams and columns by confinement using FRP wrapping; and

(f) Intervening into the beam-column joint region with beams framing into it from 4, 3 or 2 sides is not feasible. Increasing tensile cracking strength is possible of joint panel zones of beam-column joints with a beam framing on one side only provided there is no slab in the structure, using the FRP composites with fibers placed along the principal tensile stresses.

The primary objectives of using external FRP strengthening by wrapping the RC members in earthquake retrofitting are:

(1) All potential brittle failure mechanisms should be eliminated, such as shear failure, lap splice failure, buckling of bars in compression, and joint shear (if possible).
(2) The global deformation capacity of the structure should be enhanced to the extent possible, by increasing the ductility of the potential plastic hinge zones either: (a) without changing their position, or (b) relocating the potential plastic hinge zones, by using the concept of capacity design concept, wherein columns are strengthened in flexure with the goal of transforming the frame structure into an energy dissipating mechanism with strong columns and weak beams.
(3) Evaluate the increase in the ductility of the RC structure when the brittle modes of failure (in (1) above) are precluded.

Enhancement of axial strength and ductility using FRP wrapping in RC columns with rectangular or square cross-sections is less than that in RC columns with circular cross-sections, because confinement by FRP jacketing is uniform in circular cross-sections but localized along the diagonals of the square or rectangular cross-sections. In wall-like RC columns (with cross-sectional plan aspect ratio more than 3), the effectiveness of FRP jacketing is lesser than that in RC columns with cross-sectional plan aspect ratio less than 3. In wall-like columns or structural walls, the failure is affected by premature failure mechanisms, such as buckling of compression bars and spalling of cover concrete. Also, confinement models based on regression analysis are sensitive to the ultimate rupture strain of FRP. The ultimate tensile strain of FRP estimated from experiments (through flat coupon tests) is not reached even when FRP jacket ruptures in confined concrete column during compression tests [11, 12]. The ratio between this strain at which FRP rupture during confined compression tests and the ultimate strain from flat coupon test is called *efficiency factor* β.

In this chapter, an approach is presented to design the FRP jacketing based earthquake retrofitting of rectangular RC column; a design example is presented to demonstrate the procedure. Expressions are presented for evaluating the load capacity of RC columns strengthened with textile reinforced concrete and FRP strengthened concrete.

3 Retrofitting of RC Columns Using FRP for Axial Strength

The axial load capacity F_u of the textile-reinforced concrete strengthened column is given by:

$$F_u = F_c + F_{fc} = f_{cm} A_g + f_{fc,eff} A_{fc,eff} \quad (1)$$

where

f_{cm} Mean compressive strength of existing concrete core $= \alpha_c f_{cm,cube}$,
A_g $b_c d_c$, and.
$A_{fc,eff}$ $A_{eff} - A_c$,

wherein

α 0.665,
A_{eff} $\left(b_c + 2t_{eff}\right)\ \left(d_c + 2t_{eff}\right)$,
t_{eff} $t_{fc} - n_1$, and.
t_{fc} $2(1 + n_2)$.

Here, n_1 refers to the number of layers or plies with 1 mm fine-grained concrete cover per textile ply, and n_2 to the number of layers or plies with 2 mm fine-grained concrete cover per textile ply. The effective fine-grained concrete area is calculated from the 2 mm cover of the fine-grained concrete per textile layer minus the textile reinforcement with no load bearing capacity. In Eq. (2), due to different compressive stress–strain relations of existing and new fine-grained textile reinforced concrete, it is assumed that the fine-grained textile reinforced concrete has not reached the ultimate strain. Hence assuming rigid bonds, the stress in the fine-grained textile reinforced concrete is taken corresponding to ultimate strain of existing concrete.

For an existing concrete column with steel reinforcement, Eq. (1) is modified considering contribution of steel reinforcement with stress corresponding to the ultimate strain of existing concrete (Fig. 1).

$$F_u = F_c + F_s + F_{fc} = f_{cm} A_{c,net} + f_{s,\varepsilon_{cu}} A_s + f_{fc,\varepsilon_{cu}} A_{fc,eff}, \quad (2)$$

where

$A_{c,net}$ $A_g - A_s$,
$f_{fc,\varepsilon_{cu}}$ Stress in fine-grained tensile reinforced concrete corresponding to ultimate strain ε_{cu} of existing concrete, and.
$f_{s,\varepsilon_{cu}}$ Stress in steel reinforcement corresponding to ultimate strain ε_{cu} of the existing concrete.

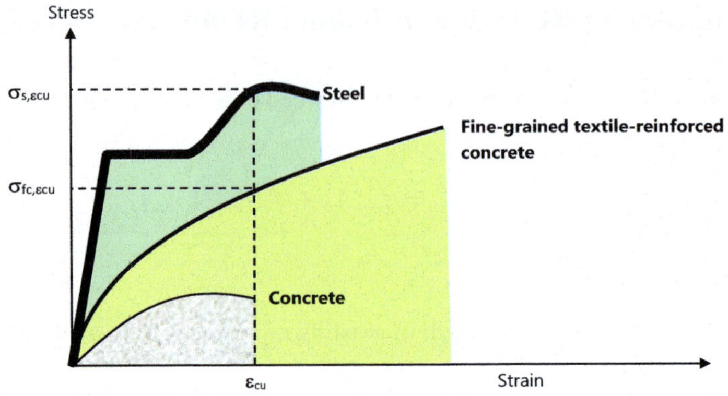

Fig. 1 Stress–strain relation of concrete and steel reinforcements bars

3.1 Confinement Models

By wrapping the column with FRP sheets, the strength of column concrete increases owing to enhancement in the axial compressive strain capacity of concrete. But, the level of increase of axial strain capacity depends on the section. FRP wrapping does not change the behaviour of the column in compression under low eccentric axial load, but delays the buckling of longitudinal bars in compression and restrains the concrete cover from spalling. Together, these actions lead to larger axial strain capacity, higher axial compression load carrying capacity and enhancement of the axial strain ductility of the RC column. ACI 440 code [13] provides an expression to estimate the confinement of concrete to evaluate the cylindrical triaxial confined concrete strength f_{cc} for a given uniform confining pressure f_l'.

$$\left(\frac{f_{cc}}{f_c'}\right) = 2.25\sqrt{1 + 7.9\left(\frac{f_l'}{f_c'}\right)} - 2\left(\frac{f_l'}{f_c'}\right) - 1.25 \tag{3}$$

where f_c' is cylinder compressive strength of concrete. For a column of rectangular cross-section (depth h less than width b), effective lateral confining pressure f_l' is given by [6]:

$$f_l' = \left(\frac{2t}{h}\right) E_f \varepsilon_{fu} \tag{4}$$

where

t Thickness of FRP jacket,

E_f Modulus of elasticity of FRP jacket material, and.

ε_{fu} Rupture strain of FRP.

For $\left(\frac{f_l'}{f_c'}\right) < 1.3$, confined compressive strength f_{cc} of concrete for a given thickness of FRP jacket [10] is given by:

$$\left(\frac{f_{cc}}{f_c'}\right) = 1 + 1.42\left(\frac{f_l'}{f_c'}\right) - 1.40\left(\frac{f_l'}{f_c'}\right)^2 + 0.30\left(\frac{f_l'}{f_c'}\right)^3 \quad (5)$$

In general, confined compressive strength f_{cc} of concrete can be determined by [14]:

$$\left(\frac{f_{cc}}{f_c'}\right) = 1 + 2.6\left(\frac{f_l'}{f_c'}\right)^{2/3} \quad (6)$$

As per Italian standard [14], the confinement is effective only when $\left(\frac{f_{l,eff}'}{f_c'}\right) > 0.05$. The ultimate axial strain ε_{ccu} of FRP-confined concrete is estimated using [14]:

$$\varepsilon_{ccu} = 0.0035 + 0.015\sqrt{\left(\frac{f_{l,eff}'}{f_c'}\right)} \quad (7)$$

$f_{l,eff}'$ Effective lateral confinement pressure $= k_{eff} f_l'$.

wherein $k_{eff}(\leq 1)$ is coefficient of effectiveness, which depends on the cross-section shape and FRP configurations (continuous or discrete).

3.1.1 Effective Coefficient Factor k_{eff}

The effectiveness coefficient k_{eff} is defined as the ratio between the volume of the effectively confined concrete $V_{c,eff}$ and that of the concrete member V_c, neglecting the area of existing internal steel reinforcement. It is given by:

$$k_{eff} = k_H k_V k_\alpha \quad (8)$$

where

k_H Coefficient of Horizontal Efficiency

$$= \begin{cases} 1 & \text{Circular cross} - \text{sections} \\ 1 - \left(\frac{b'^2 + h'^2}{3A_g}\right) & \text{Rectangular cross} - \text{sections} \end{cases},$$

k_V Coefficient of Vertical Efficiency

$$= \begin{cases} 1 & \text{Continous FRP Wrapping} \\ \left(1 - \frac{p_f'}{2d_{min}}\right)^2 & \text{FRP strips at clear spacing of } p_f' \end{cases}, \text{ and}$$

k_α Coefficient of Efficiency for fibers installed at an angle α_f to the transverse to the longitudinal axis of the Member.

$\frac{1}{1+tan^2\alpha_f}$,

wherein

b'	$b - 2r_c$,
h'	$h - 2r_c$,
b	Width of rectangular section.
h	Depth of rectangular section.
r_c	Corner radius of rectangular section.
A_g	Gross cross-sectional area of the section.
d_{min}	Minimum cross-section dimension of the member (equal to diameter in a circular section).

Also, the following inequality should be satisfied.
$p'_f \leq d_{min}/2$.

3.1.2 Lateral Confinement Pressure f_l'

The lateral confinement pressure f_l' is given by:

$$f_l' = \frac{1}{2}\rho_f E_f \varepsilon_{fd,rid} \tag{9}$$

where

ρ_f	Geometric strengthening ratio which depends on shape of sections (i.e., rectangular or circular) and FRP configurations (i.e., continuous or discontinuous wrapping),
E_f	Modulus of elasticity of the FRP in the fiber direction, and.
$\varepsilon_{fd,rid}$	Reduced design strain of FRP $min[\eta_a, \varepsilon_{fk}/\gamma_f, 0.004]$,

The Italian standard [14] does not consider confinement in slender rectangular sections with $b/h > 2$ or maximum dimension $b > 900mm$. In this situation, Eqs. (3)–(5) could be used or equations given by Spoelstra and Monti [10] could be used for such slender sections which are presented later in Sect. 4.3.

wherein

η_a Environmental conversion factor (as per Table 3–2 of Italian standard [14]).

$$\begin{cases} 0.95 \text{ Internal Environment} \\ 0.85 \text{ External Environment} \\ 0.85 \text{ Aggressive Environment} \end{cases} \text{for carbon/epoxy system.}$$

$$\begin{cases} 0.85 \text{ Internal Environment} \\ 0.75 \text{ External Environment} \\ 0.70 \text{ Aggressive Environment} \end{cases} \text{for aramid/epoxy system.}$$

$$\begin{cases} 0.75 \text{ Internal Environment} \\ 0.65 \text{ External Environment} \quad \text{for glass/epoxy system.} \\ 0.50 \text{ Aggressive Environment} \end{cases}$$

γ_f Partial safety factor for material.

$$\begin{cases} 1.0 \quad \text{Serviceability Condition} \\ 1.1 \quad \text{Ultimate Condition} \quad \text{for glass/epoxy system} \\ 1.2 - 1.5 \text{ Failure due to Debonding} \end{cases}$$

ε_{fk} Characteristic rupture strain of FRP = Specified ultimate rupture strain ε_{fu}.

3.1.3 Geometric Strengthening Ratio ρ_f

The geometric strengthening ratio, ρ_f to be used in Eq. (9) depends on shape of the sections (circular and rectangular sections). ρ_f for circular and rectangular sections as per Italian standard [14] are:

$$\rho_f = \begin{cases} \frac{4t_f\, b_f}{D\, p_f} \text{ Strip FRP Wrapping} \\ \frac{4t_f}{D} \text{ Continuous FRP Wrapping} \end{cases} \text{for circular sections, and}$$

$$\rho_f = \begin{cases} \frac{2t_f\,(b+h)\, b_f}{b\, h\, p_f} \text{ Strip FRP Wrapping} \\ \frac{2t_f\,(b+h)}{b\, h} \text{ Continuous FRP Wrapping} \end{cases} \text{for rectangular sections,}$$

where t_f, b_f and p_f are thickness, width and center to center spacing of FRP strips, respectively, D, the diameter of circular section, and b and h are width and depth, respectively, of rectangular section.

4 Member Deformations and Plastic Hinge Length

The yielding rotation θ_y, ultimate rotation θ_u and plastic hinge length L_{pl} of members can be computed as per Eurocode 8 [13] as:

$$\theta_y = \beta_{flex}\, \varphi_y\, L_v + \beta_{shear} + \beta_{slip}\, \frac{d_{bL}\, f_y}{\sqrt{f_c'}}, \tag{10}$$

$$\theta_u = \gamma_u \left[\theta_y + \left(\varphi_u - \varphi_y \right) L_{pl} \left(1 - \frac{0.5\, L_{pl}}{L_v} \right) \right] \tag{11}$$

$$L_{pl} = \alpha_{flex}\, L_v + \alpha_{shear}\, h + \alpha_{slip}\, d_{bL}\, f_y \tag{12}$$

where

L_v Shear span = $0.5L$ [15],
d_{bL} Diameter of longitudinal bars,

f_c' Cylinder strength of concrete, and.
f_y Yield strength of steel,
φ_y Yield curvature,
φ_u Ultimate curvature,

$$\alpha_{flex} = 0.1; \alpha_{shear} = 0.17; \alpha_{slip} = \frac{0.24}{\sqrt{f_c'}};$$

$$\beta_{flex} = \frac{1}{3}; \beta_{shear} = 0.0013\left(1 + 1.5\frac{h}{L_v}\right); \beta_{slip} = 0.13\phi_y;$$

$$\gamma_u = \frac{1}{\gamma_{el}},$$

wherein

$$\gamma_{el} = \begin{cases} 1.5 \text{ Primary Members} \\ 1.0 \text{ Secondary Members} \end{cases}.$$

4.1 *FRP Earthquake Retrofit Approach without Relocalization of Plastic Hinges*

For structures designed for gravity load only, the overall deformation capacity is governed by the limited rotation capacity in the plastic hinges at beam and column ends due to small member dimensions and low amount of longitudinal reinforcements. Thus, the FRP confinement of columns can help increase the overall structure deformation capacity, because column wrapping enhances the usable concrete compressive strain capacity, which, in turn, leads to increase of curvature ductility. The enhancement of curvature ductility will lead to proportional increase in the plastic hinge rotation capacity, assuming that FRP wrapping does not significantly affect the plastic hinge length. Also, the FRP confinement of columns at their ends induces considerable increase in section ductility, but does not lead to significant increase in strength and hence this retrofit approach using FRP wrapping does not change the strength hierarchy of the structure. Hence, global deficiency like the *weak column—strong beam* system cannot be converted to *strong column—weak beam* system by FRP wrapping.

To design the earthquake retrofit in terms of number of plies and layers of FRP fabrics required for critical structural elements, the following steps should be followed:

(1) Estimate through pushover analysis the maximum deformation capacity of designed structure, corresponding to the first plastic hinge reaching $0.75\theta_u$; and
(2) Determine the amount of FRP plies to be used to provide the required ductility increase of plastic hinges at columns ends by:
    (a) Determining the maximum ratio γ of demand chord rotations θ_{demand} and ultimate capacity chord rotations $\theta_{u,capacity}$, using capacity spectrum approach (i.e., design response spectrum and capacity curve (Fig. 2), as:

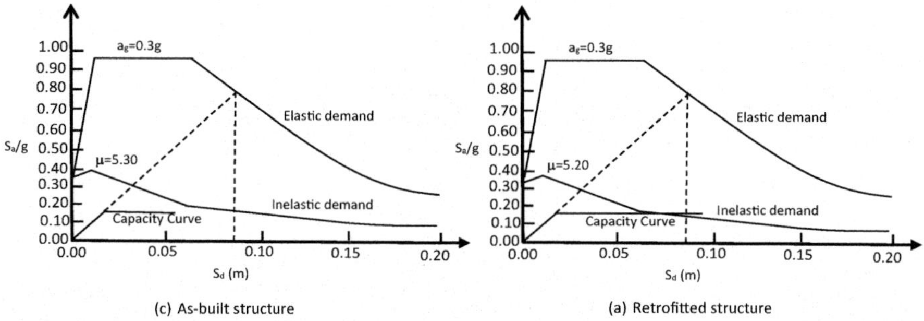

Fig. 2 Typical theoretical seismic performance of as-built and retrofitted structure for 0.3 g PGA

$$\gamma = \frac{\theta_{u,demand}}{\theta_{u,capacity}},$$

(b) Determining target rotation capacity $\gamma\ \theta_{u,capacity}$ and corresponding design ultimate curvature $\phi_{u,target}$ at the cross-section;
(c) Estimating (by cross-section analysis) the ultimate concrete strain $\varepsilon_{cu,target}$ to achieve target curvature $\phi_{u,target}$, and
(d) Determining the amount of FRP plies required to reach the target ultimate concrete strain $\varepsilon_{cu,target}$.

In earthquake retrofit design, it should be ensured that structure meets the ductility requirement by retrofitting structural elements, such as columns as described above. But, it should be ensured that all brittle failure modes are precluded as discussed above in the primary objectives of external FRP strengthening. Also, the structural global performance before and after strengthening should be assessed by nonlinear static pushover analysis in both the positive and negative directions of global X and Y-axes. A typical capacity diagram for as-built and retrofitted structure (Fig. 3) demonstrates that how deformation capacity of the structure can be enhanced by retrofitting (i.e., top displacement of a framed structure) without significantly enhancing the load carrying capacity as expressed by base shear. Thus, to pursue the objectives of retrofitting, two main aspects should be focused on, namely:

(1) Increasing the global deformation capacity of structure to dissipate its global response, and
(2) Fully exploiting the increased deformation capacity by avoiding brittle collapse modes.

The design principles of the rehabilitation strategy for framed building structures consists of:

(a) design of column confinement, and
(b) design of exterior beam-column joints and shear strengthening of wall-type columns (columns having aspect ratio, b/h > 3).

Fig. 3 Typical base shear versus top displacement curves of as-built and retrofitted structures

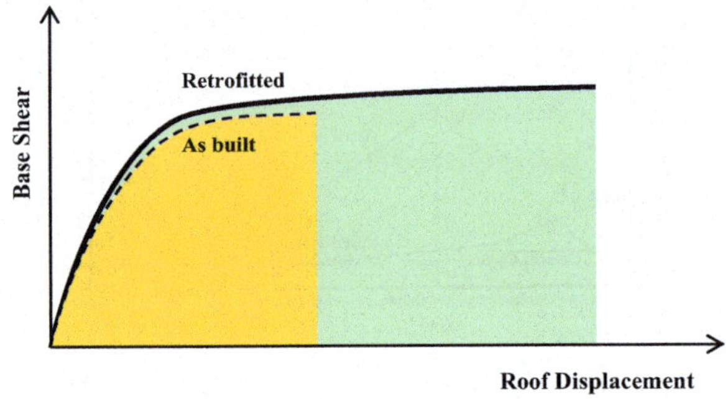

4.2 *Brittle Collapse Mechanisms*

For earthquake retrofit of structures, the following brittle collapse mechanisms [14] should be prevented using FRP strengthening:

(1) Shear failure,
(2) Failure due to loss of bond in steel overlapping areas into columns,
(3) Failure due to buckling of longitudinal bars into the columns, and
(4) Failure due to tensile stresses on the beam-column joint.

The shear capacity can be increased by providing FRP wrapping with fibers oriented in direction perpendicular to the axis of the member. The loss of bond in steel overlapping areas into columns can be reduced by confining the member cross-section with FRP wrapping. In members with circular cross-section (diameter D), the thickness of FRP sheets needed for confining the section is given by:

$$t_f = \frac{D(f_l - \sigma_{sw})}{0.002\, E_f} \tag{13}$$

where

σ_{sw} Smaller of tensile stress in stirrups corresponding to a strain of 0.001 and the mortar injection pressure between the FRP reinforcement and the RC column, if present,

$f_l =$ Confinement pressure at the lap splice location with a length L_s. $\frac{A_s f_{yd}}{\left[\frac{u_e}{2n}+2(d_b+c)\right] L_s}$, and.

wherein f_{yd} is the design yield strength of longitudinal steel reinforcement, u_e the perimeter of the cross-section within the polygon circumscribing the longitudinal bars having average diameter d_b, n the number of bars spliced along u_e, and c the cover to concrete.

In addition to the above brittle failure modes, the buckling of vertical steel reinforcement bars in RC columns can be delayed by confining the member cross-section with FRP of thickness t_f given by:

$$t_f = \frac{0.45\, n\, f_{yd}^2\, d}{4\, E_{ds}\, E_f} \approx \frac{10\, n\, d}{E_f} \tag{14}$$

where

- n Total number of existing steel longitudinal bars subjected to buckling.
- f_{yd} Design yield strength of longitudinal steel reinforcement.
- d Size of cross-section parallel to the bending plane.
- E_f Modulus of elasticity of FRP reinforcement in the direction of existing steel vertical bars.
- E_{ds} Reduced modulus of steel reinforcing bar $= \frac{4\, E_s\, E_i}{\left(\sqrt{E_s}+\sqrt{E_i}\right)^2}$,

in which E_s is are initial modulus of elasticity and E_i the tangent modulus of vertical steel bars after yielding.

4.3 Concrete Confinement Models

In this section, uniaxial models are presented of concrete confined with fiber-reinforced polymers, steel jackets and conventional transverse reinforcement. Especially, equations are presented to estimate the ultimate strength and strain of concrete confined with FRP, which have been validated by experimental studies. Two sets of models are available, namely exact expressions for ultimate strength and ultimate strain, and expressions based on regression analysis of test results.

(a) **Exact Expressions**

The ultimate confinement pressure f_{lu} induced by FRP is given by [10]:

$$f_{lu} = \frac{1}{2}\, \rho_j\, f_{ju} = \frac{2\, t_j\, f_{ju}}{d_j}, \tag{15}$$

where

- ρ_j Volumetric jacket ratio $= 4t_j / d_j$.
- t_j Thickness of FRP jacket,
- d_j Diameter of transverse reinforcement bars, and.
- f_{ju} Ultimate tensile strength of FRP layer.

The confined stress–strain curve of RC column confined by transverse steel is given by [16]:

$$1 + 7.94 \frac{f_{lu}}{f_{co'}} \tag{16}$$

$$\varepsilon_{cc} = \varepsilon_{co}\left[1 + 5\left(\frac{f_{cc'}}{f_{co'}} - 1\right)\right]. \tag{17}$$

The secant modulus of elasticity $E_{sec,u}$ at ultimate strain of ε_{cu} is given by:

$$E_{sec,u} = \frac{E_c}{1 + 2\beta\,\varepsilon_{ju}} = \frac{E_c}{1 + 2\beta\, f_{ju}/E_j} \tag{18}$$

where

$$\beta = \frac{E_c/E_{seco} - 1}{2\varepsilon_{lo}},$$

wherein
$E_{seco} = \frac{f'_{co}}{\varepsilon_{co}}$, and

$$\varepsilon_{lo} = -\nu\,\varepsilon_{co} - \left(\frac{1}{2} - \nu\right)\alpha\,\varepsilon_{co}\left(\frac{\varepsilon_{\lim} - \varepsilon_{co}}{\varepsilon_{\lim} - \alpha\,\varepsilon_{co}}\right)^2.$$

Here, ε_{lo} is concrete lateral strain at unconfined compressive strain ε_{co} corresponding to the peak unconfined compressive stress f'_{co}. β is dependent only on unconfined concrete properties f'_{co} and $\varepsilon_{co}(= 0.002)$, initial elastic modulus E_c, Poisson's ratio ν and α (0.9–1.0). ε_{lim} is the limiting axial strain beyond which micro-cracking starts, taken as 0.001. Considering $\alpha = 1.0$ and $\varepsilon_{lo} = -0.05\,\varepsilon_{co}$:

$$\beta = \frac{1}{\varepsilon_{co}} - \frac{E_c}{f'_{co}} = \frac{E_c}{|f'_{co}|} - \frac{1}{|\varepsilon_{co}|} \tag{19}$$

β can be expressed as a function of only the unconfined concrete strength f'_{co} assuming $\varepsilon_{co} = 0.002$ and $E_c = 5700\sqrt{|f'_{co}|}\ MPa$, as:

$$\beta = \frac{5700}{\sqrt{|f'_{co}|}} - 500, \tag{20}$$

where f'_{co} is in MPa. Equation (20) is reasonably accurate, if E_c is close to the actual elastic modulus.

The ultimate compressive strain ε_{cu} and stress f'_{cu} can be estimated as:

$$\varepsilon_{cu} = \varepsilon_{cc}\left(\frac{E_{sec}(E_c - E_{sec,u})}{E_{sec,u}(E_c - E_{sec})}\right)^{1-E_{sec}/E_c}, \text{ and} \tag{21}$$

$$f'_{cu} = E_{sec,u}\,\varepsilon_{cu}, \tag{22}$$

where

$$E_{sec} = \frac{f'_{cc}}{\varepsilon_{cc}},$$

wherein f'_{cc} and ε_{cc} are confined compressive strength and strain of concrete.

(b) **Regression Expressions**

The approximate ultimate compressive strength and ultimate compressive strain of confined concrete can be estimated as:

$$f'_{cu} = f'_{co}\left(0.2 + 3\sqrt{\overline{f_{lu}}}\right), \text{ and} \tag{23}$$

$$\varepsilon_{cu} = \varepsilon_{co}\left(2 + 1.25\overline{E_c}\ \varepsilon_{ju}\sqrt{\overline{f_{lu}}}\right), \tag{24}$$

where

$\overline{f_{lu}} = \frac{f_{lu}}{f'_{co}}$,

$\overline{E_c} = \frac{E_c}{f'_{co}}$, and.

$\varepsilon_{ju} = \frac{f_{ju}}{E_j}$, the ultimate strain in confining jacket.

When there is no confinement, i.e., $\overline{f_{lu}} = 0$, the expressions reduce to those of unconfined concrete, with ultimate strain $\varepsilon_{cu} = 0.004$ and ultimate strength equal to 20% of the peak strength, which are the values adopted usually. Also, ultimate strain in the confining jacket should be estimated as $\varepsilon_{ju} = f_{ju}/E_j$ and not be taken from the manufacturers sheet.

4.4 *Design Example*

4.4.1 Problem Statement

A 3-storey residential RC frame building (height 9 m with storey height 3 m) has plan dimension 12 m × 10 m (Fig. 4a). The clear column height L is 2.7 m. The upper two storeys have masonry infills, while the ground storey is open (Fig. 4b). The each column is rectangular (350 mm × 250 mm) and reinforced with 6 longitudinal bars of 14 mm diameter and 6 mm diameter steel stirrups @150 mm centers, with clear cover of 20 mm (Fig. 5). Each beam is rectangular (200 mm × 400 mm) with 200 m thick slab as flange thickness (Fig. 5b) and reinforced with 6 mm diameter stirrups@250 mm centers and anchored with 90° hook ends.

The effective depth of column section about the major (X) axis is $d = 350 - 20 - 0.5 \times 14 = 323$ mm. The shear reinforcement is less and is anchored poorly.

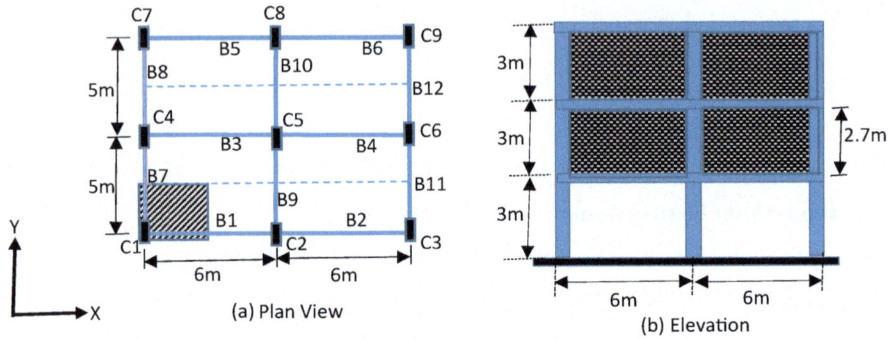

Fig. 4 Plan view and elevation view of a typical 3-storey building: **a** Plan, and **b** Elevation

The longitudinal bar is lap spliced at the bottom of the column; the lap length is $50d_b = 50 \times 14 = 700$ mm with end hooks as shown in Fig. 5a. Hence, the effective splice length is taken as $62.5d_b = 62.5 \times 14 = 875$ mm considering the positive influence of hooks under tension.

The mean in-situ cylindrical concrete compressive strength $f_{cm} = f_c' = 16$ MPa. The longitudinal reinforcement is ribbed and has average yield strength $f_{ys} = 500$ MPa. The transverse reinforcement is smooth steel and has average yield strength of $f_{yw} = 240$ MPa. Owing to presence of an open ground storey, the lateral deformation imposed by the earthquake shaking is localized in this storey, because all of the large stiffness of the upper storeys arising from the infill walls. Hence, it is critical to assess the ground storey columns (quantify their deficiencies) and suggest suitable retrofitting measures using FRP external wrapping and plates; the earthquake hazard considered is that corresponding to a peak ground acceleration of 0.24 g. Also, thickness of carbon fiber reinforced polymer (CFRP) plates $t_0 = 1.4$ mm, modulus of elasticity $E_f = 205$ GPa, ultimate tensile stress $f_{fu} = 3{,}200$ MPa, and ultimate tensile strain $\varepsilon_{fuk} = 0.017$, and partial safety factor $\gamma_f = 3$ for FRP anchored on brittle substrate.

4.4.2 Solution

The step by step evaluation of columns are given below to assess their structural weakness under earthquake loading events as the ground floor columns are more critical than beams under seismic loading situations.

Step 1: Check Slenderness of Columns

Consider dead of reinforced concrete slab ($\gamma = 25$ kN/m^3) of thickness 200 mm, additional dead load (of columns and beams) of 2 kN/m^2, masonry walls ($\gamma = 12$ kN/m^3) of thickness 0.12 m, and live load q of 3.5 kN/m^2. For the plan area of building of $12 \times 10 = 120$ m^2, the gravity load on each floor is given by:

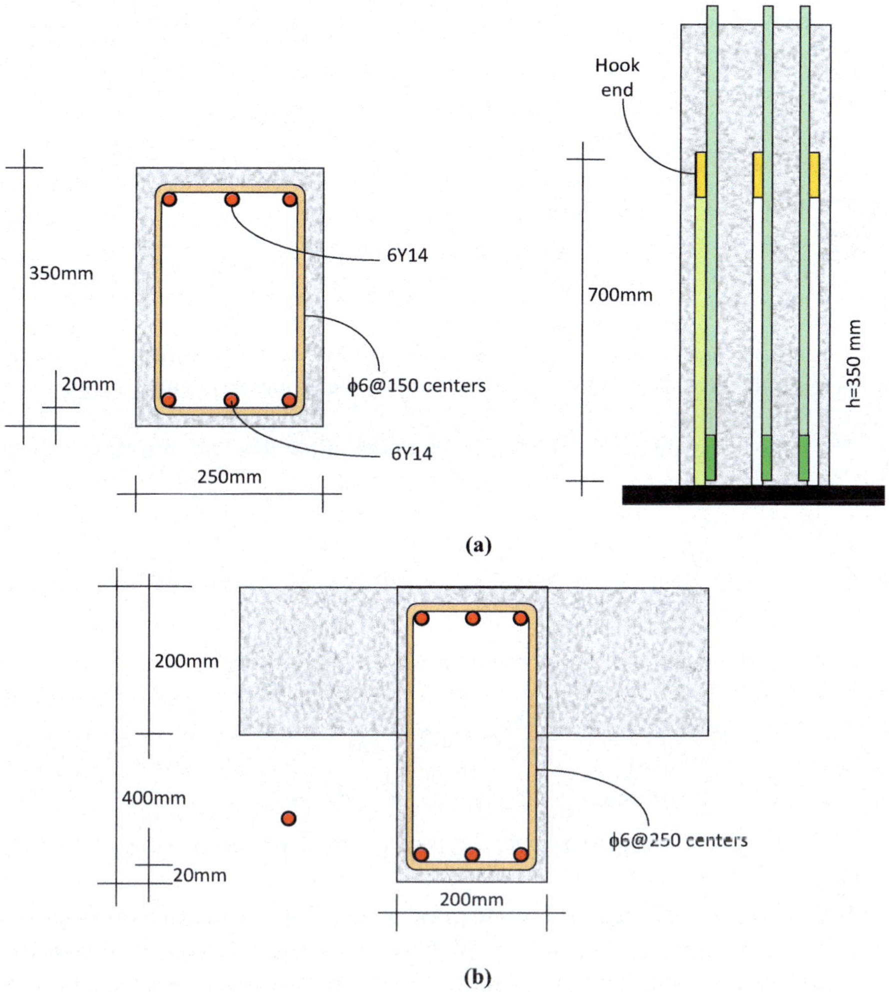

Fig. 5 Member cross-sections and reinforcement details: **a** column, and **b** beam

$$W_{Floor,RC} = (\gamma_f) \times \text{Floor Plan Area} = (25 \times 0.2 + 2) \times 120 = 840 \text{ kN}$$

$$W_{Floor,LiveLoad} = (0.25q) \times \text{Floor Plan Area} = (0.25 \times 3.5) \times 120 = 105 \text{ kN}$$

$$\begin{aligned} W_{Floor,Masonry} &= \gamma_m \times \text{Thickness} \times \text{Length} \times \text{Height} \\ &= 12 \times 0.12\text{m} \times 2 \times (8.95 + 11.25) \times 2.7 \cong 154 \text{ kN} \end{aligned}$$

Total Axial Load carried by all ground storey columns $= 3 \times 945 + 2 \times 154 = 3,143$ kN.

The central column has the largest tributary area, and hence carries the largest load, given by:

$$\mathrm{N_{ED}} = \frac{6 \times 5}{12 \times 10} \times 3,143 = 786\ \mathrm{kN}.$$

The axial load ratio of central column

$$\nu_{ED} = \frac{N_{ED}}{b\,h\,f_{cm}} = \frac{786 \times 10^3}{250 \times 350 \times 16} = 0.56$$

This is high, and to reduce it to below 0.4 for better flexural response, there is need to increase the cross-sectional area of column by using the procedures suggested in ACI 318–19 [17].

The radius of gyration r_{yy} of column about the minor axis, i.e., y-y axis is given by:

$$r_{yy} = \sqrt{\frac{I_{yy}}{A_g}} = \sqrt{\frac{\frac{hb^3}{12}}{bh}} = \frac{b}{\sqrt{12}} = 0.29\mathrm{b} = 0.29 \times 250 = 72.5\,\mathrm{mm}$$

Hence, the slenderness ratio for bending about the minor axis is:

$$\begin{aligned}\lambda &= \frac{L}{r_{yy}} = \frac{2700}{72.5} = 37.2 > \lambda_{\mathrm{lim}} \\ &= \max\{25;\ 15/15\sqrt{\nu_{ED}} - \sqrt{\nu_{ED}}\} \\ &= \max\left\{25;\ 15/15\sqrt{0.56} - \sqrt{0.56}\right\} = 20.0\end{aligned}$$

Hence, the slenderness ratio needs to be increased by increasing cross-section dimensions by concrete jacketing with high quality concrete. Let the increased cross section be b = 360 mm and h = 460 mm. Then, the revised axial load ratio of central column is:

$$\nu_{\mathrm{ED,new}} = \frac{\mathrm{N_{ED}}}{\mathrm{b\,h\,f_{cm}}} = \frac{786 \times 10^3}{360 \times 460 \times 16} = 0.30$$

and the revised radius of gyration

$$\mathrm{r_{yy,new}} = 0.29b = 0.29 \times 360 = 104.4\,\mathrm{mm}$$

$$\begin{aligned}\lambda_{new} &= \frac{L}{r_{yy,new}} = \frac{2700}{104.4} = 25.9 < \lambda_{\mathrm{lim}} \\ &= \max\{25;\ 15/15\sqrt{\nu_{ED}} - \sqrt{\nu_{ED}}\}\end{aligned}$$

$$= \max\left\{25;\ 15/15\sqrt{0.30} - \sqrt{0.30}\right\} = 27.4.$$

Thus, the increasing the dimensions of the central column satisfies the maximum slenderness ratio requirement of the column to ensure dominant flexural response.

Step 2: Determine Strength

By evaluating the strength and deformation capacity of building in y-y direction, i.e., bending about the strong x-x axis, the corner columns are expected to yield first due to lower axial load while the central column is susceptible to crushing owing to high axial load. For a RC section with low reinforcements, the curvature ϕ_y at onset of yielding of tension reinforcement is given by:

$$\varphi_y = \frac{2\,\varepsilon_{sy}}{h} = \frac{2 \times 0.0025}{350} = 1.43 \times 10^{-5} = 0.0143/\text{m}.$$

The moment at yielding can be approximated [18] as:

$$M_y = A_{sl,1}\ f_{ym}\ jd + N_{ED}(0.5h - 0.4 \times 0.25d),$$

where

jd = 0.85d is lever arm between tensile force of bottom steel reinforcement and concrete compressive force and $A_{sl,1} = 3 \times \pi \times 14^2/4 = 462\,\text{mm}^2$ (i.e., cross-sectional area of tensile reinforcement).

Yield moment of corner columns

$$\begin{aligned} M_y &= 462 \times 500 \times 0.85 \times 323 + 200.4 \times 10^3 \times (0.5 \times 350 - 0.4 \times 0.25 \times 323) \\ &= 92.02 \times 10^6\text{Nmm} = 92.02\ \text{kNm} \end{aligned}$$

Yield moment of peripheral columns

$$\begin{aligned} M_y &= 462 \times 500 \times 0.85 \times 323 + 400.8 \times 10^3 \times (0.5 \times 350 - 0.4 \times 0.25 \times 323) \\ &= 120.6 \times 10^6\ \text{Nmm} = 120.6\ \text{kNm} \end{aligned}$$

Using Fig. 6 [18, 19], the ultimate curvature of the corner column with longitudinal reinforcement ratio ρ_l is estimated as a function of normalized depth of compression zone $\xi = x/d$ for different axial load ratios υ_{ED} for symmetrically reinforced cross-section. For $\upsilon_{ED} = 0.14$ and $\rho_l = \frac{2\times 462}{250\times 350} \times 100 = 1.06\%$, $\xi = 0.24$ (Fig. 6). Taking the concrete crushing strain $\varepsilon_{cu} = 0.0035$, the ultimate curvature φ_u is:

$$\varphi_u = \frac{\varepsilon_{cu}}{\xi\,d} = \frac{0.0035}{0.9 \times 0.24 \times 323} = 5.01 \times 10^{-5}/\text{mm} = 0.05/\text{m},$$

including a reduction factor of 0.9 on ξ. Thus, the secant effective flexural rigidity to yield:

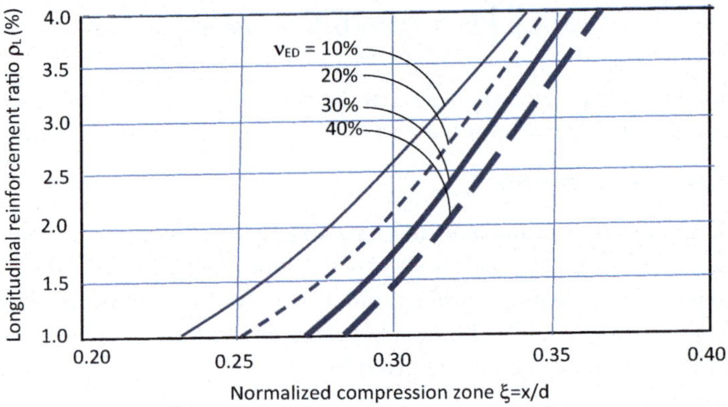

Fig. 6 Longitudinal reinforcement ratio *versus* normalized compression zone depth for different axial load ratios

$$EI = \frac{M_y}{\varphi_y} = \begin{cases} \frac{92.02}{0.0143} = 6,435\,\text{kNm}^2 \text{ corner columns} \\ \frac{120.6}{0.0143} = 8,434\,\text{kNm}^2 \text{ peripheral columns} \end{cases},$$

For modulus of elasticity of concrete of $E = 5000\sqrt{f_{ck}} = 5000\sqrt{16/0.80} \times 10^{-3} = 22.4$ GPa, the moment of inertia of uncracked section

$$I = \frac{bh^3}{12} = \frac{250 \times 350^3}{12} = 8.93 \times 10^8\,\text{mm}^4,$$

and the elastic flexural stiffness

$$EI_{el} = 22.4 \times 8.93 \times 10^8 \times 10^{-6} = 20,003\,\text{kN m}^2.$$

As in pushover analysis, cross-sections are assumed to be cracked. Hence, 50% of the elastic value is used as flexural stiffness at cracking, i.e.,

$$EI_{cr} = 0.50EI_{el} = 0.5 \times 20003 \approx 10,002\,\text{kN m}^2$$

Thus, the secant effective stiffness of columns under pure lateral translation is:

$$K = \frac{12EI}{L^3} = \begin{cases} \frac{12\times6435}{2.7^3} = 3,923\,\text{kNm}^2 \text{ corner columns} \\ \frac{12\times8434}{2.7^3} = 5,142\,\text{kNm}^2 \text{ peripheral columns} \end{cases},$$

The translational flexural stiffness under pure lateral translation are:

$$K_{el} = \frac{12EI_{el}}{L^3} = \frac{12 \times 20003}{2.7^3} = 12,195\ \text{kN/m},$$

$$K_{cr} = \frac{12EI_{cr}}{L^3} = \frac{12 \times 10002}{2.7^3} = 6,097\ \text{kN/m},$$

There is significant difference in K estimated by different approaches. K has high impact on natural period of the structure, and thereby the maximum displacement during earthquake shaking. Here, the secant to yield translational stiffness K of 3,923 kN/m is used.

The chord rotation θ_y at yield as per literature is [18] is:

$$\theta_y = \frac{\varphi_y H}{6} = \frac{0.0143 \times 2.7}{6} \times 100 = 0.64\%,$$

With reference to Eq. (10), we have

$$L_v = 0.5L = 0.5 \times 2.7 = 1.35\,\text{m} = 1350\,\text{mm}$$

$\beta_{flex} = \frac{1}{3}$; $\varphi_y = 0.0143/\text{m}$; $\beta_{shear} = 0.0013\left(1 + 1.5\frac{h}{L_v}\right) = 0.0013 \times \left(1 + 1.5 \times \frac{350}{1350}\right) = 1.81 \times 10^{-3}$

$$\beta_{slip} = 0.13\,\varphi_y = 0.13 \times 0.0143/\text{m} = 0.13 \times 0.0143 \times 10^{-3} = 1.859 \times 10^{-6}/\text{mm}$$

$$\begin{aligned}\theta_y &= \beta_{flex}\varphi_y L_v + \beta_{shear} + \beta_{slip}\frac{d_{bL} f_y}{\sqrt{f_c'}} \\ &= \frac{1}{3} \times 0.0143 \times 1.35 + 1.81 \times 10^{-3} + 1.859 \times 10^{-6} \times \frac{14 \times 500}{\sqrt{16}} \\ &- 0.0115 = 1.15\%\end{aligned}$$

Thus, Eq. (10) gives the yield rotation almost twice of the approximate value. Again, the conservative value of 0.64% is used.

The ultimate chord rotation θ_u using Eq. (11) is: *shear*.

Using, $\gamma_u = \frac{1}{\gamma_{el}} = \frac{1}{1.5} = 0.67$, $\theta_y = 0.64\%$, $\varphi_y = 0.0143/\text{m}$ and $\varphi_u = 0.05/\text{m}$, the ultimate chord rotation θ_u as:

$$\begin{aligned}\theta_u &= \gamma_u\left[\theta_y + (\varphi_u - \varphi_y)L_{pl}\left(1 - \frac{0.5L_{pl}}{L_v}\right)\right] \\ &= 0.67 \times \left[0.64 \times 10^{-2} + (0.05 - 0.0143) \times 10^{-3} \times 614.5 \right. \\ &\left. \times\left(1 - \frac{0.5 \times 614.5}{1350}\right)\right] \times 100 = 1.6\%\end{aligned}$$

Thus, the rotational and displacement ductility are given by:

$$\mu_\theta = \mu_\Delta = \frac{1.6}{0.64} = 2.5$$

Alternatively, the curvature and rotational ductility can be estimated as per literature [18] as:

$$\mu_\varphi = \begin{cases} 0.45\frac{\varepsilon_{cu}}{\varepsilon_{sy}\,\nu_{ED}} & \nu_{ED} \geq 0.2 \\ 0.45\frac{\varepsilon_{cu}}{\varepsilon_{sy}}\left(\frac{h}{\xi d}\right) & \nu_{ED} < 0.2 \end{cases}, \text{ and } \mu_\theta = \frac{\theta_u}{\theta_y} = \mu_\Delta = 0.5(\mu_\varphi + 1).$$

A reduction factor of 0.9 is applied for ξ. So for corner column, $\xi = 0.24$ and $\nu_{ED} = 0.14 < 0.2$. Hence,

$$\mu_\varphi = 0.45\frac{\varepsilon_{cu}}{\varepsilon_{sy}}\frac{h}{\xi d} = 0.45 \times \frac{0.0035}{0.0025} \times \frac{350}{0.9 \times 0.24 \times 323} = 3.16$$

To consider the contribution of reinforcement pull-out to the rotation capacity μ_θ should be multiplied by a factor of 1.5. Hence,

$$\mu_\theta = \mu_\Delta = 1.5[0.5(\mu_\varphi + 1)] = 1.5 \times [0.5 \times (3.16 + 1)] = 3.1$$

Thus,

$$\theta_u = 3.1\theta_y = 3.1 \times 0.64 = 2\%$$

The displacement ductility of 3.1 is more conservative than 2.5 calculated earlier for determining the required shear strength of column sections.

Step 3: Check for Brittle Mechanisms of Shear and Lap-Splice Failures

In Step 2, shear capacity of column was ignored when determining the flexural response of ground storey column. Thus, calculations of Step 2 are meaningful provided the columns are able to develop the lateral force V_{fl} corresponding to flexural capacity, i.e.,

$$V_{fl} = \frac{M_y}{L_v} = \begin{cases} \frac{92.02}{1.35} = 68.2\,\text{kN corner column} \\ \frac{120.6}{1.35} = 89.3\,\text{kN peripheral column} \end{cases}.$$

V_{fl} should be multiplied by a safety factor more than 1 to account for over-strength of columns. As per Eurocode [21], for medium ductility reinforced concrete buildings, the over-strength in beams and columns can be found by product of the flexural capacity with: (a) 1.15 the material safety factor for steel, (b) 1.2 due to design-action effects at the end sections of critical regions, and (c) 1 and 1.1 due to steel strain hardening for beams and columns, respectively. Hence, for columns, the over-strength factor is equal to $1.15 \times 1.2 \times 1.1 \approx 1.5$.

The shear capacity of columns is:

$$V_{RD,o} = \frac{1}{\gamma_{el}}\left\{(h - x)/(2L_v) \cdot \min(N, 0.55A_c f_c') + \left[1 - 0.05\min\left(5, \mu_\theta^{pl}\right)\right](V_{Rd,c} + V_{RD,s})\right\}$$

$$V_{RD,c} = 0.41\sqrt{f_c'}\,b\,x$$

$$V_{RD,s} = \rho_{sw}\,b_o\,h_o\,f_{y,st} = \frac{A_{sw}}{(s_h\,b_o)}b_o\,h_o\,f_{y,st}$$

For Corner Column,

$$V_{RD,c} = 0.41\sqrt{f_c'}\,b\,x = 0.41\times\sqrt{16}\times 250\times 0.9\times 0.24\times 323\times 10^{-3} = 28.6\,\text{kN}$$

$$\begin{aligned} V_{RD,s} &= \rho_{sw}b_o h_o\,f_{y,st} = \frac{A_{sw}}{(s_h b_o)}b_o h_o\,f_{y,st} \\ &= 2\times\frac{\pi}{4}\times\frac{6^2}{(150\times 210)}\times 210\times 310\times 240\times 10^{-3} \\ &= 28.05\,\text{kN} \end{aligned}$$

Thus, taking a conservative value of displacement ductility factor of 3.1,

$$\mu_\theta^{pl} = \frac{(\theta_u-\theta_y)}{\theta_y} = \mu_\theta - 1 = 3.1 - 1 = 2.1.$$

Hence,

$$1 - 0.05\ \min\left(5,\ \mu_\theta^{pl}\right) = 1 - 0.05\times min(5, 2.1) = 0.90$$

$$\gamma_{el} = 1.15$$

Thus,

$$\begin{aligned} V_{RD,o} &= \frac{1}{\gamma_{el}}\{\{(h-x)/(2L_v)\cdot\min(N, 0.55A_c f_c') \\ &+\left[1-0.05\min\left(5,\mu_\theta^{pl}\right)\right]\left(V_{Rd,c}+V_{RD,s}\right)\} \\ &= \frac{1}{1.15}\times\{(350-0.9\times 0.24\times 323)/(2\times 1350) \\ &\times\min\left(200.4, 0.55\times 250\times 350\times 16\times 10^{-3}\right)+0.90\times(28.6+28.05)\} \\ &= 62.4\,\text{kN} < 1.5\,V_{fl} = 1.5\times 68.2 = 102.3\,\text{kN} \end{aligned}$$

For Peripheral Columns,

$\xi = 0.27$; $\mu_\theta^{pl} = \frac{(\theta_u-\theta_y)}{\theta_y} = \mu_\theta - 1 = 2.6 - 1 = 1.6$, and $1 - 0.05\ \min\left(5,\ \mu_\theta^{pl}\right) = 1 - 0.05\times\min(5, 1.6) = 0.92$.

$$V_{RD,c} = 0.41\sqrt{f_c'}\,b\,x = 0.41\times\sqrt{16}\times 250\times 0.9\times 0.27\times 323\times 10^{-3} = 32.2\,\text{kN}$$

$$V_{RD,s} = \rho_{sw} b_o h_o f_{y,st} = \frac{A_{sw}}{(s_h b_o)} b_o h_o f_{y,st}$$
$$= 2 \times \frac{\pi}{4} \times \frac{6^2}{(150 \times 210)} \times 210 \times 310 \times 240 \times 10^{-3} = 28.05\text{kN}$$
$$V_{RD,o} = \frac{1}{\gamma_{el}} \{(h-x)/(2L_v).\min(N, 0.55A_c f_c')$$
$$+\left[1 - 0.05\min\left(5, \mu_\theta^{pl}\right)\right](V_{Rd,c} + V_{RD,s})\}$$
$$= \frac{1}{1.15} \times \{(350 - 0.9 \times 0.27 \times 323)/(2 \times 1350)$$
$$\times \min\left(400.8, 0.55 \times 250 \times 350 \times 16 \times 10^{-3}\right) + 0.92 \times (32.2 + 28.05)\}$$
$$= 83.2\,\text{kN} < 1.5V_{fl} = 1.5 \times 89.3 = 134$$

Thus, both the corner columns and peripheral columns will fail in shear on yielding at about the same drift (i.e.,θ_y $V_{RD,o}/V_{fl}$). The drift is $0.64\% \times 62.4/68.2 = 0.59\%$ in corner columns and $0.64\% \times 83.2/89.3 = 0.60\%$ in peripheral columns.

The other brittle failure mechanism is splice failure before development of yield stress in the reinforcement bars. Thus we need to check whether available splice length l_0 is sufficient for reinforcement bars to develop their yield strength. The available bond strength is:

$$\tau_b = \frac{2\mu_{fr}}{\pi D_b}\left[2c\, f_{ctk} + \frac{0.33A_{st}\, f_{y,st}}{N_b\, s}\right].$$

Taking $\mu_{fr} = 1$, $c = 20$ mm, $D_b = 14$ mm, $A_{st} = 2 \times \pi \times 6^2/4 = 56.5\,\text{mm}^2$, $f_{yst} = 240$ MPa, $N_b = 3$ bar pairs, s = stirrup spacing of 150 mm centers, and $f_{ctk} = 0.33\, f_{cm}^{0.5} = 0.33 \times (16)^{0.5} = 1.3\,\text{MPa}$,

$$\tau_b = \frac{2\mu_{fr}}{\pi D_b}\left[2c\, f_{ctk} + \frac{0.33A_{st}\, f_{y,st}}{N_b s}\right]$$
$$= \frac{2 \times 1.0}{\pi \times 14} \times [2 \times 20 \times 1.3 + 0.33 \times (56.5 \times 240)/(3 \times 150)] = 2.82\text{MPa}$$

The limiting value of bond strength to ensure yielding of longitudinal bars before splice failure with hook length l_0 of 875 mm,

$$\tau_{b,lim} = \frac{\gamma_{el}\, D_b\, f_{sy}}{4\, l_o} = \frac{1.15 \times 14 \times 500}{4 \times 875} = 2.3\,\text{MPa}.$$

As available bond strength $\tau_b = 2.82\,\text{MPa}$ is more than $\tau_{b,lim} = 2.3\,\text{MPa}$, the lap splice is sufficient to ensure yielding of longitudinal reinforcement. But, immediately after yielding of longitudinal reinforcements corresponding to displacement ductility $\mu_\Delta = 1$, the cover may crack which implies that the term $2c\, f_{ctk} \approx 0$, and bond strength becomes equal to 0.45 MPa much less than $\tau_{b,lim} = 2.3\,\text{MPa}$. Hence, splice failure is anticipated immediately after the yielding of longitudinal reinforcements.

Thus, from the above assessment, the ground columns may fail in shear before any yielding of longitudinal reinforcements, at a drift of about 0.6% corresponding to lateral displacement at the top of the ground storey, i.e., $\Delta = 0.6\% \times 2.7\,\text{m} = 0.006 \times 2700 = 16.2\,\text{mm}$. In the same way, in straight splices without hooks, the lap splice failure occurs in both the peripheral and the corner columns before yielding.

Step 4: Check for Global Strengthening Requirements

The available effective stiffness K_{eff} of existing building is examined to check whether it is sufficient to resist the earthquake demand or if there is need to reduce the drift demand. To reduce the drift demand, the retrofit solution should increase the global lateral stiffness K_{eff}. Local strengthening (like FRP wrapping) may be employed to preventing brittle modes of failure in the columns. By increasing K_{eff}, the demand may be reduced in two different ways, namely: (i) reducing the predominant period, and (ii) reducing deformation demand on individual members.

Here, total building mass M is 320.6 tonnes. Assuming that the building has a soft storey, wherein all lateral translation is concentrated in the ground storey, i.e.,

$$K_{eff} = \sum_{i}^{n} K_i \,\Delta\Phi_i^2 = K_1 \times 1^2 = K_1$$

Thus, the translational stiffness against earthquake lateral sway of the ground storey about major axis, i.e., x-x axis (in y-y direction) is given by:

$$\begin{aligned} K_{eff} &= 4(corner\ columns) \times 3923 + 4(Peripheral\ columns) \times 5142 \\ &\quad + (central\ column) \times 3923 = 40183\,\text{kN/m} \end{aligned}$$

Effective Natural Period $T_{eff} = 2\pi\sqrt{\frac{M}{K}} = 2\pi\sqrt{\frac{320.6}{40183}} = 0.56\ \text{s}$

Empirical reference natural period, $T_{ref} = 0.075 H_{tot}^{0.75} = 0.075 \times 9^{0.75} = 0.39\ \text{s}$

$$\frac{T_{eff}}{T_{ref}} = \frac{0.56}{0.39} = 1.44$$

T_{eff} is more than T_{ref} by more than 25%, and hence K_{eff} should be increased to reduce T_{eff} to an acceptable value of 0.4 s – 0.5 s for 3-storey building.

For the earthquake assessment of the building, the peak ground acceleration PGA of 0.24 g is considered as per problem statement. For $T_C \le T \le T_D$ [21] with $S = 1.2$ for Soil Class B, $T_C = 0.50s$, $T_D = 2s$, $a_g = 0.24g$, $\eta = 1$ $(for\ \xi = 5\%)$ and $\beta_o = 2.5$, the elastic spectral displacement demand for T_{eff} of 0.56 s is:

$$S_d(T) = a_g\, S\, \eta\, \beta_o (T_C\, T)/40 = 0.24g \times 1.2 \times 1 \times 2.5 \times (0.5 \times 0.56)/40 = 0.049\,\text{m}.$$

The lateral drift Δ in the ground storey

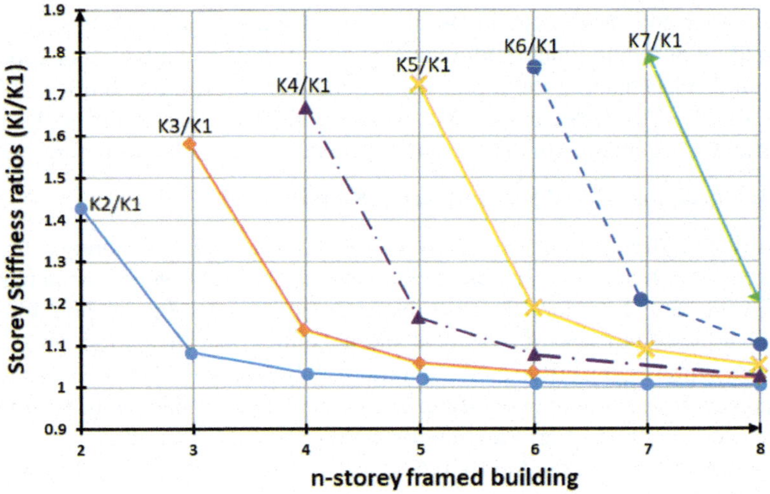

Fig. 7 Storey Stiffness ratios (K_i/K_1) for n-storey frame building [18]

$$\Delta = \frac{S_d}{H} = \frac{0.049}{2.7} \times 100 = 1.8\%.$$

Corresponding drift or displacement ductility demand, $\mu_\Delta = \frac{1.8}{0.64} = 2.8$. Thus, displacement ductility demand of 2.8 for PGA of 0.24 g is more than drift supplied by the columns (i.e., 2.5) estimated earlier assuming there is no brittle failure due to shear failure or lap splice failure. The global intervention measures are required to reduce the drift demand along with improvement of the deflection shape of stilt type of building structures (i.e., with open ground storey] before considering any application of FRP jacketing.

Figure 7 shows the storey stiffness $K_2/K_1 = 1.1$ and $K_3/K_1 = 1.6$ of a 3-storey frame building.

Let the target natural period is 0.45 s (value taken between reference period and initial effective period). Then from Fig. 8 and using the first storey mass m_1 of 96.6 tons,

$$\frac{K_1}{m_1} = 0.950 \times 10^3 = 950$$
$$\Rightarrow K_1 = 950 \times 96.6 = 91770\,\text{kN/m},$$

$$\frac{K_2}{K_1} = 1.1$$
$$\Rightarrow K_2 = 1.1\,K_1 = 1.1 \times 91770 = 100947\,\text{kN/m}$$

$$\frac{K_3}{K_1} = 1.6$$

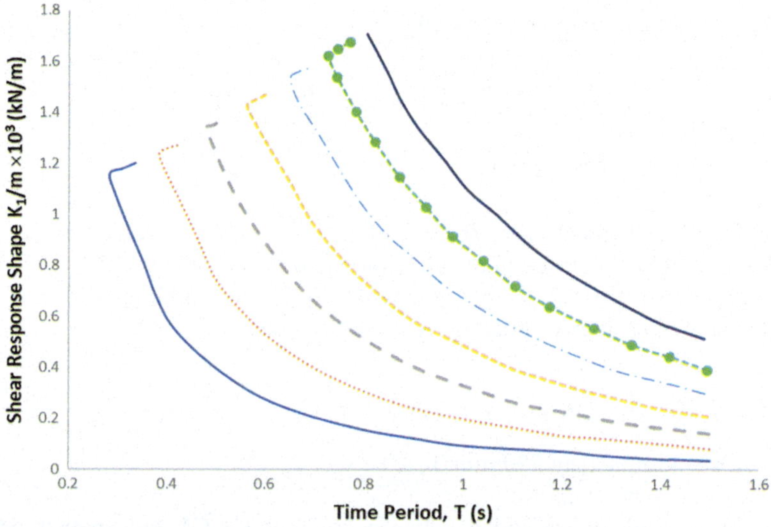

Fig. 8 Reduction of stiffness to mass ratio K_1/m of first storey with Natural Period [18]

$$\Rightarrow K_3 = 1.6\,K_1 = 1.6 \times 91770 = 146832\,\text{kN/m}$$

$K_{eff} = 40183\,\text{kN/m}$ as calculated earlier.

Thus, the target stiffness $K_1 = 91770\,\text{kN/m}$ of the ground storey is much more than available $K_{eff} = 40183\,\text{kN/m}$. The increase in ground storey stiffness is achieved by adding metallic cross-bracings in the considered direction of earthquake shaking, i.e., y-y direction.

The increased stiffness of the central column due increased dimensions of 360 mm × 460 mm can be estimated assuming the column to be pinned at its base. Hence,

$$K_{c,new} = \frac{3\,EI}{H^3} = \frac{3 \times 29 \times \left(\frac{1}{12} \times 360 \times 460^3\right) \times 10^{-6}}{2.7^3} = 12907\,\text{kN/m}$$

The initial value of K_c is 3,923 kN/m, and stiffnesses are 3923 kN/m and 5142 kN/m of corner and peripheral columns, respectively.

Required cross-bracing stiffness,

$$\begin{aligned}\sum K_x &= K_1 - K_{c,new} - 4(corner\,columns) \times 3923\ \text{kN/m} \\ &\quad - 4(peripheral\,columns) \times 5142\ \text{kN/m} \\ &= 91770 - 12907 - 4 \times 3923 - 4 \times 5142 = 42,603\ \text{kN/m}\end{aligned}$$

Length of each brace is given by

$$L = 5 - 1.5 \times 0.350 = 4.475\,\text{m}$$

Angle of inclination of cross bracings

$$\varphi = tan^{-1}\left(\frac{h_{st,cl}}{L}\right) = tan^{-1}\left(\frac{2.7}{4.475}\right) = 0.6$$
$$\Rightarrow \varphi = 31.1^{\circ}$$

Length of the diagonal brace $D_x = \sqrt{(4475)^2 + (2700)^2} = 5226\,\text{mm}$.
Taking modulus of elasticity of steel bracing as $E_s = 150000\,\text{MPa}$.

Providing two cross bracings symmetrically in the building, required cross-sectional area of each cross bracing, A_{br} is estimated as:

$$\sum K_X = 2\,E_s\,A_{br}\,/D_x\,\cos^2\varphi = 42603 \Rightarrow A_{br} = \frac{42603 \times 5226 \times \cos^2 31.1^o}{2 \times 150000}$$
$$= 544\ \text{mm}^2(\text{say } 550\ \text{mm}^2)$$

One brace could be provided between columns C1-C4 and another between columns C6-C9.

Considering the stiffness of the exterior infill walls in the upper storeys, total cross-sectional area A_w of side-walls in y-y direction for each of the upper two storeys, is:

$$A_w = 2 \times (10 - 3 \times 0.350) \times 0.12 = 2.15\,\text{m}^2$$

Thus,

$$\rho_{mw} = \frac{A_w}{A_{floor}} = \frac{2.15}{12 \times 10} = 0.018$$

For $h_i = 2.7$ m, $f_{bc} = 5$ MPa, $f_{mc} = 5$ MPa, $\gamma_m = 1.5$, $\mu_y^{mw} = 2.5$ and $\theta_y^{mw} = 0.15\%$,

$$\text{Design}\quad f_{bc} = \frac{f_{bc}}{\gamma_m} = \frac{5}{1.5} = 3.33\,\text{MPa}$$

$$\text{Design}\quad f_{mc} = \frac{f_{mc}}{\gamma_m} = \frac{5}{1.5} = 3.33\,\text{MPa}$$

Masonry wall stiffness K_{mw} in each floor is:

$$K_{mw} = \rho_{mw,i}\frac{A_{floor}}{h_i}\left(\frac{0.1\,f_{bc}^{0.7}\,f_{mc}^{0.3}}{\mu_y^{mw}\,\theta_y^{mw}}\right) \quad (25)$$

Thus,

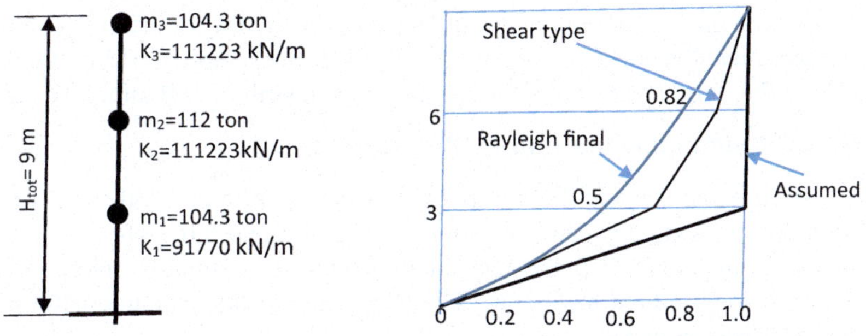

(a) Stiffness distribution (b) Lateral displacement profile after Rayleigh iteration

Fig. 9 Stiffness distribution and lateral displacement profile of the3-storey building

$$K_{mw} = 0.018 \times \frac{120}{2.7} \times 10^3 \times \left(0.1 \times 3.33^{0.7} \times 3.33^{0.3}\right)/(2.5 \times 0.0015)$$
$$= 71040\text{kN/m}$$

The available effective stiffness of the 2nd and 3rd storeys is:
$K_2 = K_3 = K_{eff} + K_{mw} = 40183 + 71040 = 111223$ kN/m (Fig. 9a)

Since the available magnitudes of K_2 and K_3 are higher than or close to the required value for chosen shear type response of the building (i.e., $K_2 = 100{,}947$ kN/m and $K_3 = 146832$ kN/m), there is no need of any global retrofit measures in these floors. In Fig. 9a, floor mass has been increased by adding ¼ of the masonry weight of 30.8 ton at top and bottom while the half the masonry weight has been added at 2nd floor.

Taking the target natural period T_{trg} as 0.45 s ($<T_c = 0.5$ s), $a_g = 0.24g$, S = 1.2 (Soil Class B), $\eta = 1$ ($\xi = 5\%$) and $\beta_o = 2.5$, the elastic spectral displacement demand is estimated as [21].

$$S_d(T) = a_g\, S\, \eta\, \beta_o\, T^2/40 = 0.24 \times 9.81 \times 1.2 \times 1 \times 2.5 \times 0.45^2/40 = 0.036\,\text{m}.$$

Considering that the displacement will be increased by about 20% when transferring from the spectrum (based on single degree of freedom system) to the actual structure (multi-degree of freedom system), the relative drift demand at the ground storey for $\Delta\varphi = 0.5$ (Fig. 9b) is:

$$\theta_{dem} = 1.2\Delta \frac{S_d}{H} = 1.2 \times 0.5 \times \frac{0.036}{2.7} \times 100 = 0.8\%.$$

The corresponding displacement ductility

$$\mu_\Delta = \frac{0.80\%}{0.64\%} = 1.25$$

After placing the bracing in the ground storey, the ductility demand is reduced to a value lower than that corresponding to the plastic hinge length of 615 mm, i.e., 2.5, and close to that corresponding to the plastic hinge length of 160 mm, i.e., 1.2.

Step 5: Local Strengthening with FRP Wrapping

The local strengthening using FRP wrapping is needed to prevent brittle failure modes in columns when flexural capacity of columns is developed fully. The FRP confinement helps in enhancing displacement ductility and shear strength, and preventing failure due to longitudinal steel bars. The properties of the available carbon fiber fabric system used in jacketing are:

Thickness $t_o = 0.12$ mm.
Modulus of elasticity $E_f = 165,000$ MPa.
Ultimate tensile stress $f_{fu} = 2970$ MPa.
Ultimate strain $\varepsilon_{fuk} = 0.018$.
Slip strength $s_o = 0.5$ MPa.
Bond strength $\tau_a = 5$ MPa.

As per approach given in literature [2], the design tensile strain of FRP layer is:

$$\varepsilon_{fu,h} = \eta_1\,\eta_2\,\eta_3\,\varepsilon_{fuk}/\gamma_f$$

where η_1 accounts for the radius of chamfer, $R = c{+}0.5D_b = 20{+}0.5{\times}14 = 27$ mm and straight part of the largest cross-section side $h' = 350 - 2 \times 27 = 296$ mm, and is given by:

$$\eta_1 = 0.25 + 2(2R + D_b)/h' = 0.25 + 2 \times (2 \times 27 + 14)/296 = 0.71$$

(1) Factor $\eta_2 = 1$ which accounts for the sufficiency of the wrap development length, i.e., straight parts of the cross section, $h' = 350 - 2 \times 27 = 296$ mm and $b' = 250 - 2 \times 27 = 196$ mm are sufficient to accommodate the minimum anchorage length of the external FRP layer, $l_b^{\min} = 0.5\pi\sqrt{E_f\,t_o\,s_o/\tau_a} = 0.5\pi \times \sqrt{165000 \times 0.12 \times 0.5/5} \approx 70$ mm. So external layer of the FRP jacket can be anchored over the column's shorter side.
(2) Factor $\eta_3 = 1$ for fully wrapped jacket and considers the redundancy of the jacket against debonding.

The design tensile strain $\varepsilon_{fu,h}$ is:

$$\varepsilon_{fu,h} = 0.71 \times 1 \times 1 \times 0.018/1.5 = 0.0085$$

(i) ***Confinement Required for Design Ductility Demand***

For the required displacement ductility μ_Δ of 1.25 (estimated in Step 4), the associated curvature ductility μ_φ is:

$$\mu_{\varphi} = 2\mu_{\theta} - 1 = 2 \times 1.25 - 1 = 1.5.$$

The maximum compressive strain demand $\varepsilon_{cu,c}$ is given by:

$$\varepsilon_{cu,c} = 2.2\mu_{\varphi}\varepsilon_{sy}\nu_{ED} = \begin{cases} 2.2 \times 1.5 \times 0.0025 \times 0.14 = 0.0012 < 0.0035 \text{ corner columns} \\ 2.2 \times 1.5 \times 0.0025 \times 0.28 = 0.0023 < 0.0035 \text{ periperal columns} \end{cases}$$

Thus, from the displacement ductility consideration, there is no need to confine concrete using closed FRP jackets to meet its strain demands in the compression zone.

Shear Strengthening

As mentioned earlier, the flexural response dominates when shear strength V_{RD} exceeds *1.5* V_{fl}.

Maximum design shear force $1.5\,V_{fl} = \max\left(1.5\,V_{fl,corner}, 1.5\,V_{fl,peripheral}\right) =$ 134 kN.

Minimum design shear strength available $V_{RD,o} = \min\left(V_{RD,o,corner}, V_{RD,o,peripheral}\right) = 62.4$ kN.

The shear force to be resisted by FRP closed-type jacketing is $\Delta V = 1.5V_{fl} - V_{RD,o} = 134 - 62.4 = 71.6$ kN.

Using the relation $\Delta V = V_{RD,f} = \left(\frac{2t_f}{b}\right) bh\, E_f\, \varepsilon_{fu,h} = 71.6$ kN,

$$t_f = \frac{71.6 \times 10^3 \times 250}{2 \times 250 \times 350 \times 165000 \times 0.0085} = 0.07\,\text{mm} < t_o = 0.12\,\text{mm}.$$

Thus, a single layer FRP jacket is sufficient for shear strengthening along the column's height.

(ii) **Strengthening of Splice Region**

As estimated in Step 2, for the lap splice with hook of 875 mm, the available bond strength $\tau_b = 2.82$ MPa is more than $\tau_{b,lim} = 2.3$ MPa, the lap splice is sufficient to ensure yielding of longitudinal reinforcement. But, immediately after yielding of longitudinal reinforcements, the cover may crack which implies that the term $2c\,f_{ctk} \approx 0$ for calculation of bond strength τ_b, which becomes equal to 0.45 MPa much less than $\tau_{b,lim} = 2.3$ MPa. Hence, splice failure is anticipated immediately after the yielding of longitudinal reinforcements.

Thus, the contribution of FRP jacketing in restoring the bond strength to the yield limit, $\tau_{b,lim} = 2.3$ MPa is given by:

$$\tau_b = \frac{2\mu_{fr}}{\pi D_b}\left[\frac{0.33\, A_{st}\, f_{y,st}}{N_b\, s} + \frac{2\, t_f\, E_f\, \varepsilon_{f,sl}}{N_b}\right].$$

Thus,

$$2.3 = \frac{2 \times 1}{\pi \times 14}\left[\frac{0.33 \times 56.5 \times 240}{3 \times 150} + \frac{2n \times 0.12 \times 165000 \times 0.0015}{3}\right]$$

$$\Rightarrow 19.8\text{n} = 40.636$$

Hence, 2 layers are needed to wrap the splice region of $l_o = 700$ mm at the bottom of column.

(iii) **Prevention of Buckling of Longitudinal Columns**

The required thickness t_f of FRP jacket for preventing the buckling of existing vertical longitudinal reinforcements, is:

$$t_f = \frac{0.45\, n\, f_{yd}^2\, d}{4\, E_{ds}\, E_f} \approx \frac{10\, n\, d}{E_f} = \frac{10 \times 6 \times 350}{165000} = 0.127\,\text{mm} \Rightarrow 1\, layer\ of\ FRP\ jacket$$

where

n Total number of existing steel longitudinal bars subjected to buckling = 6,
f_{yd} Design yield strength of longitudinal steel reinforcement = 500 MPa,
d Size of cross-section parallel to the bending plane = 350 mm, and.
E_f Modulus of elasticity of FRP reinforcement in the direction of steel vertical bars = 165,000 MPa.

But, considering the relation between strain ductility and spacing of stirrups given in literature [18], to eliminate buckling of bars in compression for required curvature ductility $\mu_\varphi = 1.5$, the compression strain demand $\varepsilon_{cu,c}$ calculated previously is less than 0.0035 for corner and peripheral columns. From Fig. 5 [18], for $\frac{s}{D_b} = \frac{150}{14} = 10.7 \approx 11$, the strain μ_{ec} at which the bars become unstable is:

$$\mu_{ec} \approx 2 = \frac{\varepsilon_{s,crit}}{\varepsilon_{s,y}}$$

$$\Rightarrow \varepsilon_{s,crit} = 2\varepsilon_{sy} = 2 \times 0.0025 = 0.005$$

When estimating the required jacket confinement, it must be ensured that:

$$\varepsilon_{cu,c} \geq \max\left(\varepsilon_{cu} = 0.0035,\ \varepsilon_{s,crit} = 0.005\right) = 0.005$$

where

$$\varepsilon_{cu,c} = \varepsilon_{cu} + 0.075\left[\varsigma\left(\alpha_f\, \rho_{fv} E_f\, \varepsilon_{fu,h} + \alpha_w \rho_{sv} f_{y,st}\right)/f_c - 0.1\right] \geq \varepsilon_{cu},$$

$a_f \approx 1 - \left((b - 2R)^2 + (h - 2R)^2\right)/(3bh)$, and

$$\rho_{fv} = 2t_f(h + b)/hb.$$

Thus,

$$a_f \approx 1 - \left((b-2R)^2 + (h-2R)^2\right)/(3bh)$$
$$= 1 - \left((250 - 2\times 27)^2 + (350 - 2\times 27)^2\right)/(3\times 250\times 350) = 0.52$$

Volumetric ratio of transverse reinforcement $\rho_{sv} = \frac{(2\times 210 + 2\times 310)\times \frac{\pi}{4}\times 6^2}{(150\times 210\times 310)} = 0.0030.$
For

$$\varepsilon_{cu} = 0.0035,\ \varepsilon_{cu,c} = 0.005,$$
$$\alpha_w = 0.15(confinement\,effectiveness\,factor\,defined\,for\,stirrups),$$
$$\xi = 1(for\ \varepsilon_{cu,c} \le 0.01)$$

Thus,

$$\varepsilon_{cu,c} = \varepsilon_{cu} + 0.075\left[\varsigma\left(\alpha_f\rho_{fv}E_f\varepsilon_{fu,h} + \alpha_w\rho_{sv}f_{y,st}\right)/f_c - 0.1\right] \Rightarrow 0.005$$
$$= 0.0035 + 0.075\times\left[1\times\left(0.52\times\rho_{fv}\times 165000\times 0.0085 + 0.15\times 0.0030\times 240\right)/16 - 0.1\right] \Rightarrow \rho_{fv}$$
$$= 0.0025 = \frac{2t_f(250+350)}{250\times 350} \Rightarrow t_f$$
$$= 0.18\text{mm} = nt_o = n\times 0.12 \Rightarrow n = 2FRP\,layers$$

Thus, provide 2 layers of FRP along the critical regions of all columns at top and bottom for $l_{pl} = 615$ mm, which is rounded to 600 mm. Thus, for eliminating premature buckling and splice failure, 2 layers of FRP jacketing (0.24 mm thickness with $\rho_{fv} = 0.0033$) provides a limited displacement ductility as:

$$\mu_\Delta = 1.3 + 12.4\left(0.5\alpha_f\,\rho_{fv}\,E_f\,\varepsilon_{fu,h}/f_c\right) - 0.1) \ge 1.3$$

$$\mu_\Delta = 1.3 + 12.4\times(0.5\times 0.52\times 0.0033\times 165000\times 0.0085/16 - 0.1) \ge 1.3$$
$$\Rightarrow \mu_\Delta = 0.99 \ge 1.3,$$
$$\therefore \mu_\Delta = 1.3 > required\ \mu_\Delta = 1.25$$

So, 2 layers of FRP jacketing are appropriate at top and bottom of columns and 1 layer along the whole height of columns (corners and peripheral columns) as shown in Fig. 10. A single CFRP layer wrapped along the height of each column alleviates the shear failure. One additional CFRP layer at the lower 700 mm of each column strengthen the splice region while the additional CFRP layer wrapped at the upper 600 mm of each column prevents the compressive bar buckling. The central column C5 does not require any FRP strengthening as its cross-section size is increased through reinforced concrete jacketing.

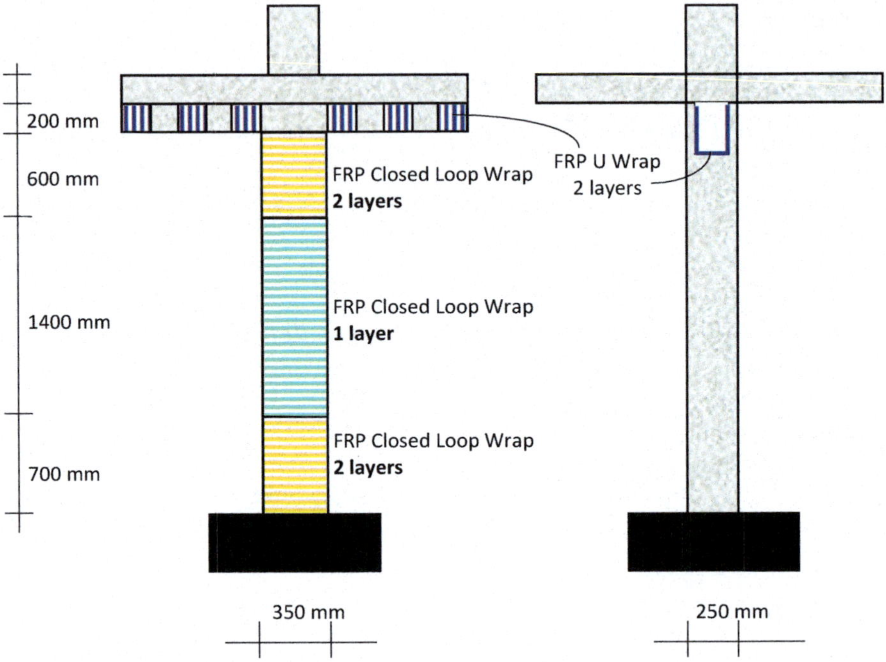

Fig. 10 Details of FRP Jacketing in ground floor columns and beam

5 Conclusions

In this chapter, an approach is presented for assessment and seismic retrofit design using FRP materials of deficient columns in RC buildings. Details of confinement models for FRP wrapped reinforced concrete columns are presented, which help to estimate the increase in the ductility of the column members; the FRP wrapping also reduces the likelihood of brittle failure modes (e.g., shear failure, lap splice failure and buckling of longitudinal reinforcements) before the flexural failure mode manifests. Basics of global and local strengthening are discussed.

FRP jacketing are recommended for local strengthening and the method of quantifying the retrofit intervention is demonstrated through a design example; it shows assessing and retrofitting 3-storey frame building towards seismic retrofitting of buildings using FRP materials.

References

1. ACI 440 (2012) Guide for the design and construction of exteranlly bonded FRP systems for strengthening concrete stuctures, No. ACI 440.2R-12, Farmington Hills, Michigan, USA
2. fib (2001) Technical report on the by EBR working party of fib TG 9.3. fib Bull 14. ISBN 2–88394–054–1
3. Pantazopoulou S, Tatsani S, Thermou G, Triantafillou T, Monti G, Bournas D, Guadagnini M (2016) Background to the European seismic design provisions for retrofitting RC elements using FRP material. Struct Concr 17(2):194–219
4. Sarker P, Begum M, Nasrin S (2011) Fiber reinforced polymers for structural retrofitting: a review. J Civil Eng 39(1):49–57. Institution of Engineers Bangladesh
5. Kam WY, Pampanin S (2011) Displacement based seismic retrofit design for non-ductile RC frame structures using local retrofit interventions at beam-column joints. In: Proceedings of ninth pacific conference on earthquake engineering building an earthquake-resilient society. Auckland, New Zealand
6. Prota A, Manfredi G, Monti G, Ludovica MD, Lignola GP (2023) Retrofitting of existing RC buildings with FRP. https://www.academia.edu/22931902/Retrofitting_of_Existing_RC_Buildings_With_FRP,10pp
7. Ortlepp R, Ortlepp S (2017) Textile reinforced concrete for strengthening of RC columns: a contribution to resource conservation through the preservation of structures. Constr Buid Mater 132:150–160
8. Di Ludovico M, Prota A, Manfredi G, Cosenza E (2008) Seismic design of an under-designed RC structure with FRP. Earthq Eng Struct Dyn 37:141–162
9. Di Ludovico M, Manfredi G, Mola E, Negro P, Prota A (2008) Seismic behavior of a full-scale RC structure retrofitted using GFRP laminates. J Struct Eng ASCE 134(5):810–821
10. Spoelstra MR, Monti G (1999) FRP-confined concrete model. J Compos Constr 3(3):ASCE, 143–150
11. Lignola GP, Prota A, Manfredi G, Cosenza E (2008) Effective strain in FRP jackets on circular RC columns. In: Procedings of 4th international conference on FRP composites in civil engineering (CICE 2008). Zurich, Switzerland, Paper 2.A.5
12. Lignola GP, Prota A, Manfredi G, Cosenza E (2009) Analysis of FRP confinement on prismatic RC columns. In: Procedings of 9th international symposium on fiber reinforced polymer reinforcement for concrete structures (FRPRCS9). Sydeney, Australia
13. EN 1998–3 (2003) Design of structures for earthquake resistance, Part 3: strengthening and repair of buildings, European Standard, EN 1998–3. Eurocode 8, Doc CEN/TC250/SC8/N343, Draft No.3
14. CNR-DT (2004) Guide for the design and construction of externally bonded FRP systems for strengthening, CNR-DT 200/400, Available on http://www.cnr.it/sitocnr/IlCNR/Attivita/NormazioneeCertificazione/NormazioneeCertificazionefile/IstruzioniCNR DT200 2004 eng.pdf
15. Park R, Paulay T (1975) Reinforced concrete structures. Wiley, New York, USA
16. Mander JB, Priestley MJN, Park R (1988) Theoretical stress-strain model for confined concrete. J Struct Eng ASCE 114(8):1804–1826
17. ACI 318 (2019) building code requirements for structural concrete and commentary, ACI 318–19, American Concrete Institute, Farmington Hills, MI, USA
18. Tatsani S, Thermou G (2022) An example-guide for rapid assessment and FRP strengthening of substandard RC buildings. Appl Sci 12(12950):1–20. https://doi.org/10.3390/app122412950
19. Thermou GE, Pantazopoulou SJ, Elnashai AS (2012) Global interventions for seismic upgrading of substandard RC buildings. J Struct Eng ASCE 138:387–401
20. Dai KY, Yu XH, Lu DG (2020) Phenomenological hysteretic model for corroded RC columns. Eng Struct 210:110315

21. EN 1998-1 (2004) Design of Structures for Earthquake Resistance-Part 1: General Rules, Seismic Actions and Rules for Buildings, EN 1998-1: Eurocode 8, European Committee for Standardization (CEN), Brussels, Belgium
22. GSARC (2017) Earhquake planning and protection organization. Greek Seismic Assessment and Retrofit Code, Official Government Gazette 2984 B, 30–8–2017. Available online: www.oasp.gr

Carbon Fibre Reinforced Polymer Composite Retrofitted Steel Profiles Using Automated Fibre Placement

Ebrahim Oromiehie, Feleb Matti, Fidelis Mashiri, and Gangadhara B. Prusty

Abstract Traditional methods for repairing impaired structures such as concreting, steel jackets, or timber splicing are impractical because of the inherent constraints associated with these materials. They would be susceptible to the same deterioration as the existing structure, leading to an ongoing cycle of repairs. Fibre-reinforced polymer (FRP) composite jackets offer a wide range of advantages including superior corrosion resistance, lightweight properties, and long-lasting durability. These characteristics make FRP composite jackets highly advantageous compared to conventional repair systems. Additionally, they can be effectively utilized for repairing various types of structures, including those made of timber, steel, and concrete. FRP composite jackets can be implemented through several techniques. However, in the experimental investigation presented in this chapter, automated fibre placement (AFP) was used to overwrap and reinforce two sets of thin-walled square hollow sections (SHS), columns and beams, with thermoplastic carbon fibre reinforced polymer (CFRP). The results obtained were then compared with the control samples without CFRP reinforcement. For the control columns, a good agreement was observed between the predicted and experimental ultimate compressive loads. The ultimate loads of CFRP reinforced columns exceeded the ultimate loads of the control columns. Inward and outward buckling was observed in each column. Debonding, tearing, and snapping of the CFRP plies was observed in column specimens with thermoplastic CFRP reinforcement. For the control beams, there was a comparable agreement between the predicted ultimate load and the experimental ultimate load. It was found that the ultimate loads for some strengthened beams were higher

E. Oromiehie (✉) · G. B. Prusty
ARC Training Centre for Automated Manufacture of Advanced Composites (AMAC), School of Mechanical and Manufacturing Engineering, UNSW Sydney, Kensington, NSW 2052, Australia
e-mail: e.oromiehie@unsw.edu.au

G. B. Prusty
Sovereign Manufacturing Automation for Composites CRC Ltd. (SOMAC CRC), Sydney, NSW, Australia

F. Matti · F. Mashiri
School of Engineering, Design and Built Environment, Western Sydney University, Penrith, NSW 2751, Australia

S. B. Singh and C. V. R. Murty (eds.), *RC Structures Strengthened with FRP for Earthquake Resistance*, Composites Science and Technology,
https://doi.org/10.1007/978-981-97-0102-5_3

than that of the control beams. For all beams, there was inward deformation on the upper surface of each beam, and outward deformations were observed on the two side walls of the SHS beams. This experimental investigation showed that the current strengthening processes using AFP is not comparable to traditional CFRP strengthening methods which use epoxy and FRP plies and that further research is required in this space. Due to the failure modes observed, future research is planned to improve the reinforcement method using AFP. The planned improvements are surface preparation, AFP processing conditions, number of CFRP layers and orientation of the CFRP.

Keywords Automated fibre placement · Thermoplastic composites · Steel beam retrofitting · Fibre reinforced polymer jackets

1 Introduction

1.1 Retrofitting in Construction

Retrofitting the critical component of structures has become more vital in the progress of civil engineering in present days. This involves the process of strengthening existing structures, both in the short and long-term, through the addition of new materials to protect them from natural events such as flooding, earthquakes, and high winds as well as maintenance of structural cracks and damages, correction of errors in design or construction, ensuring the security and the safety of the structure, reconstructing the structure for the advantages of excessive loading, etc. Retrofitting is mainly used for changing, repairing, and modifying the structures after they have been constructed [1]. In the construction industry, retrofitting is classified into global and local with subclassifications [2] which are depicted in Fig. 1.

1.1.1 Jacketing Methods

The focus of this chapter is on the local jacketing method for steel structures/profiles that are mainly used in civil structures. This technique is employed to enhance the ability of concrete, steel, and timber structures to bear heavier loads when the structural members lack sufficient capacity and also to safeguard them from further deterioration [3]. In general, jacketing is classified into (a) reinforced concrete jacketing, (b) steel jacketing, and (c) fibre reinforced polymer composite jacketing.

Reinforced Concrete Jacket (RCJ)

To reinforce concrete elements in buildings, retrofitting methods are employed on structural elements such as walls, columns, and beams using various combinations such as surface, web and flange strengthening. They aim to improve the axial and

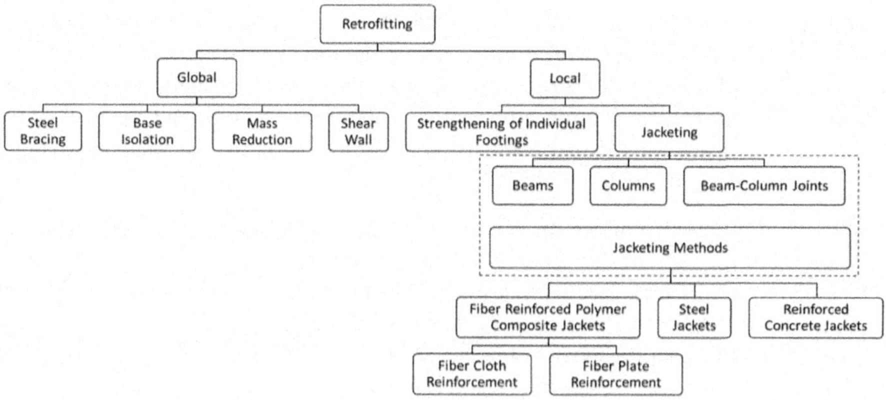

Fig. 1 Retrofitting classification

shear strength of columns while minimizing the need for extensive foundation reinforcement. In this retrofitting technique for reinforced cement concrete (RCC) buildings a cage-like structure made from steel reinforcement, or a composite material wrap is constructed around the deteriorated section. This cage serves as a foundation onto which cast-in-place concrete or shotcrete is applied. Jacketing is also effective for repairing collapsed columns, piers, and piles can even be utilized in underwater applications [2–4].

Steel Jacket (SJ):

The steel jacketing reinforcement technique involves wrapping section steel, typically angle steel, around the corners of a structural member to enhance the load-bearing capabilities, ductility, and rigidity of the component without altering its size, resulting in more reliable structural performance. It is particularly suitable for reinforcing structures where increasing the cross-sectional size of the original member is not feasible but a substantial boost in load-bearing capacity, ductility, and rigidity is required. The steel jacketing technique requires applying structural adhesive to affix steel onto the reinforced component, creating a unified and integrated system. This method is suitable for strengthening structures where enlarging the original member's cross-sectional size is not possible, but there is a significant need to enhance the cross-section's load-bearing capacity, as well as improve ductility and rigidity. The main advantages of steel jacketing include its simplicity, ease of construction, ability to significantly enhance the load-carrying capacity of the structure, low onsite workload, and short construction time [2–5].

Fibre Reinforced Polymer Composite Jacket (FRPCJ):

FRP composite jackets, provide enhanced characteristics like corrosion resistance, lightweight, and durability, which make them superior to traditional repair systems. They are also compatible with steel, concrete, and timber structures.

Fibre reinforced polymer composite (FRPC) jacketing is classified into fibre cloth reinforcement and fibre plate reinforcement technology, which are relatively new for repair and reinforcement of structures. This method involves using resin bonding materials to adhere fibre-reinforced polymer (FRP) products onto the surfaces of the building components that are expected to experience high stress. The main advantages of FRPC jacketing are listed below [2, 3, 6],

- Provide a distinct reinforcing effect on the narrow working surface because of their flexibility and ability to be wrapped around structures.
- Lightweight yet high-strength, resulting in minimal changes to the weight and appearance of the reinforced structure. Additionally, their low thickness has a negligible impact on shear moment.
- The construction process is quick and efficient.
- Carbon fibres with their high elastic modulus, offer improved control over factors such as temperature, cracks, and expansion due to rust after reinforcement.
- The original structure remains undamaged as no fasteners such as bolts, anchors, or external compression are necessary.
- Reduced manpower, materials, and construction time [4].

The effectiveness of an FRP jacket system is influenced primarily by three material parameters: thickness, fibre type, and fibre orientation. The thickness of the FRP jacket plays a significant role in the strength and ductility of repaired columns. It is directly related to the confinement pressure exerted by the FRP jacket, with higher thickness leading to increased confinement effectiveness [7, 8].

FRP wraps with higher thickness significantly enhance the strength and ductility of wrapped concrete columns [9]. The same effect is observed in both steel and timber structures since the exerted confining pressure is the crucial factor [10]. Nevertheless, for hollow steel tubes, once the jacket thickness reaches a specific threshold, where the dominant behaviour becomes inward buckling deformation of the jacket, further increasing the jacket thickness does not provide significant additional benefit as the jacket's resistance to inward buckling deformation is limited [11]. There is also a limitation on the thickness of a multilayer FRP laminate strengthening system due to the increased potential failure modes introduced by additional layers. Failure can occur in the adhesive between each layer, increasing the risk of failure within the FRP. For example, two layers for pultruded plates and three layers for FRP fabrics [12].

The magnitude of the confining stress applied by the prefabricated FRP jacket is the main factor influencing the effectiveness of the repair system and is highly influenced by the type and orientation of the fibres, regardless of the core material (concrete, steel, or timber) [13, 14]. Glass fibres are more cost-effective than carbon fibres. However, carbon fibres exhibit superior characteristics. Aramid fibres, on the other hand, have lower compressive load capacities compared to other fibres [15]. However, in prefabricated FRP jackets, fibres are oriented in the circumferential direction, in order to produce higher lateral stresses. This will result in an increased axial load capacity. Also, additional fibres with various angles relative to the hoop and longitudinal directions are used to provide resistance against multi-axial strains, enhance the structural integrity of the entire FRP shell, and exhibit more ductile behaviour upon failure [16]. Finally, increasing the confining pressure significantly increases the ductility enhancement ratio [7, 17].

1.2 Reinforcement Techniques

1.2.1 Wet Lay-Up

The wet layup process involves grinding the surface of a column or structure and applying resin to the surface. This is followed by layering on reinforcing fibre and using a roller to impregnate the fibres with resin and compact the layers. Proper application and impregnation of the reinforcement is essential for effective FRP strengthening, as it ensures the transfer of tangential stress from the existing structural member to the FRP retrofitting. Additionally, an external resin layer contributes to the impregnation of the fabric while providing protection to the fibres against potentially harmful environments, such as marine environments. The wet lay-up method is commonly employed in the production of large structures such as ship hulls. Historically, it has been a dominant manufacturing technique due to its low initial setup and tooling costs, as well as the relatively low skill level required for component manufacturing [18]. It is a labor-intensive process that involves applying each layer and ensuring proper wetting of the fibre reinforcement. This approach has several limitations, resulting in highly inconsistent components in terms of weight, fibre volume fraction, and strength. To mitigate these variations and minimize excess resin, vacuum bagging is commonly employed in conjunction with the wet layup procedure [6, 19].

1.2.2 Interposition of a Separating Film (ISF)

The ISF technique involves the use of a polymer film placed between the column and the FRP. This method prevents the substrate from being wetted by the thermosetting matrix, thereby avoiding irreversible permanent bonding. It is particularly suitable for confinement applications where the interaction is based on contact rather than bond. In cases of confinement, the FRP is activated passively by the lateral deformation of the material, generating stress in the FRP jacket in the hoop direction. In this scenario, the contact between the substrate and the confinement jacket remains intact until either the fibres rupture or adhesion is lost in the overlap region. For successful implementation of the ISF technique, the interposed polymer film should possess several characteristics, including low roughness to minimize friction between the FRP and the substrate, thereby facilitating potential future removal of the retrofitting. Also, it should exhibit low stiffness to simplify and expedite installation. Moreover, the film should be resistant to the heat generated by the exothermic crosslinking reaction of the epoxy matrix and be cost-effective to avoid significant cost increases in interventions that already involve expensive materials like carbon fibre [20].

1.2.3 Liquid Adhesion Release-Agent (LAR)

This method involves applying an inhibitor to the masonry surface. The objective is to create a transparent and protective layer that prevents adhesion between the epoxy matrix of the FRP and the masonry. When implementing the LAR technique, it is essential to choose a suitable adhesion-inhibiting material that ensures the detachment of the resin matrix without causing chemical alterations (maintaining breathability) or aesthetic changes (chromaticity) to the substrate. Adhesion between two materials can be primarily categorized as either chemical or mechanical. Chemical adhesion involves various types of bonds such as pure or polar primary covalent bonds, metallic, ionic, polar (or hydrogen) bonds, and Van der Waals forces. On the other hand, mechanical adhesion relies on the surface roughness, enabling one material to grip onto another. The LAR technique belongs to chemical adhesion. The primary chemical adhesion mechanism when using epoxy resin (e.g., for FRP) is hydrogen bonding. Liquid release agents, which exhibit water repellent properties, are employed to prevent hydrogen bonding. Additionally, these materials also serve as protectors for the masonry substrate, safeguarding it from weathering effects [20].

1.2.4 Filament Winding (FW)

In FW, fibre tows are impregnated with resin prior to being wrapped around a rotating mandrel. This process is commonly employed for cylindrical and spherical components, such as pipes and pressure vessels. The winding process is inherently automated, enabling cost-effective production of consistent components. This method is constrained to tubular parts and is not suitable for manufacturing open structures. Nevertheless, with a sufficiently flexible machine, it can be adapted to produce slightly more complex shapes. One challenge is achieving uniform fibre distribution and resin content across the entire thickness of the laminate. The filament winding process demonstrates excellent repeatability, making it a cost-effective choice. By maintaining a fibre to matrix volume fraction of 60:40. This technique can produce components with relatively high strength. An additional advantage of this process is the enhanced torsional strength it imparts to the final beam [18, 20].

1.2.5 Pultrusion

This is a fully automated method for producing composite materials. It is commonly employed to create fibre-reinforced sections with diverse shapes and sizes. This process is widely available and predominantly utilized in the civil engineering sector.

Its primary purpose is to fabricate composite components with a consistent cross-section. In this method, tows or fabric are saturated with resin and then passed through a heated die for curing. Additionally, resin can be directly injected into the die during the process. Pultrusion stands out as the most efficient composite manufacturing technique, boasting high productivity when compared to other methods. However, a limitation of this approach is that the fibres tend to be oriented primarily in the longitudinal direction of the beam element [18].

1.2.6 Automated Fibre Placement (AFP)

Automated fibre placement (AFP) is a high-speed manufacturing process that combines the laying, curing/melting, and consolidation steps into a single apparatus. This integration significantly improves the efficiency and output of this approach [21–23]. The AFP machine comprises of a placement head and a robotic arm, both of which are controlled by advanced software packages, Fig. 2 [24, 25].

During a lay-up, the placement head brings the surfaces of the prepreg tape together under heat and pressure to form the bonding. To apply this pressure and eliminate air from the composite structure, a consolidation roller is utilized, while a

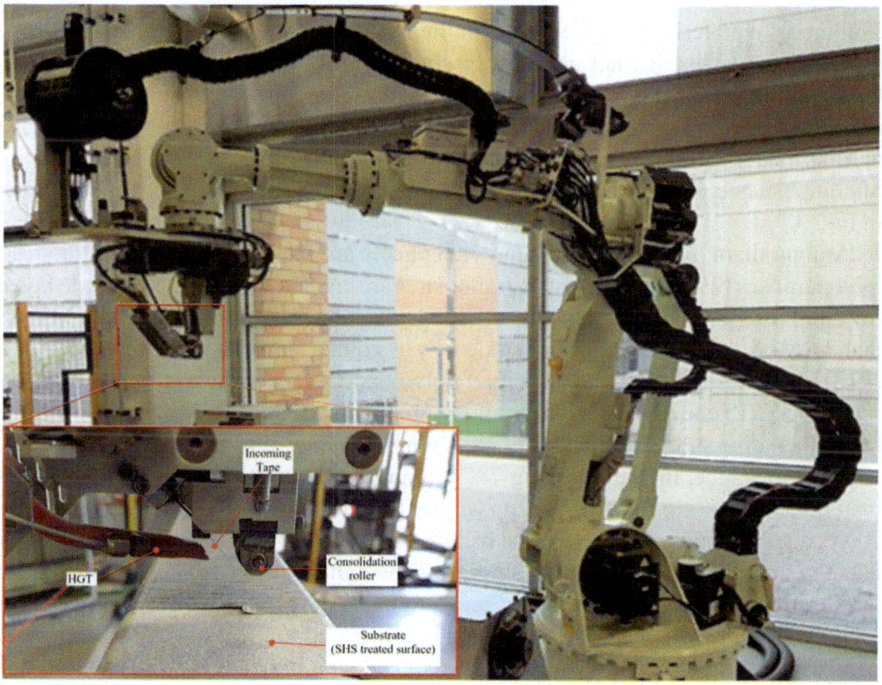

Fig. 2 Automated fibre placement (AFP) machine with thermoplastic head. *Photo courtesy* UNSW Sydney

heat source such as a hot gas torch (HGT) or laser system is employed. The cooling process is done at room temperature. The quality of the end product is influenced by various factors such as temperature, pressure, and the deposition rate of materials. Hence, it is important to ensure that these factors remain within the allowed limits or tolerances to maintain the desired quality [25]. In recent decades, the aerospace industry has witnessed a revolutionary transformation in the production of composite structures through automated manufacturing techniques such as AFP [26]. These methods offer various advantages, including enhanced precision in fibre placement, automated debulking, and the elimination of post-processing steps like oven or autoclave curing [27]. Furthermore, they contribute to reductions in material and labor costs [28]. Consequently, major manufacturers like Boeing, Airbus and NASA have adopted these techniques extensively to manufacture highly precise components like wing skins, nose cones, and fuselages [29–32]. Additionally, ongoing studies are exploring the utilization of these methods into other sectors and industries like marine, automotive and wind energy for making composite propellers, wheels and turbine blades [29–31, 33–37].

1.3 Composite Materials: Structures and Types

A composite material is formed by combining two or more different materials, each of which possesses distinct physical or chemical properties. When these materials are combined, they create a new material with unique characteristics. Composites are widely used in many applications across multiple industries, including aerospace, marine, aviation, transportation, sports, and civil engineering [38]. The categorization of composites based on the types of reinforcement and matrix materials is depicted in Fig. 3.

Among them, laminated and sandwich panels are increasingly utilized for structural purposes [39, 40]. Laminated composites are created by bonding multiple layers (referred to as ply or lamina) together. Each layer consists of strong, continuous fibres that provide reinforcement, surrounded by a comparatively weaker matrix material. The use of carbon fibre reinforced polymers (CFRP) and glass fibre reinforced plastic (GFRP) composites in structural components has greatly expanded in various fields including aircraft, sports vehicles, wind turbines, and infrastructure projects [26]. For instance, the new Boeing 787 Dreamliner aircraft incorporates composite materials accounting for ~52% of its weight [41], and the new Airbus A380 utilizes composites that makeup ~25% of its weight [42]. The exceptional properties of laminated composites, such as a high strength-to-weight ratio, resistance to creep, high tensile strength at high temperatures, and toughness, have sparked increasing interest across different industries [38, 43]. Another key driver for the industry's adoption of composite materials is the advantage of reduced weight, leading to substantial cost savings, energy efficiency improvements up to 18%, and lower CO_2 emissions [44]. Figure 4 shows the application of FRPC in different industries.

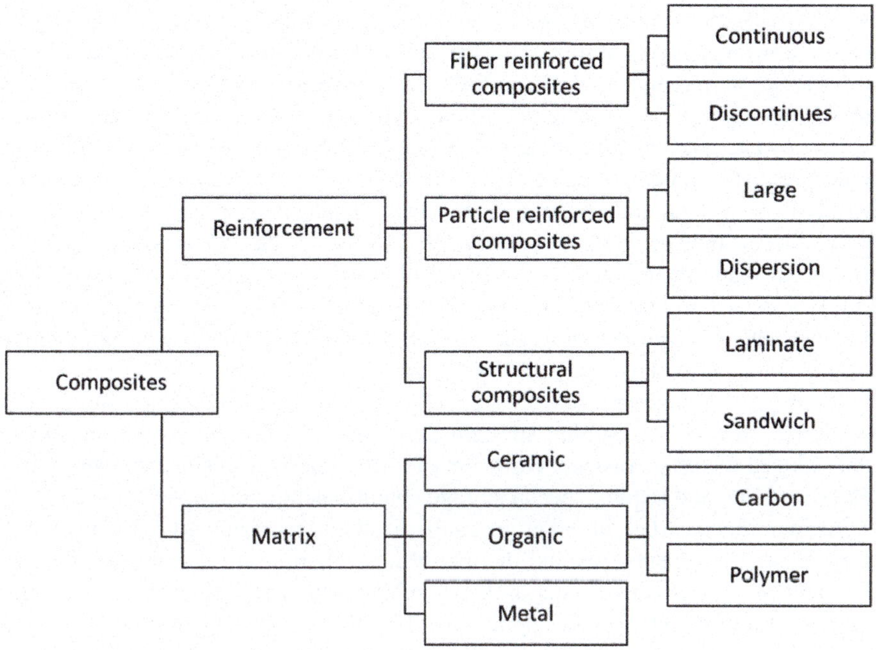

Fig. 3 Classification of composites based on reinforcement and matrix materials

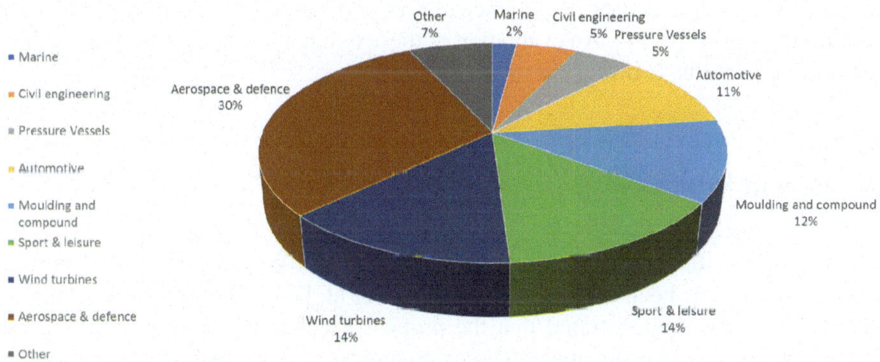

Fig. 4 Application of fibre reinforced polymer composites in Europe for different industry sectors [45]

1.3.1 Composite Materials for AFP

Prepreg tapes are commonly utilized in AFP. They consist of carbon or glass fibres impregnated with a thermosetting or thermoplastic resin matrix. They are available in rolls of different widths and are characterized by factors such as resin content, self-adhesive strength, drapeability, shelf life, and gel time. In prepregs, the reinforcing fibres are pre-impregnated with a partially cured resin to a specific level. This allows the prepreg tapes to be directly placed into a mould without the need for additional resin. However, heat and pressure must be applied during the manufacturing process to ensure a strong bond. In the case of thermoplastic composites, when the heated surfaces of the tape and substrate come into contact under pressure, the polymer chains diffuse between the matrix layers of each surface through thermal vibrations, forming a bond. The bonded area is then cooled under pressure at room temperature [28]. Thermoset prepregs, on the other hand, require controlled heat during the curing process. This heat allows the resin to flow within the laminate and polymerize, ultimately reaching the cured state [46].

Prepregs offer several advantages over non-impregnated tapes and fibres, including enhanced maximum strength properties, uniformity, consistent quality, and reduced material wastage. However, they are relatively expensive, require heat for processing, and require special storage conditions [47, 48].

As mentioned earlier, prepregs are categorized into two main types based on their polymer matrix: thermosets and thermoplastics. Thermoset matrices are the most used and are cost-effective materials that exhibit resistance to solvents. They undergo irreversible curing through the application of heat or suitable irradiation [49]. Epoxy resin is the most prevalent thermoset materials, offering superior performance at a relatively low cost. On the other hand, thermoplastic matrices can be softened by heating them to their processing temperatures. The process is reversible since no cross linking occurs. The choice of polymer matrix depends on the specific application requirements [44, 50, 51]. An evaluation of these two matrix types, considering their processing benefits, is discussed in [52].

2 Recent Studies

The viability of three novel approaches, including ISF, FW, and LAR, for retrofitting stone columns has been examined by Cascardi et al. [20]. Through axial compressive tests, it was demonstrated that the proposed techniques did not compromise the effectiveness of confinement. The examination of the substrates after the tests confirmed the reversibility of the proposed solutions, as evidenced by the absence of detachment of the stone surface. The investigated retrofitting techniques demonstrated their potential to serve as a valuable and efficient methods for increasing the seismic protection of columns in heritage buildings.

Mohammed et al. [8] conducted a comprehensive review of the current practices and potential applications of prefabricated composite jackets for structural repair. According to their findings, prefabricated FRP composite repair systems are considered preferable to wet lay-up methods due to their ease of installation, improved safety measures, higher quality, minimized on-site workforce needs and minimized resource wastage. However, the design of effective joints, that facilitate easy, quick, and safe installation, plays a crucial role in ensuring structural continuity when using prefabricated FRP composite jackets. The effectiveness of prefabricated FRP composite jackets depends on factors such as fibre thickness, orientation, the properties of the infill grout, and the extent and shape of existing structural damage. Thus, a thorough understanding of these design parameters is essential for achieving an optimal and safe design for prefabricated FRP jacket repair systems. Additionally, the existing models for predicting the strength and behaviour of structures strengthened with FRP composite jackets often overlook the level of damage present in the original structures. Hence, it is necessary to develop numerical and/or analytical models that systematically account for the influence of key parameters on the overall response of repaired structures. This will result in a more reliable and secure repair system.

Zeng et al. [53] have investigated the behaviour of fibre reinforced polymer (FRP) confined elliptical concrete-filled steel tube (ECFST) columns and the effects of various factors including the FRP jacket thickness and the cross-sectional aspect ratio. The authors aimed to (a) examine the efficiency and benefits of FRP confinement to ECFST columns under axial compression, (b) understand the behaviour of confined concrete in FRP-confined ECFSTs, and (c) examine the capability and accuracy of existing theoretical models for concrete in FRP-confined ECFSTs. Considering the high vulnerability of high-strength steel (HSS) to buckling, Q690 steel tubes were utilized in this study. According to their findings, an FRP jacket significantly improves the elastic stiffness, yield axial load, and ultimate axial load of ECFST columns. It was also shown that, the axial load-strain behaviour of ECFST columns can be divided into two segments. An ascending linear segment, which is connected to a secondary section can exhibit either an upward trend or a downward trend.. The confinement mechanism of concrete in FRP-confined ECFST columns resembles that of FRP-confined concrete, whose axial stress–strain curve typically exhibits a bilinear behaviour. Both the ultimate axial stress and ultimate axial strain increase with increasing FRP jacket thickness at a given cross-sectional aspect ratio. However, for a given FRP jacket thickness, the confinement efficiency decreases as the cross-sectional aspect ratio increases, leading to lower ultimate axial stress and a shallower slope of the second ascending portion. FRP-confined ECFST specimens with a cross-sectional aspect ratio of 2.0 exhibit poor FRP confinement efficiency. An increase in cross-sectional aspect ratio results in a decrease in ultimate axial stress and ultimate axial strain. The enhancement index for elastic stiffness increases with the cross-sectional aspect ratio for specimens with a given FRP thickness. This indicates that the FRP jacket is more effective in improving the elastic stiffness of ECFST columns compared to circular concrete-filled steel tube (CCFST) columns[53].

In the following section, the effectiveness of a new automated technique to retrofit short columns and beams is discussed. For this purpose, the behaviour of short steel columns and beams overwrapped with thermoplastic carbon fibres using AFP is presented.

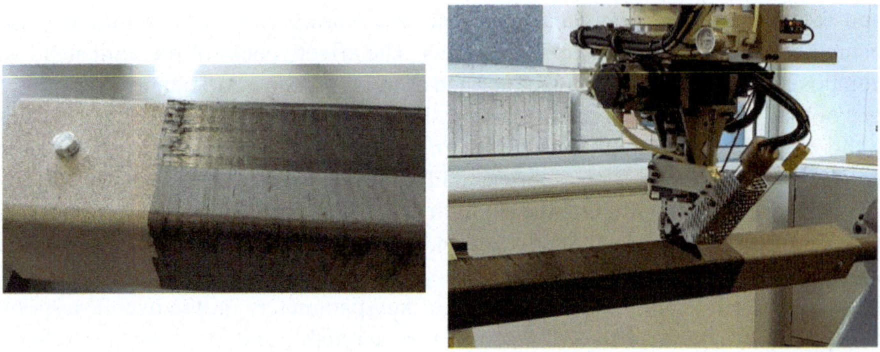

Fig. 5 Over wrapping steel profiles with thermoplastic CFRP using AFP. *Photo courtesy* Automated Composites Laboratory, UNSW Sydney

3 Experimental Program

3.1 *Sample Preparation*

The Automated Dynamics-built AFP machine at UNSW Sydney was utilized for reinforcing the SHS specimens, as shown in Fig. 5. This AFP machine is a seven-axis robot platform, including a coordinated spindle. The thermoplastic placement head comprises of a consolidation system, prepreg tape dispensing system, hot gas torch (HGT) as a heat source (Fig. 2), and a computer controller [25]. Carbon-PEEK prepreg tapes (AS4/APC2, supplied by Solvay) were used to reinforce the steel profiles. The prepreg tapes were 0.15 mm thick and 6.35 mm wide. The key material properties of the prepreg tape of density, modulus of elasticity, shear modulus, Poisson's ratio and fibre volume fraction were 1570 kg/m^3, 138 GPa, 5 GPa, 0.28 and 0.6, respectively.

The AFP deposition rate, HGT temperature and consolidation force were selected based on the authors' earlier investigations [54–59] and set as 76 mm/s, 900 °C and 200 N, respectively. The layup sequences were chosen to ensure a strong and durable bond between the prepreg and SHS for all reinforced profiles and are summarized in Tables 1 and 2. The surface of the steel profiles was sandblasted using Class 2.0 blast with a GL40 steel grit size.

3.2 *Column Specimens*

Fourteen column specimens made up of SHS were tested under axial compression for the determination of the ultimate load. The steel profiles were 3.5 mm thick with cross sectional outer dimensions of 89 mm × 89 mm according to Standards Australia Online [60]. As shown in Table 1, the length of each column was 267 mm, three times

Table 1 Details of the column specimens

Specimen		Direction (T/L)	Lay-up	Overwrapped with CF-PEEK
C	a	–	–	No
	b	–	–	No
1	a	T	[90, 90]	Yes
	b	T	[90, 90]	Yes
	c	T	[90, 90]	Yes
	d	T	[90, 90]	Yes
2	a	L	[0, 0]	Yes
	b	L	[0, 0]	Yes
	c	L	[0, 0]	Yes
	d	L	[0, 0]	Yes
	e	L	[0, 0]	Yes
	f	L	[0, 0]	Yes
	g	L	[0, 0]	Yes
	h	L	[0, 0]	Yes

Note T = Transverse, L = Longitudinal

Table 2 Details of the tested beam specimens

Specimen		Direction (H/V)	Lay-up	Overwrapped with CF-PEEK
B	a	–	–	No
	b	–	–	No
3	a	H	[0, 0]	Yes
	b	H	[0, 0]	Yes
4	a	H	[0, 0]	Yes
	b	H	[0, 0]	Yes
5	a	V	[90, 90]	Yes
	b	V	[90, 90]	Yes
6	a	V, 45°	[90, 90, 45]	Yes
	b	V, 45°	[90, 90, 45]	Yes
7	a	V	[90, 90]	Yes
	b	V	[90, 90]	Yes
8	a	HV	[90, 0]	Yes
	b	HV	[90, 0]	Yes
9	a	HV	[90, 0]	Yes
	b	HV	[90, 0]	Yes

Note H = Horizontal; and V = Vertical

that of the breadth of the specimen. The measured lengths of the columns ranged from 267 to 269 mm. Two control column specimens (C-a and C-b) were tested under axial compression (Fig. 6). The control columns did not have thermoplastic carbon fibre reinforced polymer (CFRP). Twenty-one specimens were externally reinforced with thermoplastic CFRP plies AFP machine at UNSW Sydney. Four specimens (1-a, 1-b, 1-c and 1-d) had two layers of thermoplastic CFRP laid transversally (perpendicular to axial loading) as shown in Fig. 7. Additionally, eight columns (2-a, 2-b, 2-c, 2-d, 2-e, 2-f, 2-g and 2-h) had two layers of thermoplastic CFRP plies laid longitudinally (parallel to axial loading), see Fig. 8. Two Stainless Steel Hose Clamps were installed on specimens (1-c, 1-d, 2-c, 2-d, 2-e, 2-f, 2-g and 2-h) to find out if the ultimate loads of the CFRP strengthened columns with steel hose clamps increase in comparison to the CFRP strengthened columns with no steel hose clamps. Steel hose clamps were trailed as the CFRP plies that were laid longitudinally on column specimens separated from the steel under axial compression loading. The location of the stainless-steel hose clamps is shown in Fig. 9.

Fig. 6 Control column specimens

Fig. 7 Transverse reinforced column specimens

Fig. 8 Longitudinal reinforced column specimens

3.3 Beam Specimens

Two control beams and fourteen thermoplastic CFRP overwrapped beams shown in Table 2 were tested for their bending capacity. The steel beam specimens were in the

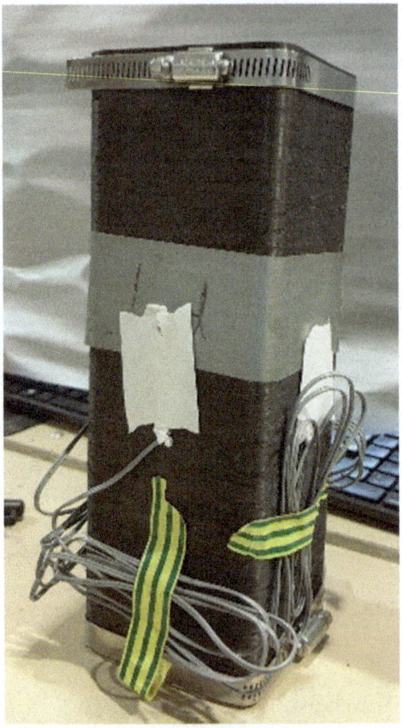

Fig. 9 Locations of stainless-steel hose clamps

form SHS with thickness of 3.5 mm and cross-sectional outer dimensions of 89 mm × 89 mm. The length of each beam was approximately 750 mm.

As shown in Fig. 10, the control beams (B-a and B-b) did not have thermoplastic CFRP wrap. Beam specimens (3-a, 3-b, 4-a and 4-b) had two layers of CFRP plies laid horizontally (parallel to length), as in Fig. 11. Specimens (5-a, 5-b, 7-a and 7-b) had two layers of CFRP laid vertically (perpendicular to length), as shown in Fig. 12. Beam specimens (6-a and 6-b) shown in Fig. 13 were made with three layers of CFRP placed vertically and at angle of 45°. Beam specimens (8-a, 8-b, 9-a and 9-b) shown in Fig. 14 had two layers of CFRP laid horizontally and vertically.

Fig. 10 Control beam specimens

Fig. 11 Longitudinal reinforced beam specimens

Fig. 12 Transverse reinforced beam specimens

Fig. 13 Reinforced beam specimens with additional layer of 45°

Fig. 14 Transverse and longitudinal beam reinforced specimens

3.4 Instrumentation and Testing of Column Specimens

Axial deformations were measured using a linear potentiometer (LP1516). This is detailed in Fig. 15. Two control columns and twelve CFRP columns were tested under axial compressive loading using a compression test machine shown in Fig. 16. The column specimens had a fixed support at the base of the column with the load being applied to the top of the column via a flat plate (see Fig. 16). For determining the peak load, the test specimens were subjected to a continuous axial compressive load applied at a rate of 0.016 mm/s. The test continued at the same rate after the peak load was obtained and then was stopped when the load reduced to a value of 300 kN for control columns and 150 kN for CFRP columns.

Fig. 15 Plan view of instrumentation locations

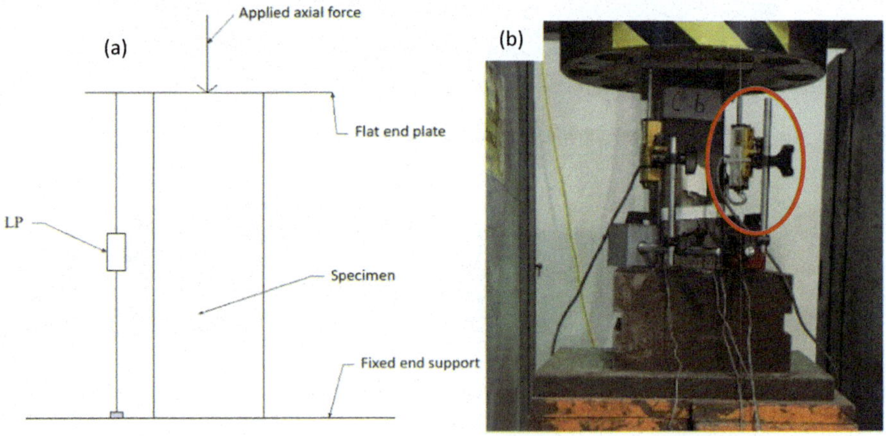

Fig. 16 Test setup: **a** Schematic diagram and **b** test setup in compression-testing machine showing LP1516

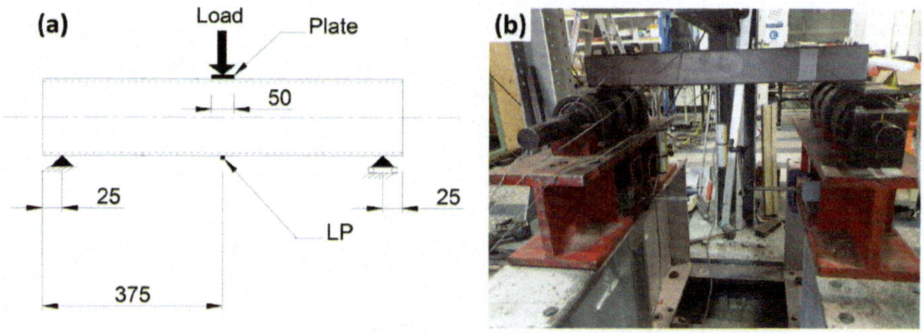

Fig. 17 Test setup: **a** schematic diagram and **b** test rig (all dimensions are in mm)

3.5 *Instrumentation and Testing of Beam Specimens*

A liner potentiometer (LP) was used at the bottom face of the beams to measure deflections, see Fig. 17a. The bending capacity of the control beams and 4 CFRP strengthened beams were investigated using the test setup shown in Fig. 17a. Loads were applied at the midspan of the beams using a hydraulic jack, as shown in Fig. 17b. The control beam behaviour was compared to the different methods of strengthening using CFRP. The beam specimens were pin connected at both ends of each beam. The pinned supports were located 25 mm away from the end edges of the beam. The load was applied to these test specimens at a rate of 0.016 mm/s. The test continued at the same rate after obtaining the peak load until the load dropped to a value of 49 kN or the specimens failed.

4 Performance Prediction of Reinforced Steel Profiles

4.1 *Strength of Steel Beams Reinforced with Thermoplastic Carbon Fibre Reinforced Polymer*

4.1.1 Behaviour of Beams

The ultimate loads for the SHS control beams and CFRP beams are shown in Table 3. There is a good agreement between the predicted ultimate compressive loads (Standards Australia Online [61] and OneSteel Market Mills [62]) and experimental ultimate compressive loads. The predicted value of the ultimate load for the control beams was 93.9 kN. The values of the experimental ultimate load of specimens B-a and B-b were 87.2 and 77.8 kN, respectively.

The ultimate load of B-b was lower than B-a. This may be because the weld line of B-b was on the bottom surface (Fig. 18) whereas the weld line of B-a was on the

Table 3 Experimental values of ultimate loads

Specimen		Direction (H/V)	Lay-up	Ultimate load (kN)
B	a	–	–	87.21
	b	–	–	77.82
3	a	H	[0, 0]	81.85
	b	H	[0, 0]	82.73
4	a	H	[0, 0]	86.24
	b	H	[0, 0]	85.99
5	a	V	[90, 90]	83.30
	b	V	[90, 90]	88.81
6	a	V, 45°	[90, 90, 45]	82.98
	b	V, 45°	[90, 90, 45]	87.26
7	a	V	[90, 90]	73.59
	b	V	[90, 90]	72.40
8	a	HV	[90, 0]	74.15
	b	HV	[90, 0]	76.21
9	a	HV	[90, 0]	84.33
	b	HV	[90, 0]	83.57

top surface. This shows SHS beams may carry higher load when the weld line is on the top surface.

Fig. 18 Weld line of specimen B-b

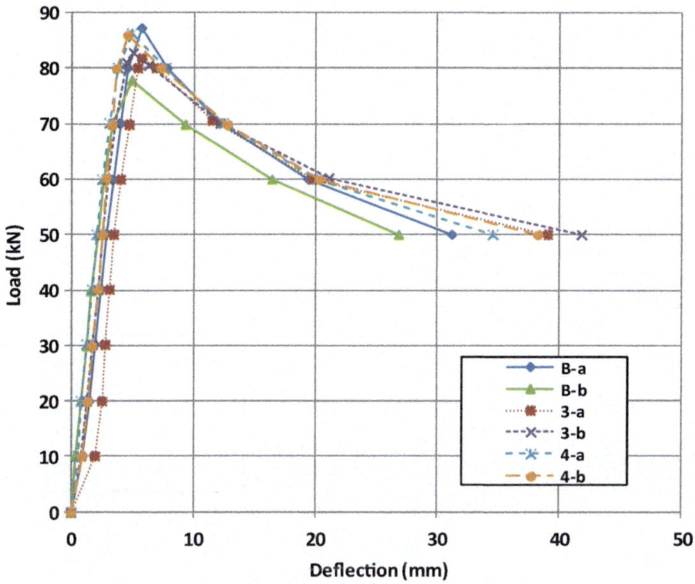

Fig. 19 Load–deflection curves for the control beams and longitudinally reinforced CFRP beams

The load–deflection curves of the control beams and specimens 3-a, 3-b, 4-a and 4-b with longitudinal CFRP reinforcement are shown in Fig. 19. The values of the experimental ultimate load of specimens 3-a, 3-b, 4-a and 4-b beams are 81.9, 82.7, 86.2 and 86.0 kN, respectively. These values are higher than the ultimate load of specimen B-b of 77.8 kN. Specimens 3 and 4 were compared with control beam B-b as the weld lines of beams 3, 4 and B-b were not located on the top surface.

The load–deflection curves of specimens 5-a, 5-b, 7-a and 7-b with transverse CFRP reinforcement are shown in Fig. 20. The values of the experimental ultimate load of specimens 5-a, 5-b, 7-a and 7-b beams were 83.3, 88.8, 73.6 and 72.4 kN, respectively. The ultimate load of CFRP beam 5-a (83.3 kN) was greater than the ultimate load of control beam B-b (77.8 kN). CFRP beam 5-a was compared with control beam B-b as the weld lines of these beams was not located on the top surface. The ultimate load of CFRP beam 5-b (88.8 kN) was greater than the ultimate load of control beam B-a (87.2 kN). The weld lines of specimens 5-b and B-a were located on the top surface. The ultimate loads of CFRP beams 7-a and 7-b were 73.6 and 72.4 kN, respectively. These values were lower than the ultimate loads of control beams.

The load–deflection curves of specimens 6-a and 6-b with CFRP reinforcement at 45° are shown in Fig. 21. The values of the experimental ultimate loads for specimens 6-a and 6-b beams were 83.0 and 87.3 kN, respectively. These values were greater than the ultimate load of control beam C-b (77.8 kN) and were lower than the ultimate load of control beam C-a (87.2 kN).

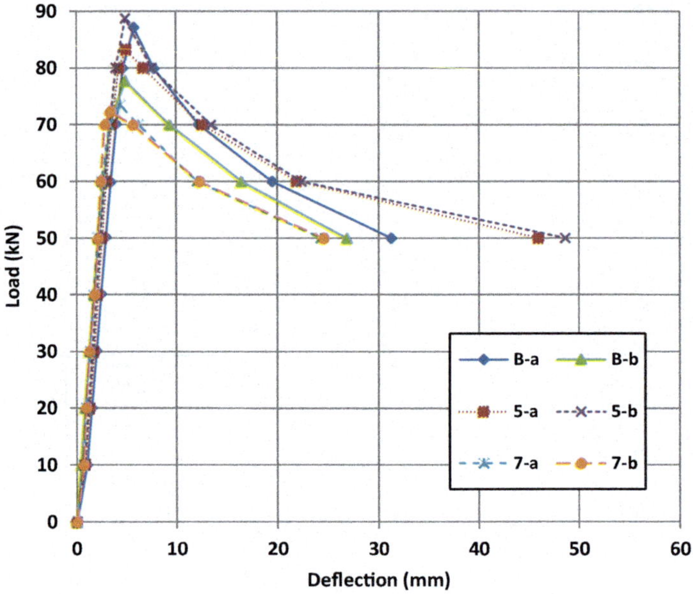

Fig. 20 Load–deflection curves for the control beams and transversally reinforced CFRP beams

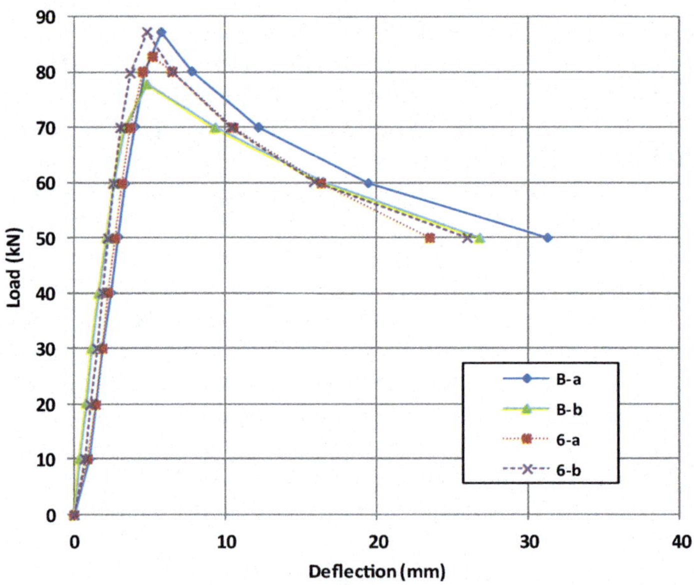

Fig. 21 Load–deflection curves for the control beams and beams with CFRP reinforcement at 45°

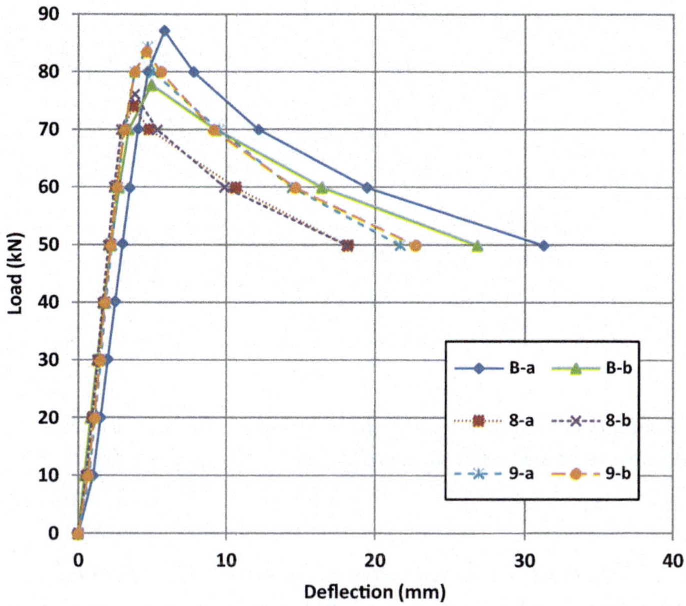

Fig. 22 Load–deflection curves of control beams and beams with CFRP laid longitudinally and transversally

The load–deflection curves for specimens 8-a, 8-b, 9-a and 9-b with longitudinal and transverse reinforcement are shown in Fig. 22. The values of the experimental ultimate load of specimens 8-a, 8-b, 9-a and 9-b beams were 74.2, 76.2, 84.3 and 83.6 kN, respectively. The ultimate loads for the strengthened beams 8-a, 8-b, 9-a and 9-b were similar to that of the control beams.

4.1.2 Failure Modes

For the SHS beams with and without thermoplastic CFRP overwrap, local deformations were observed at the location of the load application. As shown in Fig. 23, there was inward deformation (concave) on the upper surface of each beam. Outward deformations (convex) were observed on the two side walls of the beams.

For beam specimens 3-a, 3-b, 4-a and 4-b with CFRP laid longitudinally, the CFRP composites separated from steel SHS, see Fig. 23b for a typical failure mode. For beam specimens 5-a, 5-b, 7-a and 7-b whose CFRP was laid transversally, the thermoplastic CFRP layers snapped as the load was applied to the specimens, see Fig. 23c for a typical failure mode. Figure 23d shows a typical failure mode for a beam specimen whose CFRP was laid at an angle of 45°. The 45° CFRP layers de-bonded from the thermoplastic transverse CFRP at the location of the load (top and sides locations). For beam specimens 8-a, 8-b, 9-a and 9-b whose CFRP was

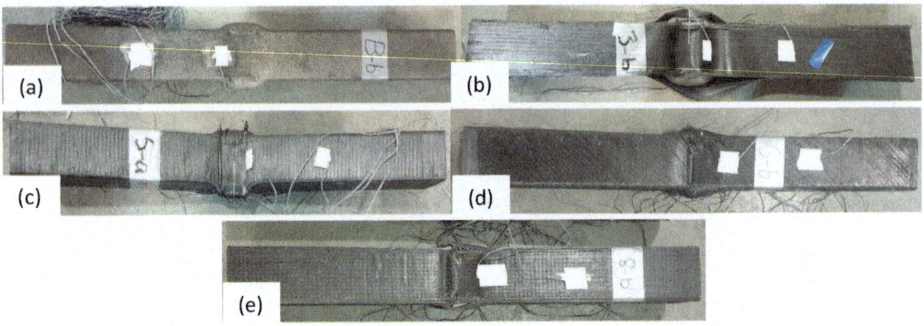

Fig. 23 Typical failure modes of the SHS beams: **a** B-b; **b** 3-b; **c** 5-a; **d** 6-b; and **e** 8-b

laid longitudinally and transversally, de-bonding of the thermoplastic CFRP from the steel tube were observed at the corners, as shown in Fig. 23e.

4.2 Axial Strength of Steel Short Columns Reinforced with Thermoplastic Carbon Fibre Reinforced Polymer (CFRP)

4.2.1 Behaviour of Columns

The ultimate loads and the deflections at ultimate load for the two SHS control columns and twelve CFRP columns are shown in Table 4. For control columns, a good agreement was observed between the predicted and experimental ultimate compressive loads. The experimental ultimate load of C-a and C-b were 527.9 and 524.1 kN, respectively. According to Standards Australia Online [61] and, OneSteel Market Mills [62], the predicted value of the ultimate load for the control columns was 517.5 kN (Table 4). As shown in Table 4, Figs. 24 and 25, the ultimate loads for the thermoplastic CFRP columns did not exceed the ultimate loads for the control columns.

4.2.2 Failure Modes

The failure modes for the steel SHS control columns were local buckling failure, as shown in Fig. 26a. Both inward and outward buckling were observed in the control columns. Bottom end buckling occurred on specimen C-a and top end buckling can be observed on specimen C-b. Inward and outward buckling are also observed in the transverse CFRP columns. As shown in Fig. 26b, bottom end buckling occurred on specimen 1-a, 1-b and 1-c whereas top end buckling can be observed on specimen 1-d. Also, the CFRP layers for the transversally reinforced CFRP columns with

Table 4 Ultimate loads and deflections for SHS columns

Specimen		Direction (T/L)	Steel Hose Clamps installed on column	Ultimate load (kN)
C	a	–	No	527.94
	b	–	No	524.08
1	a	T	No	478.10
	b	T	No	507.88
	c	T	Yes	493.00
	d	T	Yes	513.63
2	a	L	No	484.53
	b	L	No	499.46
	c	L	Yes	448.13
	d	L	Yes	404.03
	e	L	Yes	494.51
	f	L	Yes	491.50
	g	L	Yes	498.04
	h	L	Yes	504.20

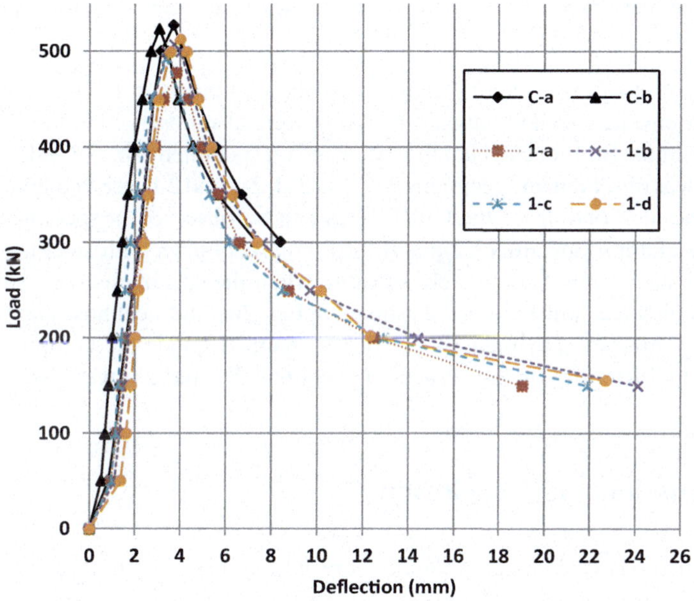

Fig. 24 Load–deflection curves for control columns and transversally reinforced CFRP columns

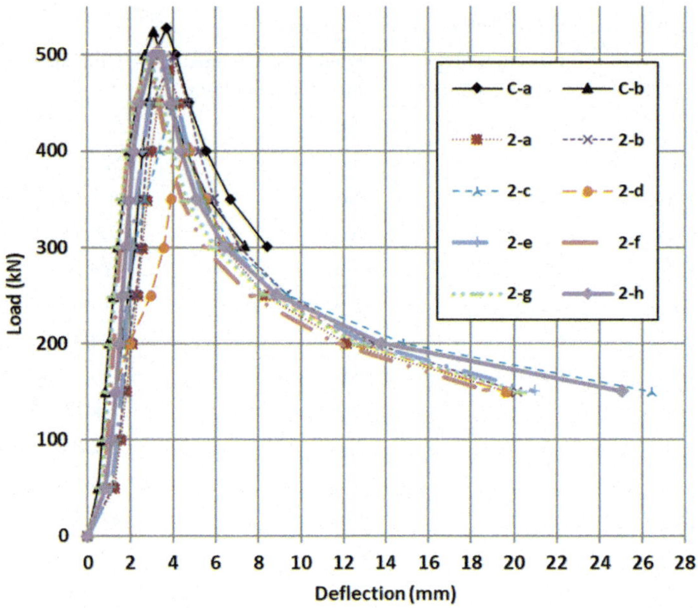

Fig. 25 Load–deflection curves for control columns and longitudinally reinforced CFRP columns

and without steel hose clamps snapped as the compression load was applied on the specimens. Inward and outward buckling were also observed in the longitudinal CFRP columns. Bottom end buckling occurred on specimen 2-b, 2-e and 2-f. Top end buckling was observed on specimens 2-a, 2-c, 2-d, 2-g and 2-h, see Fig. 26c. As shown in Fig. 26c, de-bonding of the CFRP plies can be observed in specimens without steel hose clamps (specimens 2-a and 2-b). For specimens with steel hose clamps, the CFRP steel did not completely separate when the specimens were subjected to compression loads. Furthermore, as shown in Fig. 26c, the steel hose clamps located on the top edge of specimens 2-d and 2-g snapped under the compression load. Tearing of CFRP plies on the corners of specimen 2-g and 2-h was also observed.

5 Discussion and Conclusion

Cold-formed SHS beams and columns are widely used in engineering structures such as bridges, cranes, and towers. In engineering structures, these sections are often subjected to cyclic loading from imposed loads such as vehicles and wind loads. When loads are applied to a structure, cracks may initiate and propagate, and subsequent fatigue failure may occur. Improving the mechanical behaviour of such structures can be achieved with CFRP reinforcement in the form of a jacket. The utilization of FRPC jackets provides a range of significant advantages including

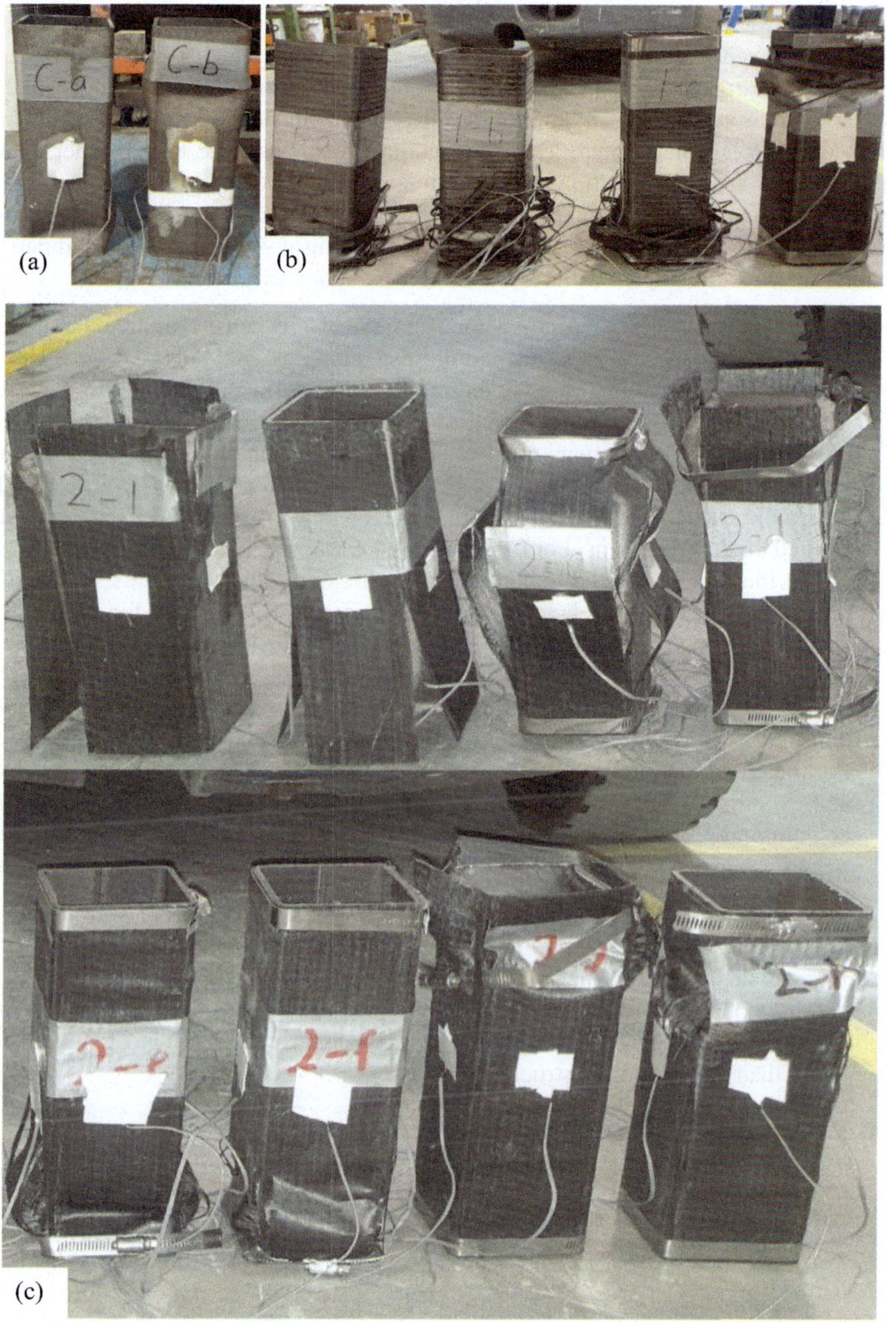

Fig. 26 Failure mode of: **a** control columns, **b** transverse CFRP columns with and without steel hose clamps, and **c** longitudinal CFRP columns with and without steel hose clamps

superior resistance to corrosion, durability and long-life compared to the limitations of conventional repair systems such as concrete and steel jackets. This experimental investigation presented in this chapter focused on reinforcing thin-walled SHS columns and beams with thermoplastic CFRP using hot gas torch assisted automated fibre placement. Fourteen SHS column specimens were tested under axial compression for determination of their ultimate loads. Two control beams and fourteen thermoplastic CFRP overwrapped beams were also tested for the determination of their bending capacity. The results obtained from this investigation were compared with the baseline samples without CFRP and are summarized below,

- For the control beams, there was a good agreement between the predicted ultimate compressive loads and experimental ultimate compressive loads.
- The experimental ultimate load of thermoplastic CFRP beams did not show significant improvement using the current AFP strengthening method due to the failure modes. For beam specimens whose CFRP was laid longitudinally, the thermoplastic CFRP plies separated from the steel. For beam specimens whose CFRP was laid transversally, some thermoplastic CFRP layers snapped as the load was applied on the specimens. For beam specimens whose CFRP was laid at an angle of 45°, the CFRP layers de-bonded from the thermoplastic transverse CFRP at the location of the load. For beam specimens whose CFRP was laid longitudinally and transversally, de-bonding of the thermoplastic CFRP from the steel tube was observed at the corners. For the SHS beams with and without thermoplastic CFRP, inward deformation was observed on the upper surface of each beam and outward deformation was observed on the two side walls of the beams.
- For the control columns, a good agreement was observed between the predicted and experimental ultimate compressive loads. The ultimate loads of the CFRP columns were lower than the ultimate loads for the control columns. De-bonding, tearing and snapping of the CFRP plies are observed in specimens with thermoplastic CFRP.

This investigation is still ongoing, as further research is being conducted to explore the effect of factors such as the number of overwrapped plies, ply orientation, and AFP processing conditions on the strength of reinforced SHS profiles. These continued efforts aim to enhance our understanding of FRP composite jacketing techniques and optimize their application in structural repairs.

Acknowledgements This project was funded by Western Sydney University (WSU) and University of New South Wales (UNSW). The authors would like to acknowledge the support received through following funding schemes of Australian Government:

ARC LIEF: An Australasian facility for the automated fabrication of high-performance bespoke components (LE140100082).

ARC ITTC: ARC Training Centre for Automated Manufacture of Advanced Composites (IC160100040).

References

1. Kumar Nigam P, Akhtar S (2021) Retrofitting practices in various categories of RCC structures—a comprehensive review. Mater Today: Proc 45:7123–7131
2. 10 Best Retrofitting Techniques for RCC Buildings [Internet]. Civil String Get the Knowledge (2021). Available from: https://civilstring.com/retrofitting-techniques-for-rcc-buildings/
3. Waghmare SPB (2011) Materials and jacketing technique for retrofitting of structures. Int J Adv Eng Res Stud 1(1):15–19
4. FRP Jacketing versus Steel Jacketing—structural strengthening. In: Horse (ed) Solutions (2023)
5. Di Trapani F, Malavisi M, Marano G, Greco R, Ferrotto M (2020) Optimal design algorithm for seismic retrofitting of RC columns with steel jacketing technique. Procedia Manuf 44:639–646
6. Samy K, Fawzy A, Fouda MA (2023) Strengthening of historic reinforced concrete columns using concrete and FRP jacketing techniques. Asian J Civil Eng 24(3):885–896
7. Ozbakkaloglu T (2013) Compressive behavior of concrete-filled FRP tube columns: assessment of critical column parameters. Eng Struct 51:188–199
8. Mohammed AA, Manalo AC, Ferdous W et al (2020) State-of-the-art of prefabricated FRP composite jackets for structural repair. Eng Sci Technol Int J 23(5):1244–1258
9. Li G, Kidane S, Pang S-S, Helms JE, Stubblefield MA (2003) Investigation into FRP repaired RC columns. Compos Struct 62(1):83–89
10. Saafi M, Asa E (eds) (2010) Extending the service life of electric distribution and transmission wooden poles using a wet layup FRP composite strengthening system. J Perform Constructed Facil
11. Teng JG, Yu T, Fernando D (2012) Strengthening of steel structures with fiber-reinforced polymer composites. J Constr Steel Res 78:131–143
12. Roads V (ed) (2018) Code of practice: FRP for strengthening of bridge structures
13. Kim H, Lee KH, Lee YH, Lee J (2012) Axial behavior of concrete-filled carbon fiber-reinforced polymer composite columns. Struct Design Tall Spec Build 21(3):178–193
14. Vincent T, Ozbakkaloglu T (2013) Influence of fiber orientation and specimen end condition on axial compressive behavior of FRP-confined concrete. Constr Build Mater 47:814–826
15. Teng J, Chen J-F, Smith ST, Lam L (2002) FRP: strengthened RC structures
16. Wu R-Y, Pantelides CP (2017) Rapid repair and replacement of earthquake-damaged concrete columns using plastic hinge relocation. Compos Struct 180:467–483
17. Lam L, Teng JG (2003) Design-oriented stress–strain model for FRP-confined concrete. Constr Build Mater 17(6):471–489
18. Mazumdar SK (2002) Composites manufacturing: materials, product, and process engineering. CRC Press LLC, United States of America
19. Performance Proven Worldwide. Wet Lay-Up Process United States of America: Performance Proven Worldwide; cited 2015. Available from: http://www.pactinc.com/capabilities/wet-lay-uphand-lay-up-method/
20. Cascardi A, Dell'Anna R, Micelli F, Lionetto F, Aiello MA, Maffezzoli A (2019) Reversible techniques for FRP-confinement of masonry columns. Constr Build Mater 225:415–428
21. Gutowski T (1997) Advanced composites manufacturing. Wiley, New York, p 600
22. August Z, Ostrander G, Hauber D (eds) (2015) Additive manufacturing with high performance thermoplastic composites. In: The second international symposium on automated composites manufacturing. Montreal, Canada
23. Werner D (ed) (2015) Multi-material-head novel laser-assisted tape, prepreg and dry-fiber placement system. In: The second international symposium on automated composites manufacturing. Montreal, Canada
24. Stokes-Griffin CM, Compston P (2015) The effect of processing temperature and placement rate on the short beam strength of carbon fibre–PEEK manufactured using a laser tape placement process. Compos A Appl Sci Manuf 78:274–283
25. Oromiehie E, Prusty BG, Compston P, Rajan G (2019) Automated fibre placement based composite structures: review on the defects, impacts and inspections techniques. Compos Struct 224:110987

26. Landry A (ed) (2015) The needs for automation in composite design and manufacturing. In: The second international symposium on automated composites manufacturing. Montreal, Canada
27. Béland S (1990) High performance thermoplastic resins and their composites: Noyes Data Corp
28. August Z, Ostrander G, Michasiow J, Hauber D (2014) Recent development in automated fibre placement of thermoplastic composites. SAMPE 50(2):30–37
29. Jonsson M, Eklund D, Järneteg A, Åkermo M, Sjölander J (eds) (2015) Automated manufacturing of an integrated pre-preg structure. In: The second international symposium on automated composites manufacturing. Montreal, Canada
30. Di Francesco M, Giddings P, Potter K (eds) (2015) On the layup of convex corners using automated fibre placement: an evaluation method. In: the second international symposium on automated composites manufacturing. Montreal, Canada
31. Girdauskaite L, Krzywinski S, Schmidt-Eisenlohr C, Krings M (eds) (2015) Introduction of automated 3d vacuum buildup in the composite manufacturing chain. In: The second international symposium on automated composites manufacturing. Montreal, Canada
32. Quddus M, Brux A, Larregain B, Galarneau Y (eds) (2015) A study to determine the influence of robotic lamination programs on the precision of automated fiber placement through statistical comparisons. In: The second international symposium on automated composites manufacturing. Montreal, Canada
33. Serubibi A, Hazell PJ, Escobedo JP et al (2023) Fibre-metal laminate structures: high-velocity impact, penetration, and blast loading–a review. Compos A Appl Sci Manuf 173:107674
34. Serubibi A, Hazell PJ, Escobedo JP, Wang H, Oromiehie E, Prusty GB (2021) Analysis of AFP manufactured fibre metal laminate structures under impact loading. Engineers Australia, BARTON, ACT
35. Maung PT, Prusty BG, Oromiehie E, Phillips AW, St John NA (2022) Design and manufacture of a shape-adaptive full-scale composite hydrofoil using automated fibre placement. Int J Adv Manuf Technol 123(11):4093–4108
36. Air A, Shamsuddoha M, Oromiehie E, Prusty BG (2023) Development of an automated fibre placement-based hybrid composite wheel for a solar-powered car. Int J Adv Manuf Technol 125(9):4083–4097
37. Maung PT, Prusty BG, Donough MJ, Oromiehie E, Phillips AW, St John NA (2023) Automated manufacture of optimised shape-adaptive composite hydrofoils with curvilinear fibre paths for improved bend-twist performance. Mar Struct 87:103327
38. Garg DP, Zikry MA, Anderson GL (2001) Current and potential future research activities in adaptive structures: an ARO perspective. Smart Mater Struct 10(4):610–623
39. Chung DDL (2010) Composite materials: science and applications, 2nd edn. Springer, London, United Kingdom, p 371
40. Bannister M (2001) Challenges for composites into the next millennium—a reinforcement perspective. Compos A Appl Sci Manuf 32(7):901–910
41. Spenser J (2006) Boeing technologies developed for commercial jetliners are now being integrated into some military products. Boeing Frontiers Online
42. Efficiency and reliability: Airbus (2023). Available from: https://www.airbus.com/en/products-services/commercial-aircraft/passenger-aircraft/a380
43. Balageas D (2006) Introduction to structural health monitoring. Wiley Online Library
44. Prusty BG, Oromiehie E, Rajan G (2016) Introduction to composite materials and smart structures. In: Iniewski K (ed) Structural health monitoring of composite structures using fiber optic methods. Devices, Circuits, and Systems. CRC Press, New York, pp 1–19
45. Elmar W, Bernhard J (2014) Composites market report. The European GRP market (AVK) TgCmC
46. Prepregs: Composites One (2017). Available from: http://www.compositesone.com/product/prepreg/
47. What are prepregs? Fibre Glast Development Corporation (n.d.) [Available from: http://www.fibreglast.com/product/about-prepregs/Learning_Center

48. Daniel IM, Ishai O (2006) Engineering mechanics of composite materials, 2nd edn. Oxford University Press, New York
49. Thermosets: Polymers International Australia (2014) [Available from: http://polymers.com.au/thermosets/
50. Common Thermoplastics and Thermoseting and there uses: United Kingdom: Stephens Plastic Mouldings (n.d.) cited 2015. Available from: http://www.stephensinjectionmoulding.co.uk/thermoplastics/
51. Chang IY, Lees JK (1988) Recent development in thermoplastic composites: a review of matrix systems and processing methods. J Thermoplast Compos Mater 1(3):277–296
52. Muzzy JD, Kays AO (2004) Thermoplastic versus thermosetting structural composites. Polymer Compos 5(3)
53. Zeng J-J, Liang S-D, Li Y-L, Guo Y-C, Shan G-Y (2021) Compressive behavior of FRP-confined elliptical concrete-filled high-strength steel tube columns. Compos Struct 266:113808
54. Gain AK, Oromiehie E, Prusty BG (2022) Nanomechanical characterisation of CF-PEEK composites manufactured using automated fibre placement (AFP). Compos Commun 31:101109
55. Oromiehie E, Gain AK, Donough MJ, Prusty BG (2022) Fracture toughness assessment of CF-PEEK composites consolidated using hot gas torch assisted automated fibre placement. Compos Struct 279:114762
56. Oromiehie E, Gain AK, Prusty BG (2021) Processing parameter optimisation for automated fibre placement (AFP) manufactured thermoplastic composites. Compos Struct 272:114223
57. Arns J-Y, Oromiehie E, Arns C, Prusty BG (2021) Micro-CT analysis of process-induced defects in composite laminates using AFP. Mater Manuf Process 1–10
58. Oromiehie E, Garbe U, Prusty BG (2019) Porosity analysis of carbon fibre reinforced polymer laminates manufactured using automated fibre placement. Compos Mater
59. Matti FN, Oromiehie E, Mashiri FR, Prusty BG (2022) Strengthening of steel beams with thermoplastic CFRP overwrapping using automated fibre placement. In: Proceedings of the 3rd International Conference on Structural Engineering Research (iCSER 2022), 20–22 November 2022, Sydney, Australia, pp 107–114
60. Online SA (2016) Steel structures, AS 1163–2016. SAI Global database 1998
61. Online SA (1998) Steel structures, AS 4100–1998. SAI Global database 1998
62. OneSteel Market Mills (2004) Cold formed structural hollow sections and profiles, 4th edn

Seismic Performance of Hybrid Structures Subjected to Extreme Earthquakes

Mohit Bhandari, Harmanpreet Singh, S. D. Bharti, and Mahendra K. Shrimali

Abstract The present study aims to the evaluation of the effectiveness of hybrid building systems to resist extreme earthquake forces. The study is conducted by considering a five-storey building frame with different variants consisting of different types of primary structural members: steel, reinforced cement concrete, concrete-filled steel tube (CFST), and truss system. The seismic performance of five variants of building frames is evaluated under the effect of near-field earthquakes also referred to as extreme earthquakes by performing nonlinear time history analysis (NTHA). The results obtained are compared for different cases considering various seismic parameters of the building model. The performance of hybrid frames was evaluated in terms of inter-storey drift, top-storey displacement, top storey acceleration and base shear. The study concludes that hybrid building systems consisting of CFST columns and steel beams perform outbound as compared to the other variants with the decreased storey drift, and displacement and the hybrid structure consisting of RCC beam produced lower accelerations.

Keywords Hybrid buildings · Near-field earthquake · Extreme earthquakes · Seismic performance · Nonlinear time history analysis

1 Introduction

Hybrid buildings are structures that combine different building systems and materials to achieve improved structural performance and sustainability. These buildings often incorporate elements of both traditional and modern construction techniques, such as using steel and concrete in combination with wood or other renewable materials.

M. Bhandari (✉) · H. Singh
University Institute of Civil Engineering, Chandigarh University, Mohali 140413, Punjab, India
e-mail: Mohit.e8967@cumail.in

S. D. Bharti · M. K. Shrimali
National Centre for Disaster Mitigation and Management, Malaviya National Institute of Technology, Jaipur 302017, Rajasthan, India

S. B. Singh and C. V. R. Murty (eds.), *RC Structures Strengthened with FRP for Earthquake Resistance*, Composites Science and Technology,
https://doi.org/10.1007/978-981-97-0102-5_4

Hybrid buildings are an emerging trend in the field of construction that has gained increasing attention in recent years [1]. These buildings are characterized by the integration of different types of building systems and materials to achieve improved structural performance and sustainability. The use of hybrid building systems can result in structures stronger, more durable, and more energy-efficient structures built using only one type of building system. Overall, hybrid buildings are becoming increasingly popular in the construction industry as they offer a balance of structural integrity, efficiency, and sustainability [2, 3].

Hybrid buildings are important for a number of reasons. Firstly, they offer improved structural performance by combining different types of building systems and materials. This can result in structures that are stronger, more durable, and better able to withstand natural disasters such as earthquakes, hurricanes, and floods. Secondly, hybrid buildings can be more energy efficient than traditional buildings due to the use of advanced materials and building technologies. This can lead to reduced energy consumption and lower operating costs for the building's occupants. Thirdly, hybrid buildings can have a lower environmental impact than traditional buildings by incorporating sustainable materials and construction techniques [4, 5].

Recent studies have shown that hybrid buildings have significant advantages over traditional buildings in terms of structural performance, energy efficiency, and sustainability. Zhang et al. [6] investigated the influence of shear connectors on the behavior of FRP-concrete-steel (FCS) hollow beams. The outcome of the study showed that the shear connectors play an important role in determining the behavior of the beams. The beams under study exhibited a ductile behavior, and the use of recycled aggregate concrete was found to be effective in enhancing their performance. Zhang et al. [7] outlined the design and development of two prototype structures using modelling techniques and the outcome of the nonlinear cyclic pushover and nonlinear dynamic analyses of the developed model. A comprehensive examination of the behavior of model structures during earthquakes, including the maximum floor movements during and after seismic activity, has been carried out. The results of the analyses showed that the self-centering hybrid-steel-frame (SC-HSF) structures exhibited a promising performance, with minimal residual drift.

Swarna and Reddy Suda [8] studied the comparison between Reinforced Cement Concrete (RCC) structures and Concrete Filled Steel Tube (CFST) structures for multi-story buildings in terms of seismic performance. The study considers multi-story structures with heights of 10, 20 and 30 storey. The results of the study showed that CFST structures have a better seismic performance compared to RCC structures. The storey drift of the CFST structure was less than the RCC structure by 47% but both remained within the permissible limit. Ghanem et al. [9] investigated the utilization of waste plastic in the construction industry in non-structural concrete beams. To investigate the effectiveness of the waste plastic sheets, the flexural behavior of concrete beams containing waste plastic meshes was studied. The study involved the preparation of beams with steel reinforcement and waste plastic sheets with varying widths and patterns, which then were subjected to increasing loads until failure. The waste plastic increased the moment capacity of the concrete beams and made them more ductile.

Poursadrollah et al. [10] presented a new approach to modular construction using hybrid systems made of lightweight steel truss girders and tubular columns connected by plug-and-play joints. The feasibility of using this system in low to moderate seismic zones is investigated through a case study that follows the requirements of EC3. The results of pushover analyses indicate that the main lateral resistance of the frame is provided by the joint's configuration, while plastic hinges tend to form in the base columns. The authors suggest that the use of this type of modular structure in low seismicity areas is feasible, as the inherent lateral resistance provided by the components' configuration is advantageous without the need for additional detailing or elements.

Gao et al. [11] evaluated the behavior of steel–concrete hybrid transfer joints through cyclic testing of four specimens. The research aimed to examine various strength properties along with the damage pattern of the joints. The results showed that all specimens exhibited stable load–displacement response, good energy dissipation capacity and ductility, with beam flexural breakdown as the prominent collapse pattern. Jiang et al. [12] studied the performance of high-rise hybrid structures in China. The authors compared the seismic behavior and construction cost of steel structures and hybrid structures and concluded that the latter system is a better choice. Further, static nonlinear (pushover) and dynamic nonlinear analyses were performed to determine the earthquake response of the hybrid structure. Further results indicated that the structure has satisfactory seismic performance and satisfies inspection requirements. Vineetha et al. [13] discussed the seismic response of hybrid RC-masonry structures. Additionally, the paper explores a framework for retrieving geometrical, structural, and material data for modeling during routine condition assessment of a chosen structure. The paper focuses on the seismic behavior of a specific hybrid RC-masonry institutional building constructed in India in the 1950s. Finally, the authors suggest that proper studies are required to capture the feasibility of modern hybrid structures, with wooden frames and RC foundations, in hilly regions.

CFST columns are a form of composite structure which combine the advantages of both concrete and steel. They form a hybrid structure that takes advantage of the inherent properties of steel and concrete. CFST columns have been widely used in various structures, including high-rise buildings, bridges, and offshore structures. They offer several benefits compared to traditional solid concrete columns or steel columns, such as improved axial strength, fire resistance, and seismic performance [14, 15].

In recent years, CFST columns have received increasing attention from researchers and engineers due to their potential for providing efficient and economical solutions for various structural applications. Despite the numerous benefits of CFST columns, their behavior under extreme loading conditions and their ability to resist lateral forces are still being studied. The continued development of design guidelines, construction techniques, and advanced computational tools has increased the understanding of the behavior of CFST columns and allowed for their greater use in structural engineering [16]. CFST columns represent a promising option for modern structures due to their combination of high strength, fire resistance, and seismic performance. Further research is necessary to further enhance our understanding of CFST columns and their behavior under extreme loading conditions, but they hold great potential for use in a variety of structural applications [17, 18].

He et al. [19] investigated the behavior of Concrete-Encased CFST (CE-CFST) and tests were performed to evaluate the strength of CE-CFST columns. Further, results captured a decrease in ultimate strength of column under evaluation with an increase in initial stress, with a peak decrease of 9.3%. Authors developed a FE model and based on the findings, proposed a methodology to determine the effect of inceptive stress on the maximum load capacity of a Circular Eccentrically Loaded Composite Filled Steel Tube Column. Swami et al. [20] investigated the influence of CFST columns building connections on the stability of composite modular buildings. Numerical models were established to assess the validity of the study. The results showed that incorporating CFST columns into a 10-story composite modular building improves resistance against buckling under high axial forces.

Zhang et al. [21] examined the behavior of composite frame structures subjected to seismic excitation, specifically those composed of multi-cellular T-shaped concrete-filled steel tubular (TM-CFST) columns and steel beams. The study involved physical testing of four specimens under both compressive loads and lateral cyclic loads, as well as a refined finite element (FE) modeling approach. A comparison between FE and experimental results showed good agreement, validating the FE modeling method. In conclusion, the study found that TM-CFST frames are suitable for low to medium-rise building structures and have good structural properties.

Guo et al. [22] evaluated the seismic capacity of (CECFST) columns connected to steel beams using bolted joints. Experimental tests were performed on bolted CECFST column joints to analyze their behavior and identify typical failure modes. An analytical model was also established using the component method. The study confirmed the accuracy of the proposed method through experimentation and found it to be useful in the practical engineering design of bolted joints. A full-scale CECFST moment resisting frame was evaluated for its seismic behavior in different seismic events. The analysis revealed that the CECFST moment frame displayed exceptional performance under earthquakes and exhibited high resilience and resistance to collapse.

Han et al. [14] performed a review study on the recent advancements in CFST research and application, particularly in China. A brief examination of the design approaches used in different countries is also presented. The study highlights the characteristics of CFST structures and considers it as an alternative to traditional steel or reinforced concrete systems, but further evaluation is needed to fully understand their feasibility and potential for wider application. The performance of CFST was analyzed in a study where seven column specimens were tested with varying internal designs and shear-span ratios. The outcomes indicated that these internal designs significantly delayed the local buckling of the steel tube, resulting in an enhancement in behavior of the CFST columns Dong et al. [23]

Despite the benefits offered by CFST columns, there are also challenges associated with their use, such as the need for specialized design and construction knowledge, and the potential for durability issues with the filled concrete. To fully assess the feasibility of CFST columns for widespread use, it is recommended that a thorough comparison of its advantages and disadvantages be conducted and that the life-cycle performance of CFST structures be evaluated.

In conclusion, the literature review highlights the development of CFST columns and the current state of research on this type of structural system. While CFST columns offer numerous benefits over traditional steel and RC systems, further research is needed to fully evaluate their feasibility and potential for widespread use in civil engineering projects.

Overall, the literature review on hybrid buildings highlights the potential benefits and challenges of this emerging trend in the construction industry. However, it is also evident that further research is needed to fully understand the implications of hybrid building design and construction and to develop practical solutions for overcoming the challenges associated with these structures.

With this background in view, this study attempts to evaluate the seismic performance of the hybrid building under extreme earthquakes considering a five-storey 2-D frame building for the analytical study.

2 Modeling and Designing of Buildings

A five-storey building frame is considered in this study. The frame is made hybrid by adopting different types of members varying in material like steel, reinforced concrete, and CFST. A total of five variants of the frame is considered as described in Table 1. The modeling of the frame is done in the commercial structural analysis and design software ETABS. The frame consists of five storey levels of equal height of 3.2 m having five bays only in the longitudinal direction of an equal span of 5 m each. The structural members, beams and columns, are modelled using line elements and material properties are assigned respectively.

The compressive strength of concrete is taken as 25 MPa; Poisson's ratio of concrete is 0.2, and the elastic modulus of concrete is 25000 MPa. The steel properties include a yield strength equal to 500 MPa, Ultimate strength is 545 MPa. For the CFST members, concrete of compressive strength 25 MPa and grade having a yield strength of 345 MPa is used.

End length offsets are considered in the modeling and the value is auto-calculated from the member sizes. A rigid zone factor at the beam-column joint is assumed to equal to 0.5 for concrete, 1 for steel and 0.8 for CFST members. The summary of the sizes of different model variants is given in Table 1. The nonlinearity in the steel and concrete members is modelled with the help of plastic hinges located at the end of members and auto hinge properties were generated by the ETABs adopting the provisions of ASCE 41–17. The elevation view of the typical model variants is shown in Fig. 1.

The designing of the building frames is carried out in the ETABS. The steel and concrete members are designed as per Indian standards following IS 456–2000, IS 1893–2016, IS 13920–2016. The CFST members are designed as per the American

Table 1 Details of members sizes of different variants of building models

Building model ID	Type of building	Column		Beam	
		Type	Size (mm)	Type	Size (mm)
B-1	Hybrid	Concrete	300 × 300	Steel	ISMB 450
B-2	Hybrid	CFST	300 × 300 × 12	Steel I-section	ISMB 450 3–4–5 storey ISMB 500 1–2 storey
B-3	Hybrid	CFST	300 × 300 × 12	Concrete	230 × 500
B-4	Conventional	Steel H	ISWB 450	Steel I	ISMB 450 3–4–5 storey ISMB 500 1–2 storey
B-5	Conventional	Concrete	300 × 300	Concrete	230 × 300

code, AISC 360–16. The assumed loadings on the frames are identical for each variant; the dead load is 12 kN/m.

3 Analytical Study

The seismic performance of five-storey hybrid building frames is evaluated under near-field earthquakes which are referred to as extreme earthquakes in this study. The Near-field earthquakes are reordered at a rupture distance of less than 20 km. The distinct characteristic is the presence of high-velocity pulses having high amplitude and long period. Due to these characteristics, near-field earthquakes are capable of inputting high seismic energy in the building system in few seconds enabling it to yield high responses. A set of real earthquake records as given in Table 2 is used to carry out nonlinear time history analysis to compute the responses of the building models. The seismic performance is evaluated in terms of inter-storey drift, storey displacement, base shear, and floor accelerations. The analysis is performed at a scaled PGA level of 0.4 g.

4 Discussion of Results

The performance of the hybrid building frame system is compared in terms of different response parameters like inter-storey drift, storey displacement, Base shear, and floor accelerations. The detailed discussion on each parameter is discussed in the below section.

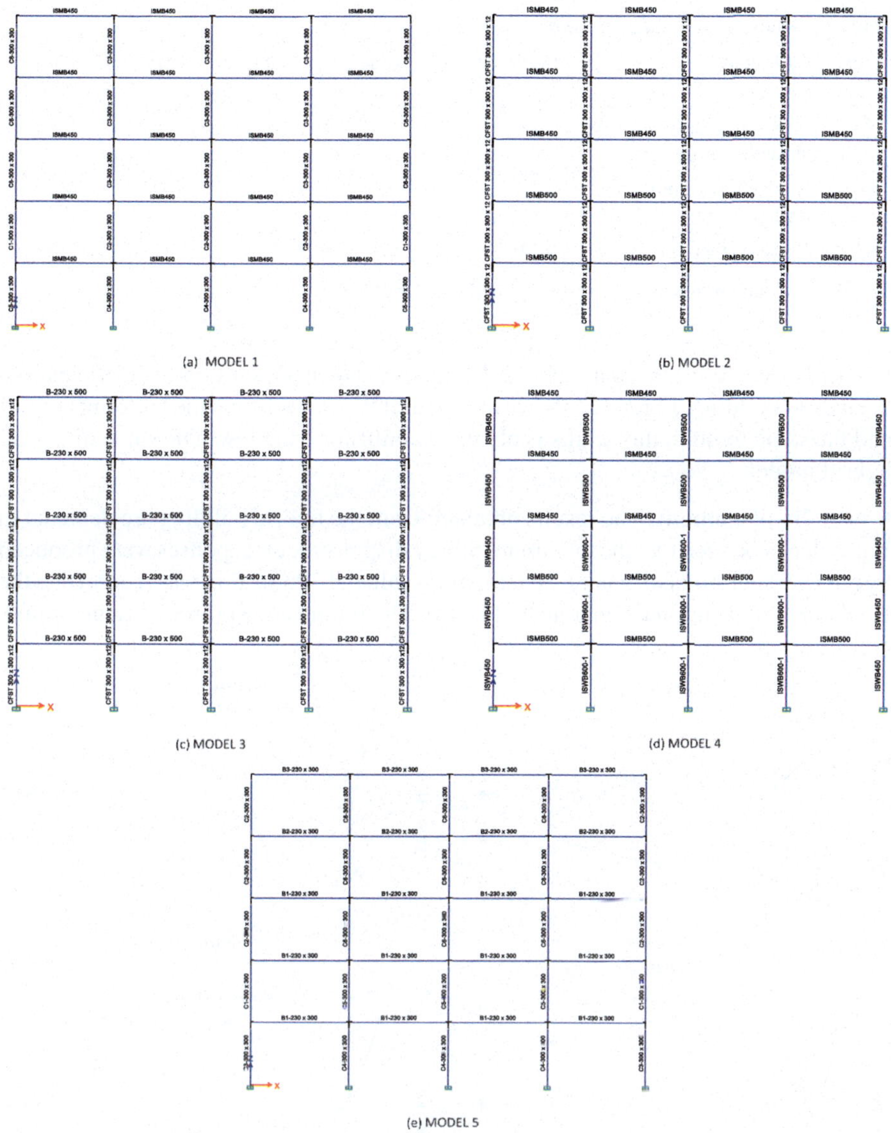

Fig. 1 Elevation view of building models

Inter-storey Drift: The trend in the results yielded by the hybrid building is presented in Fig. 2. It is evident from the pattern of the IDR of the different building frames that the maximum IDR responses have been yielded by the Kobe earthquake for which a maximum value of IDR (0.045) is yielded in the B5 model (conventional concrete buildings). The minimum value of IDR (0.0034) is produced in the B4 model (conventional steel model). The hybrid model B2 (CFST column and Steel

Table 2 Details of earthquake records

Year	Earthquake (Station)	M_w	PGA (g)	PGV (cm/s)	PGD (cm)	PGV/PGA	R_{jb}
1979	Imperial valley (Brawley airport)	6.5	0.16	36	26	225	8.54
1995	Kobe (Port island 0m)	6.9	0.35	90	39	257	3.31
1999	Kocaeli (Gebze)	7.5	0.14	33	30	235	7.57

Note Records taken from https://ngawest2.berkeley.edu/

Beam) is closely incoherent with the B4 model, which shows good performance of the hybrid building. Figure 2d shows the mean IDR values for all the building models and the same trend in the results is obtained confirming the low IDR value of the B2 hybrid model.

Storey Displacement: The results obtained from the hybrid building are depicted in Fig. 3. It can be seen that the maximum Storey displacement responses were produced by the Kobe earthquake, where the maximum value of 393.65 mm was recorded in the B5 model (conventional concrete buildings). The Indian code prescribes a maximum

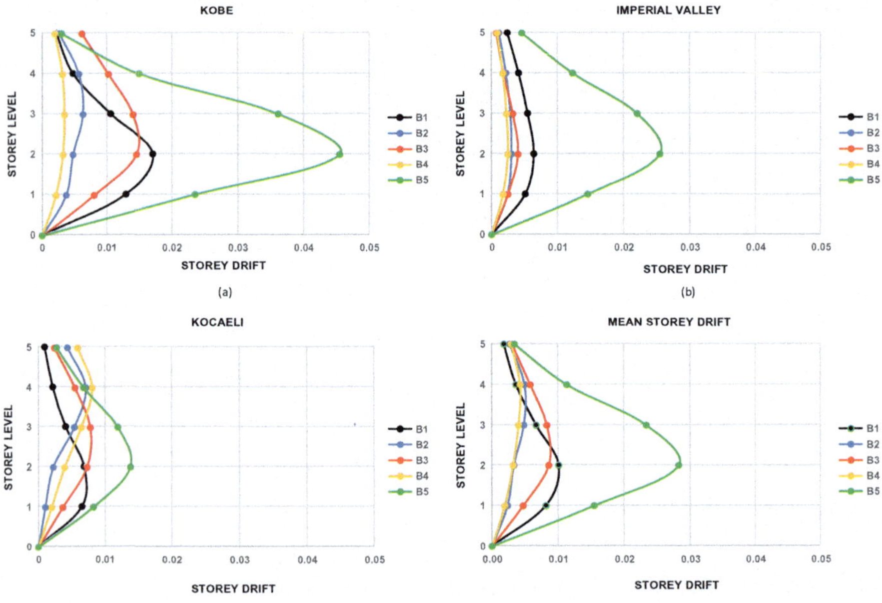

Fig. 2 Variation in the IDR along the storey level of different types of buildings subjected to different earthquakes; **a** Kobe earthquake; **b** Imperial valley; **c** Kocaeli; **d** Mean drift values of all earthquakes

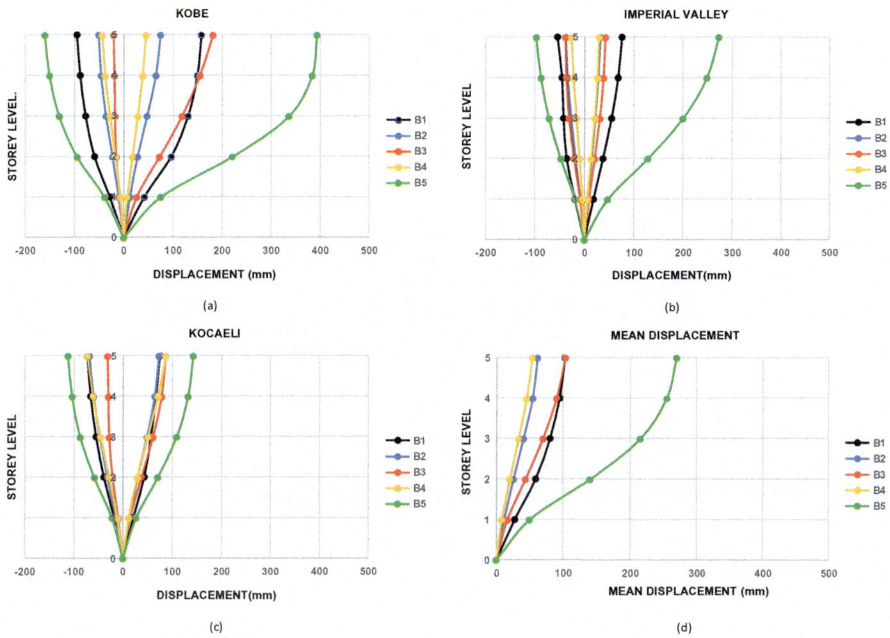

Fig. 3 Variation in the displacement along the storey level of different types of buildings subjected to different earthquakes; **a** Kobe earthquake; **b** Imperial valley; **c** Kocaeli; **d** Mean displacement values of all earthquakes

value of storey displacement as H/250 i.e. 64 mm in this case. The B4 model (conventional steel structures) recorded the lowest value of storey displacement at 44.58 mm. A similar trend as in storey drift was observed where the hybrid model B2, which features CFST columns and steel beams, exhibits results that are comparable to those of the B4 model, demonstrating the favorable behavior of the hybrid building subjected to near-field earthquakes.

Base Shear:

The base shear is a crucial factor in the design of structures, as it represents the total horizontal force acting at the base of the building or structure due to various external loads, such as wind, earthquakes, or other similar forces. The magnitude of the base shear is used to determine the necessary strength and stiffness of the structural elements and components, including the foundation, walls, columns, beams, and connections.

The results of the analysis shown in Fig. 4. indicate that the maximum base shear was experienced by the B4 model, which was anticipated as the same model displayed the minimum storey displacement and storey drift. Furthermore, the B4 model also showed the highest deviation in the base shear values, followed by the B2 model, highlighting that the response of these structural combinations is highly dependent on the type of earthquake. It is noteworthy that the B5 structure (conventional concrete)

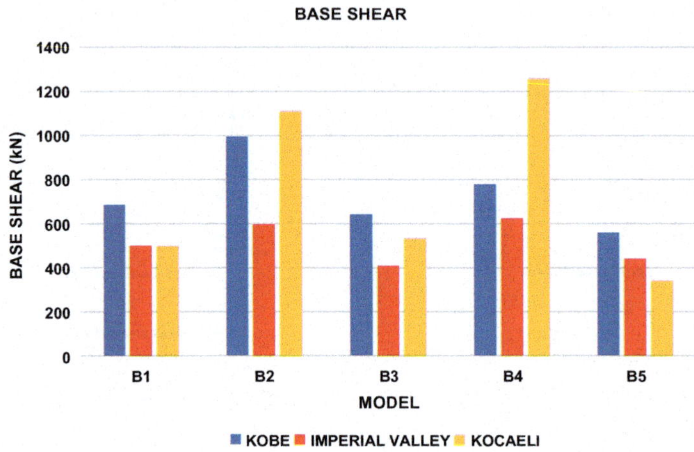

Fig. 4 Variation in the base shear yielded in the building model under three earthquakes

was unable to withstand the complete time history excitation in the case of the Kobe earthquake.

Top Floor Acceleration:

This metric measures the dynamic response of a structure to external loads, including earthquakes, wind, and other similar forces, and is an essential consideration in the design and analysis of structures. The Floor acceleration provides information on the behavior of the structure under dynamic loads and the distribution of forces throughout the structure.

The study of floor accelerations in various structures reveals the dynamic behavior and response of each system to external loads. The results of this study, as presented in Figs. 5, 6, 7, and 8, illustrate the joint acceleration experienced at the top-story corner joint. The analysis demonstrates that each system has its unique strengths and weaknesses in terms of its ability to resist joint accelerations.

In particular, the data highlights the highest joint acceleration in the B4 model during the Kocaeli earthquake, followed by the B2 model, suggesting that incorporating steel into the construction increases joint acceleration. Conversely, the B5 model, which employs conventional concrete, displayed the lowest joint acceleration for all time histories included in the study.

It is important to note that the structural system with the highest joint acceleration values does not necessarily imply weakness or a higher likelihood of collapse. Rather, this characteristic is dependent on various factors, including the type of load, the distribution of mass and stiffness, and the type of connections.

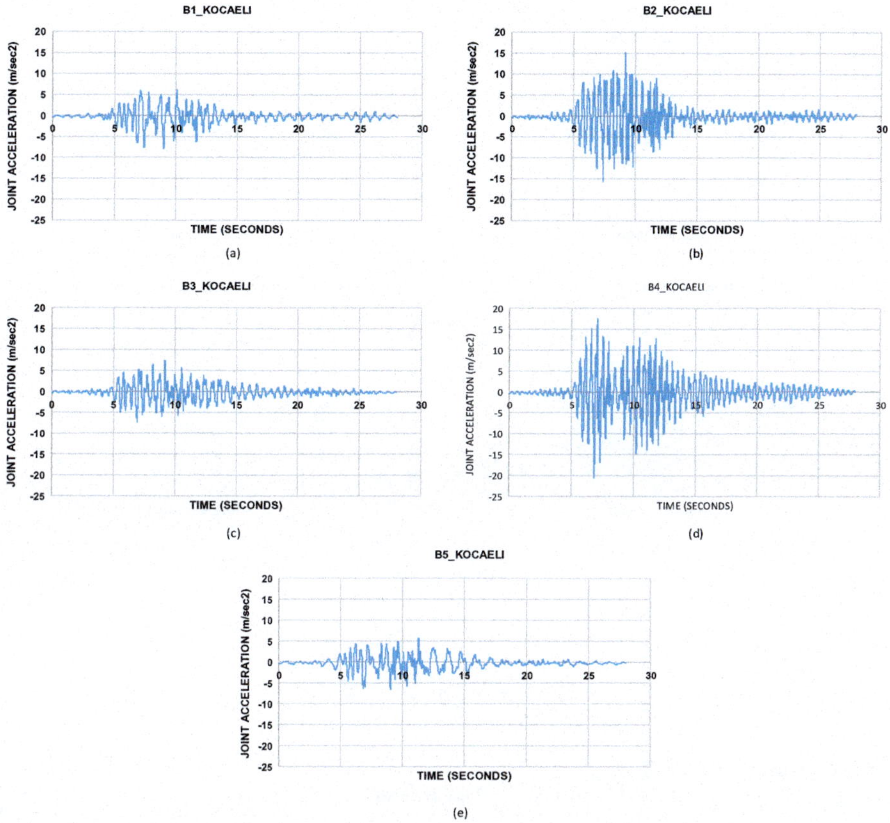

Fig. 5 Top floor acceleration time history for kocaeli earthquake for different building models

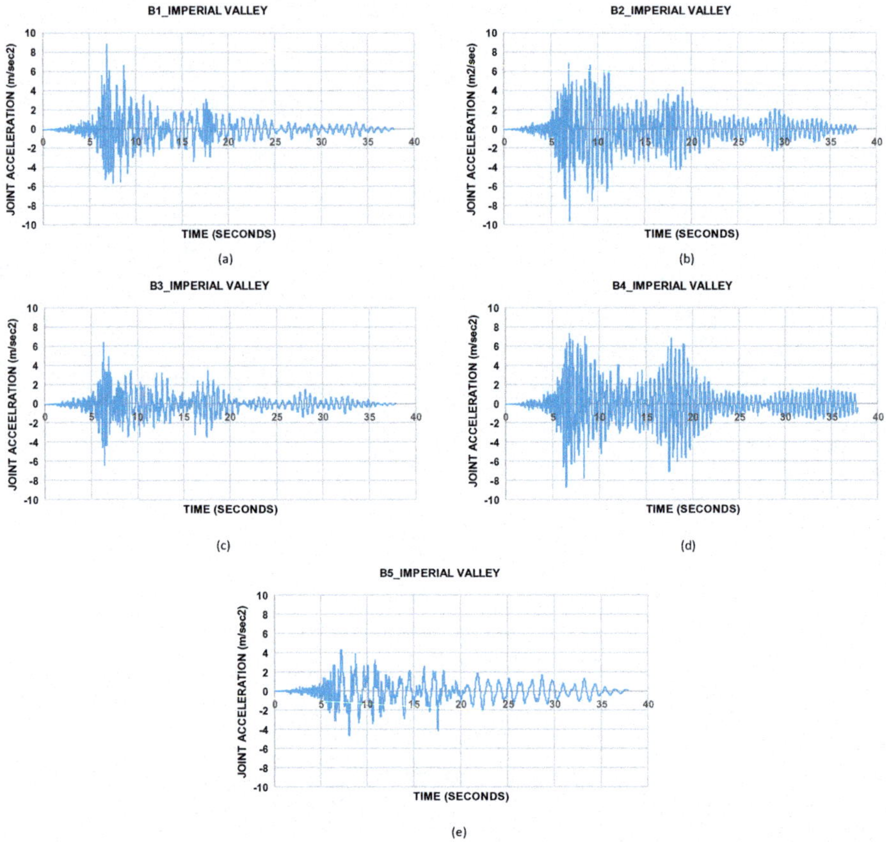

Fig. 6 Top floor acceleration time history for imperial valley earthquake for different building models

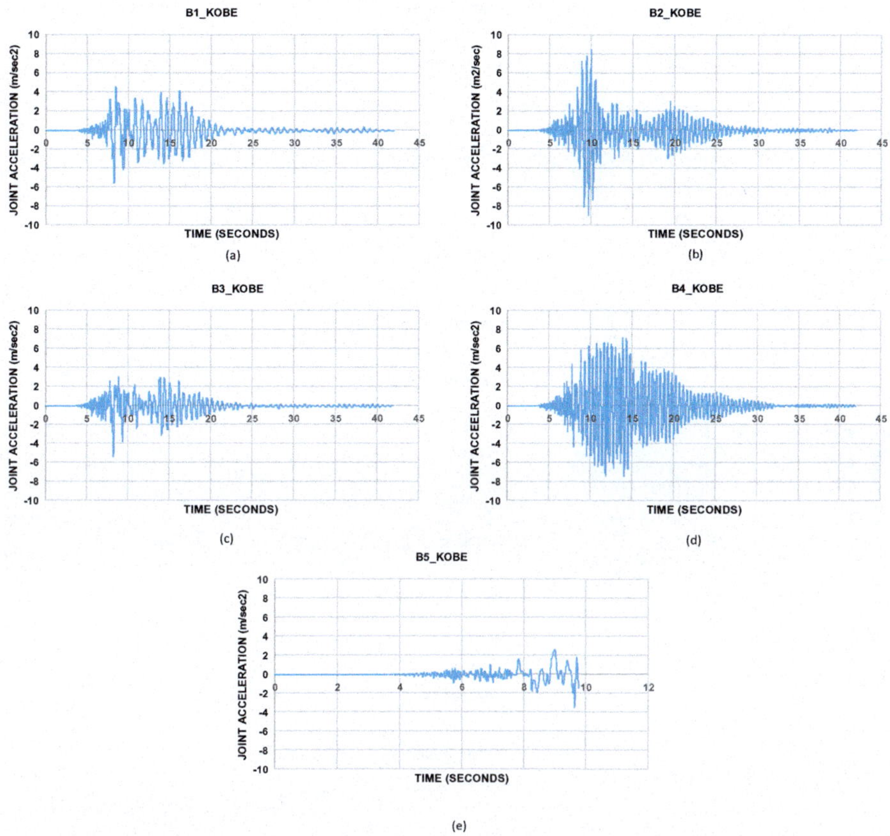

Fig. 7 Top floor acceleration time history for kobe earthquake for different building models

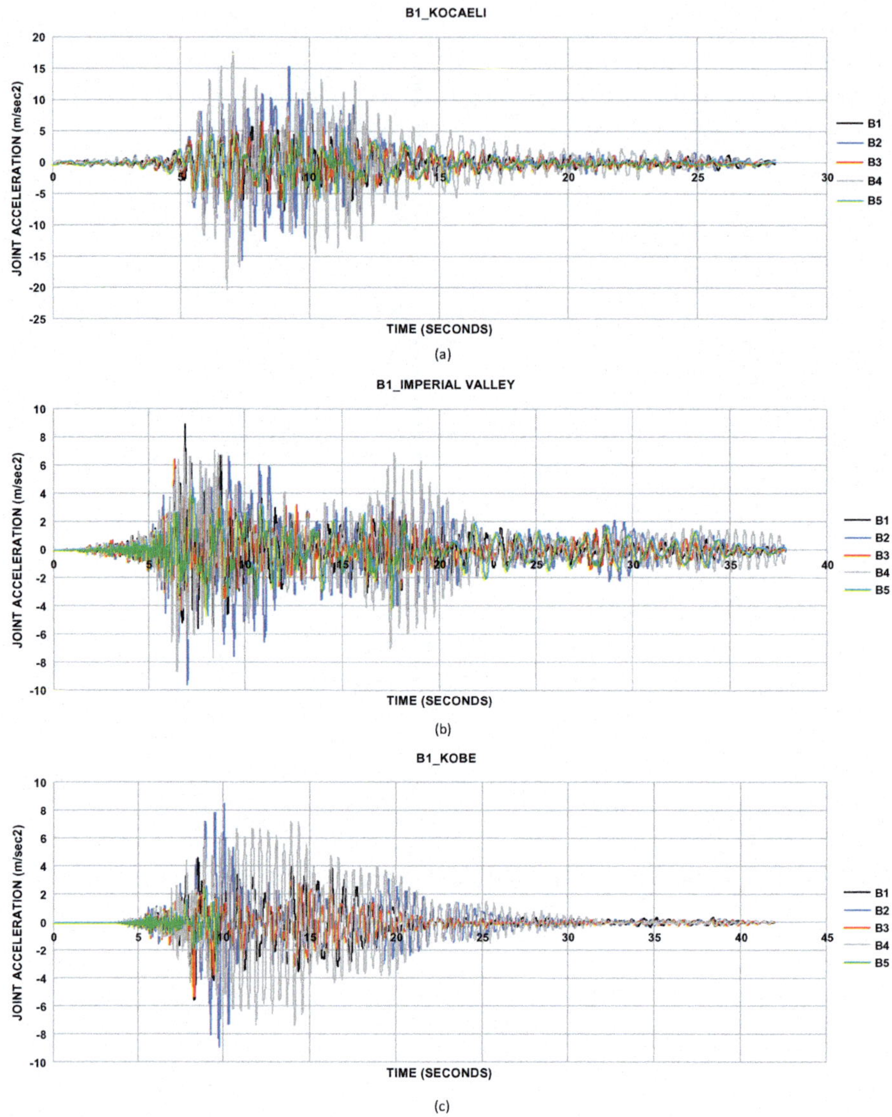

Fig. 8 Mean top floor acceleration time history for three considered earthquakes for different building models

5 Conclusion

The performance of hybrid building models is evaluated under extreme earthquakes for 5 storey building model. The Numerical analysis is conducted by performing a time history analysis employing a suite of three near-field real earthquake records.

The responses of the building models are recorded and compared for different seismic response parameters. On the basis of the assumed building models, constraints, and earthquake data, the important conclusions drawn from this study are appended below:

- The maximum IDR values were produced by the Kobe earthquake in the B5 model (conventional concrete buildings), with the minimum value recorded in the B4 model (conventional steel structures).
- The maximum storey displacement was recorded in the B5 model (conventional concrete buildings), with the lowest value in the B4 model. The hybrid model B2 exhibited results comparable to the B4 model.
- The B4 model showed the highest base shear value, with a higher deviation compared to other models. The B5 model was unable to withstand the complete time history excitation of the Kobe earthquake.
- The highest floor acceleration was recorded in the B4 model during the Kocaeli earthquake, followed by the B2 model. The B5 model (conventional concrete) had the lowest joint acceleration.
- In conclusion, the results from the analysis of five hybrid and conventional structures have shown that the hybrid structure with a steel beam and CFST Column performs similarly to a conventional steel building and significantly better than a conventional RCC structure in terms of serviceability. These findings demonstrate the potential of hybrid structures to provide superior structural performance compared to conventional structures and highlight the importance of considering hybrid structures as a viable alternative in construction projects.

References

1. Roeder CW (1998) Overview of hybrid and composite systems for seismic design in the United States. Eng Struct 20:355–363
2. Wang Z, Pan W (2020) A hybrid coupled wall system with replaceable steel coupling beams for high-rise modular buildings. J Build Eng 31:101355
3. Loss C, Piazza M, Zandonini R (2016) Connections for steel–timber hybrid prefabricated buildings. Part II: Innovative modular structures. Constr Build Mater 122:796–808
4. Stiemer S, Tesfamariam S, Karacabeyli E, Propovski M (2012) Development of steel-wood hybrid systems for buildings under dynamic loads. In: Proceedings of the 7th international specialty conference on behavior of steel structures in seismic areas (STESSA). Santiago, Chile
5. Ji X, Liu D, Molina HC (2018) Seismic performance evaluation of a high-rise building with novel hybrid coupled walls. Eng Struct 169:216–225
6. Zhang L, Zhang J, Chen G, Lin G (2023) Flexural behavior of hybrid FRP-recycled aggregate concrete-steel hollow beams. J Constr Steel Res 200:107650
7. Zhang H, Zhou X, Ke K, Yam MCH, He X, Li H (2023) Self-centring hybrid-steel-frames employing energy dissipation sequences: insights and inelastic seismic demand model. J Build Eng 63:105451

8. Swarna C, Reddy Suda VB (2022) Comparison of seismic analysis of multi storey buildings with RCC columns and CFST columns by varying heights. IOP Conf Ser: Earth Environ Sci 1086:012009
9. Ghanem H, Chahal S, Khatib J, Elkordi A (2022) Flexural behavior of concrete beams reinforced with recycled plastic mesh. Buildings 12:2085
10. Poursadrollah A, D'Aniello M, De Martino A, Landolfo R (2020) Preliminary study on the seismic performance of hybrid steel structures with truss lightweight girders and plug-and-play connections. Int J 37
11. Gao J-Y, Nie X, Ding R, Fan J-S (2020) Experimental study on seismic performance of a new transfer joint in the steel-concrete vertical hybrid structure. J Constr Steel Res 174:106259
12. Jiang J, You B, Hu M, Hao J, Li Y (2008) Seismic design of a super high-rise hybrid structure. In: The 14th world conference on earthquake engineering, pp 7–12
13. Vineetha N, Menon A, Gettu R (2012) Seismic response of hybrid buildings. 15 WCEE Lisboa
14. Han L-H, Li W, Bjorhovde R (2014) Developments and advanced applications of concrete-filled steel tubular (CFST) structures: members. J Constr Steel Res 100:211–228
15. Wang X, Fan F, Lai J (2022) Strength behavior of circular concrete-filled steel tube stub columns under axial compression: a review. Constr Build Mater 322:126144
16. Gunawardena YKR, Aslani F, Uy B, Kang W-H, Hicks S (2019) Review of strength behaviour of circular concrete filled steel tubes under monotonic pure bending. J Constr Steel Res 158:460–474
17. Wang Z-B, Tao Z, Han L-H, Uy B, Lam D, Kang W-H (2017) Strength, stiffness and ductility of concrete-filled steel columns under axial compression. Eng Struct 135:209–221
18. Li G, Hou C, Shen L, Yao G-H (2022) Performance and strength calculation of CFST columns with localized pitting corrosion damage. J Constr Steel Res 188:107011
19. He F, Li C, Chen B, Liu J, Zhang M, Briseghella B (2023) Compressive behavior of concrete-encased CFST column with initial stress. J Constr Steel Res 201:107724
20. Swami G, Thai H-T, Liu X (2023) Structural robustness of composite modular buildings: the roles of CFST columns and inter-module connections. Structures 48:1491–1504
21. Zhang J-W, Tong J-Z, Zhang L, Tong G-S, Li X-G (2022) Seismic performance evaluation on composite frame structures with T-shaped multi-cellular CFST columns. J Constr Steel Res 199:107634
22. Guo L, Wang J, Wang W, Hu Z (2022) Experimental study and analytical evaluation on seismic performance of CECFST moment resisting frame with bolted connections. Eng Struct 259:114074
23. Dong H, Qin J, Cao W, Zhao L (2022) Seismic behavior of circular CFST columns with different internal constructions. Eng Struct 260:114262

Seismic Strengthening and Retrofitting of RC Structures Using Fibre Reinforced Composites

R. Siva Chidambaram, Naveen Kumar Kothapalli, and Pankaj Agarwal

Abstract The conventional practices of strengthening RC elements like Reinforced Concrete (RC) jacketing, Ferro cement jacketing, and Steel jacketing lead to an increased cross-sectional size of the member, which interrupts the functionality and proves to be nondurable when exposed to an aggressive environment. The use of Fiber Reinforced Polymer (FRP) composite technique in strengthening work addresses these issues over the other conventional practices. The simpler execution practice, high strength-to-weight ratio and faster setting time make FRP an efficient material in RC strengthening works. Many varieties of FRP, such as Glass, Carbon, Basalt, Aramid, etc., are available in construction practice that have unique tensile properties. The properties of tensile strength, Young's modulus and rupture strain of FRP play a major role in the strength and post-yield deflection of the associated RC members. The property of adhesive used as a bonding source between FRP and concrete plays a significant role in transferring the forces and determining the failure mode. Strengthening techniques primarily depend on the demand/deficiency of structural elements. This chapter presents an overview of the strengthening techniques and experimental behaviour of FRP-strengthened RC elements.

Keywords Fibre Reinforced Polymer (FRP) · Flexural behaviour · Shear strengthening · Beam-column joint strengthening · Fracture behaviour · Retrofitting

---

R. Siva Chidambaram
Advanced Concrete, Steel and Composites Division, CSIR-Central Building Research Institute, Roorkee 247667, Uttarakhand, India
e-mail: schidambaram@cbri.res.in

N. K. Kothapalli · P. Agarwal (✉)
Department of Earthquake Engineering, Indian Institute of Technology Roorkee, Roorkee 247667, Uttarakhand, India
e-mail: pankaj.agrawal@eq.iitr.ac.in

N. K. Kothapalli
e-mail: knaveen@eq.iitr.ac.in

S. B. Singh and C. V. R. Murty (eds.), *RC Structures Strengthened with FRP for Earthquake Resistance*, Composites Science and Technology,
https://doi.org/10.1007/978-981-97-0102-5_5

1 Introduction

The strength enhancement of a deficient structural element and retrofitting of damaged RC elements in moment-resistant framed structures can be done in various ways, such as steel plate bonding, steel jacketing, RC jacketing, ferro-cement jacketing, and Fiber Reinforced Polymer (FRP) jacketing. FRP is an efficient strengthening technique composed of fibres with a high strength-to-weight ratio and resins that bind them together. A reduction in additional dead load on a retrofitted member is achieved using FRP strengthening compared to conventional techniques. FRP is non-corrosive and flexible in nature and can be moulded into any shape or configuration. Strength attainment in the case of FRP is faster and requires less setting time. The flexible nature of FRPs favours strengthening structural elements such as beams, columns, and slabs without any need for sophisticated equipment. The FRP wrapping pattern depends on the structural element's type and requirement, such as the confinement of a column being increased by wrapping it on its lateral face and the flexural strengthening of the RC beam under bending by affixing it on its soffit face. Shear strengthening is accomplished by adapting the composites in the deficient shear zone with fibres oriented parallel to the loading direction. The externally wrapped FRP effectively enhances the peak strength and stiffness rather than the deflection. Strength upgradation in the FRP technique depends on the type of FRP, the pattern in which it is established, and the chemical properties of the epoxy used. There are many types of FRP, such as CFRP (carbon fibre), GFRP (glass fibre), BFRP (basalt fibre), and AFRP (Aramid fibre), available for strengthening and retrofitting works. The mechanical properties, such as strength, modulus, and rupture strain, vary with reference to the FRP chosen. The common failure pattern observed in FRP strengthening technique is the delamination and debonding from the concrete substrate. The fire resistance of an FRP-strengthened member primarily depends on the resin rather than the temperature resistance of the fibres used. Significant degradation in stiffness has been observed when the temperature at the concrete–FRP interface of a strengthened member is greater than the glass transition temperature [1].

Flexural strengthening of RC beams:

The flexural strength capacity of RC beams can be enhanced through the application of External Bonding (EB) and Near-Surface Mounted (NSM) FRP techniques. The performance of these strengthened beams relies on various factors such as material quality, strengthening methods, conditions under which they are subjected to loads and other environmental factors. Flexural strength enhancement can effectively be achieved by affixing FRP laminates at the tension face of beams, typically located on the underside, known as the soffit face, as shown in Fig. 1a. Beams that are completely enveloped by FRP sheets demonstrate exceptional performance. However, the practical implementation of this technique can pose considerable complexities. Use of the partial wrapping technique, often referred to as the "U-wrapping," can significantly improve the flexural strength, stiffness, and resistance to deflection in reinforced concrete beams, as shown in Fig. 1b. The thickness of FRP plays a crucial role in the

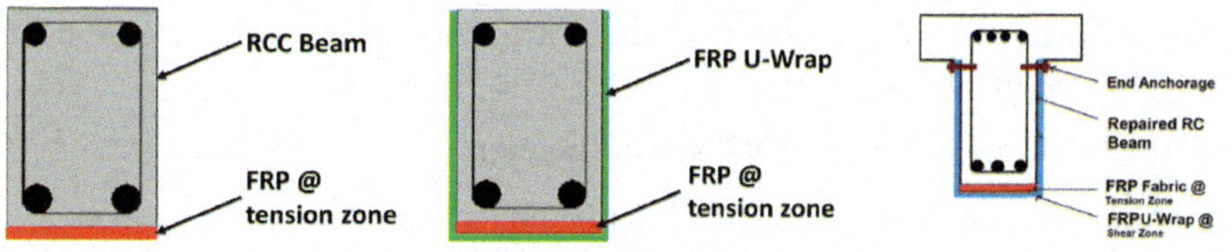

(a) Flexural strengthening (b) Flexural & U-shaped wrap (c) FRP Strengthening with end anchorage

Fig. 1 Typical FRP strengthening and anchorage in RC beams

strengthening process, as demonstrated by bending tests showing a 39% reduction in deflection for beams U-wrapped with two layers of composite as compared to those with a single layer [2]. A combination of Carbon and Glass FRP layers in a U-wrapped configuration can increase the load capacity of RC beams by 103% [3]. The ultimate effectiveness of EB wrapping is contingent upon the adhesive bond between the FRP sheet and the concrete substrate. To prevent premature bond failures, particularly those of a critical or brittle nature, it is recommended to employ additional measures such as intermediate anchorage, mechanical fastening, and strengthening with NSM FRP strips, Fig. 1c.

Shear strengthening using FRP composites:

Shear strengthening in RC beams can be executed in various ways, i.e. by adopting continuous sheets or strips in U-shape, side bonding, and Full wrap techniques. Figures 2 and 3 show the typical shear-strengthening patterns. Externally wrapped FRP reinforcement was considered similar to the internal shear reinforcement, assuming that at the ultimate limit state, the effective strain in FRP is smaller than the ultimate tensile strain of FRP. References [4, 5] investigated shear-strengthened RC beams with different fibre orientations and distributions, resulting in a conclusion that a large amount of externally bonded CFRP composites don't exhibit similar increments in failure loads. Several factors, including a moderate low FRP spacing, shear span-to-depth ratio, and transverse reinforcement, can improve the efficiency of shear strengthening. The orientation of FRP wrapping plays a pivotal role in dictating the crack pattern within the reinforced beam. For instance, beams featuring the FRP wrap along the 45° inclined direction demonstrate the ability to deter the formation of diagonal cracks. In contrast, beams with fibres wrapped along the 0 and 90° directions are less effective [6]. In cases where shear cracks occur in beams strengthened with diagonal FRP wrapping, their propagation tends to follow a downward trajectory upon reaching the ultimate load. The specific angle at which these shear cracks manifest can vary, typically falling within the range of 30–60°. This variation is influenced by factors related to the shear strength of the FRP system, as documented in references [7–9]. Prior to the shear-strengthening design measures, it is imperative to undertake a comprehensive analysis of crack size, shape, and orientation, as these factors wield a unique influence on both the design strength and the ensuing failure characteristics. Expanding the surface area covered by FRP can substantially augment the structural capacity.

Importance of anchorage in FRP strengthening:

The efficiency and efficacy of FRP strengthening schemes vary with respect to the type of fibre, orientation of fibre strands, and influence of anchorage. In shear-strengthened beams wrapped with CFRP, it is observed that the FRP wrap acts as an anchorage and improves the shear capacity of the beam, which alters the failure mode from brittle to ductile [10, 11]. Despite of the higher tensile modulus of CFRP, [12] observed a sudden failure of the strengthened RC beam after debonding of U-wrapped strips due to the absence of anchorage. A lower modulus fibre sheet has better efficiency than strengthening using multiple fibre layers and higher modulus fibre. In design, the allowable strain value of FRP is limited by codal guidelines, and it

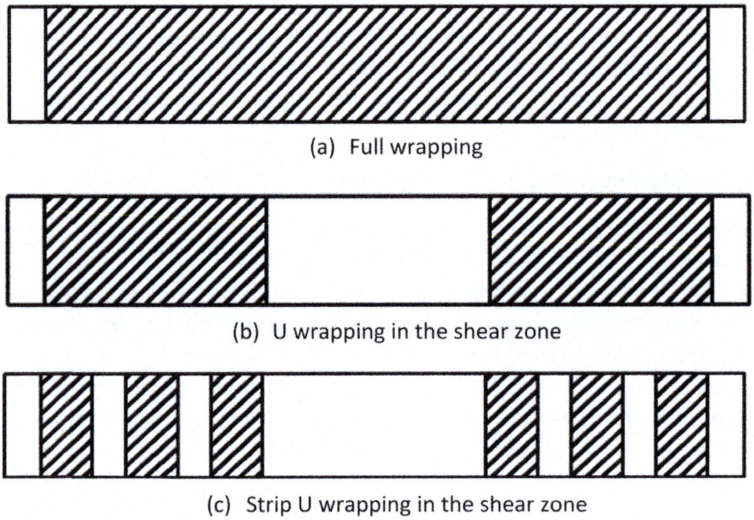

Fig. 2 Typical FRP wrapping in shear-deficient RC beams

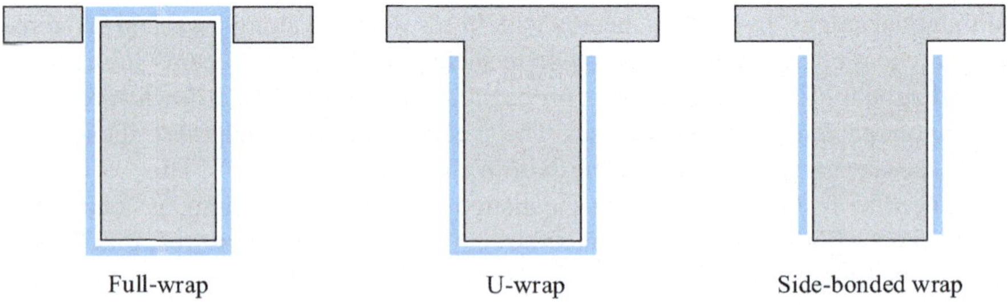

Fig. 3 Configurations for the shear strengthening of structures using FRP composites

can be as low as 10–25% of the rupture strain in cases of higher degree strengthening [13]. Adequate anchorage can be adopted to effectively use FRP strain capacity in achieving the design level and overcoming their shortcomings. Many studies have been conducted to prevent debonding failure with the use of FRP U-jackets across the span or at a specific spacing. References [14, 15] observed concrete cover separation failure rather than debonding failure when plate-end FRP U jackets were added. Compared to specimens strengthened using NSM FRP or EB FRP with no anchorage techniques, an appropriately designed end-anchorage system can contribute to better shear resistance. The use of self-end anchorage using a groove at the beam-slab corner showed 29% higher shear strength than the FRP-strengthened beam without anchorage. Further, a 2.5 times higher deflection is noticed in the anchored specimen. A specimen with sandwich anchorage showed a 12% enhancement in strength and a more than two fold increase in deflection than the specimen without anchorage [16]. These observations clearly manifest that the anchorage system in FRP strengthening plays a crucial role in enhancing strength and strain.

FRP strengthening in beam-column joints:

Beam–column joints play a vital role in resisting the lateral forces during an earthquake event. The combined forces of compression, tension and shear make the joint more susceptible in an RC-framed structure. The strong-column weak-beam design philosophy ensures that the failure is concentrated in the beam hinge regions instead of the column and joints, enhancing the structure's rotational capacity. Plastic hinge formation in the beam improves energy dissipation and displacement ductility. The reinforcement detailing in joint region plays a crucial role in achieving ductility and damage tolerance. Many joint failures during past earthquakes were caused due to improper detailing in the joint region. Inadequate confinement and anchorage lead to early rebar slip and allow the rebar to buckle, which leads to joint collapse.

The use of FRP in strengthening beam-column joints is one of the most effective solutions due to its high strength-to-weight ratio. This technique notably enhances the strength and stiffness of the joints. Many types of strengthening schemes or patterns are shown in Fig. 4, in which the feasibility of wrapping mostly relies on the hinge region rather than the joint core region due to its inaccessibility. The common practice is the use of L-shaped fabric to connect the column and beam at the hinge region, followed by confinement using FRP. The interior part of the building joint core region is not accessible compared to the corner joints of a building. The confinement in actual practice may be restricted to U-wrap due to the presence of a slab, or in some cases it may be full wrap through the holes made, as depicted in Fig. 3.

The strength and stiffness enhancement dominates the energy dissipation potential [17, 18]. The yielding phenomenon in FRP primarily depends on the area fractions, strength, modulus and anchorage used in the system, as discussed in the earlier sections. Also, flexible fabric sheets perform better than solid stirps [19]. Various factors, such as the number of layers, anchorages, joint internal retrofitting strategy, reinforcement ratio, and concrete strength play vital roles in the seismic performance of FRP-strengthened joints. A proper anchorage system enhances the post-yield behaviour of the hinge region and compels the joints to dissipate more energy than

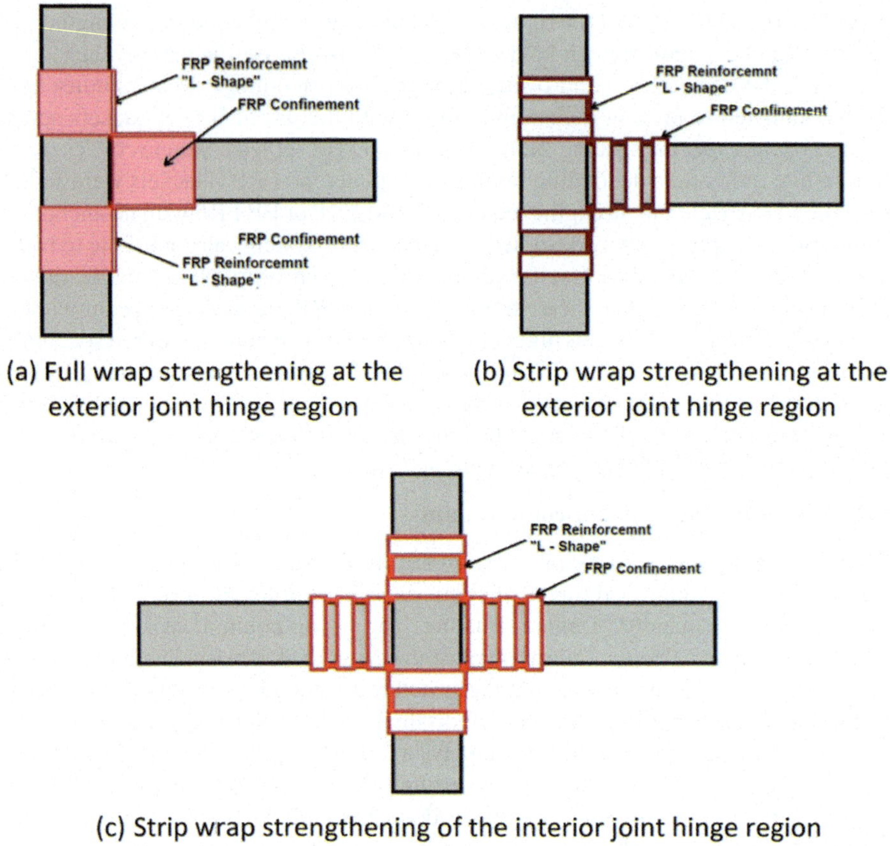

Fig. 4 Typical FRP strengthening schemes

the strengthened joints without adequate anchorage [20]. Numerous studies have been conducted on the use of glass-FRP as retrofitting materials for enhancing the performance of beam-column joints [16, 21, 22]. These investigations have demonstrated that the incorporation of GFRP anchors significantly elevates both the shear strength and ductility of RC beam-column joints. This augmentation in structural capacity is accompanied by a notable shift in failure mode from shear to flexure. Further, it results in a substantial increase in energy dissipation potential compared to control specimens. Schematic diagrams of retrofitted/strengthened beam-column joints are shown in Fig. 5.

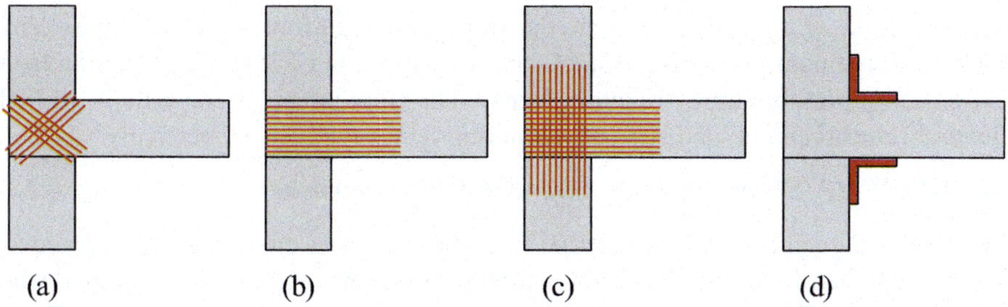

Fig. 5 Beam-column joint retrofitting/strengthening with FRP composites **a** X-shape **b** U-shape **c** T-shape **d** L-shape

2 Guidelines to Estimate the Flexural and Shear Capacity of FRP Strengthened RC Beams

The flexural and shear cracking capacity of FRP-strengthened members can be estimated using the following standard guidelines.

ACI 440.2R-17 (Flexural Strengthening with FRP Composite)

The formula for estimating the nominal flexural strength capacity '*M*' of a conventional RC frame section with FRP as per ACI 440.2R-17 is

$$M = A_s f_s\left(d - \beta_1 \frac{x}{2}\right) + \phi_f A_f f_{fe}\left(d_f - \beta_1 \frac{x}{2}\right) \tag{1}$$

A trial-and-error method can be used to estimate the neutral axis depth, x as given in Eq. (2).

$$x = \frac{A_s f_s + A_f f_{fe}}{\alpha_1 f_c' \beta_1 b} \tag{2}$$

where, α_1 and β_1 are the concrete stress block parameters. The value β_1 can be obtained from ACI-318-08 [23] and α_1 having a value of 0.85.

$$\beta_1 = \begin{cases} 0.85 f_c' \leq 28MPa \\ 0.85 - 0.007\left(f_c' - 28\right),\ f_c' > 28MPa \end{cases} \tag{3}$$

ACI model provides a simplified Intermediate crack-induced debonding strain based on the Teng [24] equation:

$$\varepsilon_{fd} = 0.41\sqrt{\frac{f_c'}{nE_f t_f}} \leq 0.9\varepsilon_{fu} \tag{4}$$

where, $A_s, A_f, f_s, f_{fe}, d, d_f, \phi_f$, are the area of steel reinforcement, area of external FRP reinforcement, the stress in steel, effective stress in FRP, depth of section from extreme compression fibre to centre of tension reinforcement, effective depth of FRP flexural reinforcement, and strength reduction factor for FRP, respectively.

ACI 440.2R-17 (Shear Strengthening with FRP Composite)

The FRP design guide ACI 440.2R 2017 [25, 26] serves as an extensive reference for shear strengthening using FRP. This guideline considers critical factors such as fibre orientation and the anticipated crack pattern [27]. By analysing the force induced due to tensile stress near the assumed crack location, the shear capacity achievable with FRP can be determined.

$$V_f = A_f f_{fe}(\sin\alpha_f + \cos\alpha_f)\frac{d_f}{s_f} \quad (5)$$

$$A_f = 2nt_f w_f \quad (6)$$

$$f_{fe} = \varepsilon_{fe} E_f \quad (7)$$

where $t_f, w_f, A_f, s_f, \alpha_f$, denote the thickness, width, area, spacing, and inclination angle of the FRP composite, respectively. The effective strain ε_{fe} depends on the wrapping scheme of FRP [5].

In accordance with the design principles outlined in ACI, the addition to the overall shear strength reduction factor ϕ, the contribution of FRP to shear strength is further adjusted by a reduction factor. φ_f, with the final shear strength capacity as follows:

$$\phi V = \phi(V_c + V_s + \varphi_f V_f) \quad (8)$$

where V_c, V_s, V_f, denotes the shear contribution of concrete, the shear contribution of steel, and the shear contribution of FRP, respectively.

Triantafillou's Model

Triantafillou [5] proposed a model to estimate the shear contribution of FRP based on the area fraction of FRP and other parameters like fibre orientation, elastic Modulus, and effective strain of FRP. The shear strength contribution of FRP can be estimated as per Eq. 9.

$$V_{frp,d} = \frac{0.9}{\gamma_{frp}}\rho_{frp}E_{frp}\varepsilon_{frp,e}b_w d(1+\cot\beta)\sin\beta \quad (9)$$

if $0 \le \rho_{frp}E_{frp} \le 1$

$$\varepsilon_{frp} = 0.0119 - 0.0205(\rho_{frp}E_{frp}) + 0.0104(\rho_{frp}E_{frp})^2$$

if $\rho_{frp} E_{frp} \geq 1$

$$\varepsilon_{frp} = -0.00065(\rho_{frp} E_{frp}) + 0.00245$$

where, ρ_{frp}, $\varepsilon_{frp,e}$, b_w, β, γ_{frp}, E_{frp}, denotes the area fraction of FRP, the effective strain of FRP, width of the beam, fibre orientation, partial safety factor for FRP in uniaxial tension, and elastic modulus of FRP, respectively.

3 Experimental Study on the Flexural and Shear Strengthening of RC Beams

The shear and flexural behaviour of FRP-strengthened RC beams was studied using scaled-down beam specimens. Beams were cast with shear deficiency, followed by strengthening using different FRPs. Ordinary Portland Cement (OPC) of 43 grade was used in the concrete mix preparation. The three FRPs adopted in this study are CFRP uni-directional, BFRP and GFRP bi-directional fibre composites. The corresponding fibre composites were adhered to the concrete surface using two types of solutions, namely the primer and saturant. Each of the two solutions consists of two components, namely resin and the hardener, which were mixed in the proportion of 1: 0.5. The main function of the primer compound was to fill the concrete pores, and that of the epoxy solution was to bind fibre composites on the primer coated concrete surface. After a day's curing of the primer coat, an epoxy coat was applied over which the FRP was adhered, immediately after which a roller was rolled down to eliminate the voids. For double-layered specimens, the epoxy coat was applied in the direction parallel to the fibres of first layer, upon which the second layer was fixed. The FRP-strengthened specimens were kept in a serene state for a period of 15 days for the epoxy to cure and gain its ultimate strength.

Fracture behaviour of the FRP strengthened concrete prisms

Plain concrete prism specimens of standard size 500 mm × 100 mm × 100 mm with a vertical groove of 20 mm depth and 2 mm width at the mid-point were used in the fracture study. Fracture specimens were strengthened by the three different FRPs to investigate the contribution of FRP in fracture strength and corresponding failure patterns. Specimens were tested in the flexural testing machine under single-point loading. FRP was fixed within a span of 400 mm between the two roller supports. The grooved portion was kept free from adhesive, and failure was anticipated to occur at the mid-region, eventually by groove widening. Details of the tested specimens are illustrated in Table 1.

The load–deflection plots of the fracture specimens are shown in Fig. 6. The Fracture strength of the conventional specimen is low compared with FRP-strengthened specimens. The fracture behaviour of the strengthened specimen depends on the property of FRP. The specimen Fr Con 20 fails with an average fracture strength of

Table 1 Details of the prism specimens fixed by FRP for fracture test

Specimens	Details of specimens	Grade of concrete
Fr Con 20	Conventional specimen	M20
Fr B 20 S	Basalt fibre single layer affixed specimen	M20
Fr C 20 S	Carbon fibre single-layer affixed specimen	M20
Fr G 20 S	Glass fibre single-layer affixed specimen	M20

3.46 MPa with brittle failure at the mid-point, breaking into two parts. The specimen Fr B 20 S exhibits 28% higher strength than conventional and possesses better post-yield deflection due to the tensile contribution of FRP. The lower modulus BFRP led the specimen to fail in a ductile manner by proper yielding of FRP at the middle of the beam, shown in Fig. 7. The fracture strength of Fr C 20 S specimen is 14.89 MPa, which is higher than all the FRP layered specimens and also shows an average increase of 330% than the conventional. The brittle failure of concrete, as shown in Fig. 7, shows lesser post-peak behaviour. Generally, the higher modulus CFRP converts the unstrengthened portion to the weak region. Thus, the failure was initiated and occurred at the unstrengthened portion by forming truss action prior to CFRP rupture. The specimen Fr G 20 S shows an average strength of 4.66 MPa, 34% higher than the conventional, along with enhanced post-peak behaviour. This infers that the influence of modulus and strength of the fibres decide the performance of FRP-affixed concrete specimens.

R.C. Beams with External FRP Full U-Wrapping

Reinforced concrete beams of 1000 mm in length having a cross-sectional size of 100 mm × 150 mm with/without FRP wrapping using three different varieties of FRP were tested under two-point loading. The reinforcement detailing was kept the

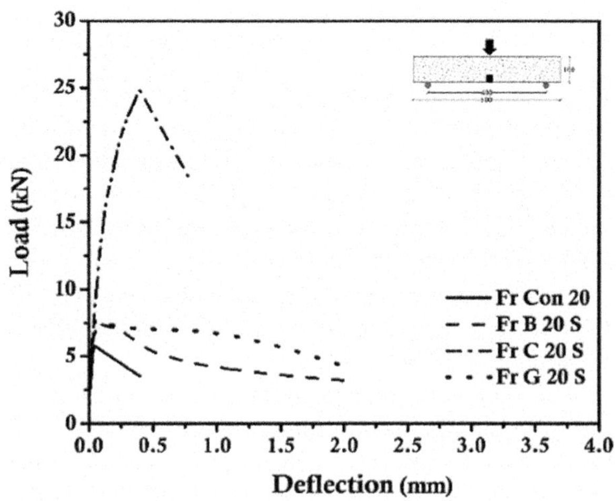

Fig. 6 Load versus deflection behavior of the fracture prisms

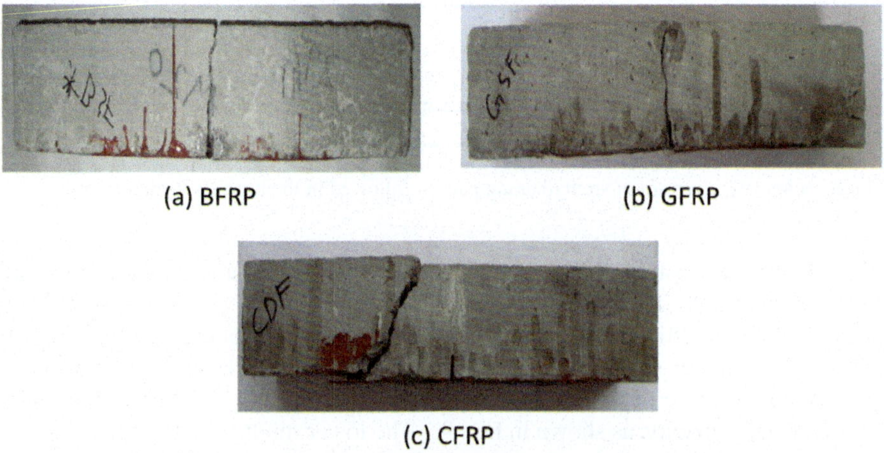

Fig. 7 Failure patterns noticed in the fracture specimens

same in all the beams with a longitudinal reinforcement of 10 mm dia. at the bottom and 8 mm dia. at the top. The shear detailing was deliberately kept high with 6 mm 2-legged stirrups placed at 220 mm center-to-centre. The configuration of wrapped FRP in RC beams is presented in Table 2. A schematic diagram of the wrapping pattern followed in all the FRP-strengthened beams is shown in Fig. 8. During the grinding process, the bottom edges of the tension zone were made smooth to avoid stress concentration on FRP during loading. After the fastening process of the FRP, a roller was used to eliminate the voids present at the FRP-epoxy interface. BFRP and GFRP, owing to their lesser thickness, were wrapped in double layers. In the sequence of strengthening for both shear and flexure, shear zone FRP was fixed over the flexure zone FRP, which acts as an anchorage to it.

The inadequate stirrup spacing fails to resist the truss action, which leads the conventional specimen B M20 0 to fail in shear. The inclined shear cracks restrict beam deflection and exhibit a sudden brittle failure. In FRP-strengthened specimens, the external confinement effectively resists the shear force and allows the formation of numerous flexural cracks in the mid-span. The strengthening measures in specimen B B M20 1 using lower modulus BFRP at the tension zone allow vertical crack

Table 2 Detailed configuration of strengthening in the R.C beams

Specimen ID	Description of specimen wrapping
Type 1	Concrete grade M20
B M20 0	Conventional specimen without any FRP wrapping
B B M20 1	Beam specimen strengthened in shear and flexure using BFRP sheets (2 layers)
B C M20 1	Beam Specimen strengthened in shear and flexure using CFRP sheet
B G M20 1	Beam specimen strengthened in shear and flexure using GFRP sheets (2 layers)

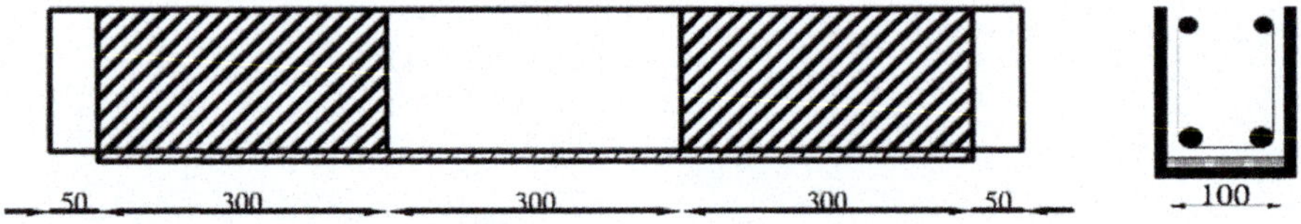

Fig. 8 Schematic diagram of strengthening pattern followed in the beams specimens

formation in the unstrengthened portion, which altered the shear failure to flexure, as shown in Fig. 9b. The higher modulus CFRP improves the load-carrying capacity of the beam by resisting its deformation. The inadequate anchorage in this specimen fails to support the shear zone CFRP, which shows FRP debonding prior to the tension zone rupture of CFRP. This led the specimen to fail in shear with a higher load-carrying capacity, as shown in Fig. 9c. The lower modulus GFRP strengthened B G M20 1 specimen's cracking behaviour was similar to B B M20 1. Vertical cracks were propagated in the unstrengthened mid region at a higher load, which led to its higher ultimate deflection. The specimen failed in flexure with rupture of FRP, followed by debonding in the tension zone, as shown in Fig. 9d.

The load–deflection behaviour of FRP-strengthened specimens shows a certain drop in load after attaining their corresponding peak loads, which proves that the trend of strength increase is evident till the active contribution of FRP, shown in Fig. 10.

Immediately after the failure of FRP, the load gets transferred to the reinforcement. The conventional specimen B M20 0 with inadequate confinement fails to restrict the formation of shear cracks in the early stage and shows a sudden drop in load after attaining a peak load of 45.83 kN with meagre post-peak behaviour. The BFRP-strengthened B B M20 1 specimen fails at a load of 77.80 kN, which is 70% higher

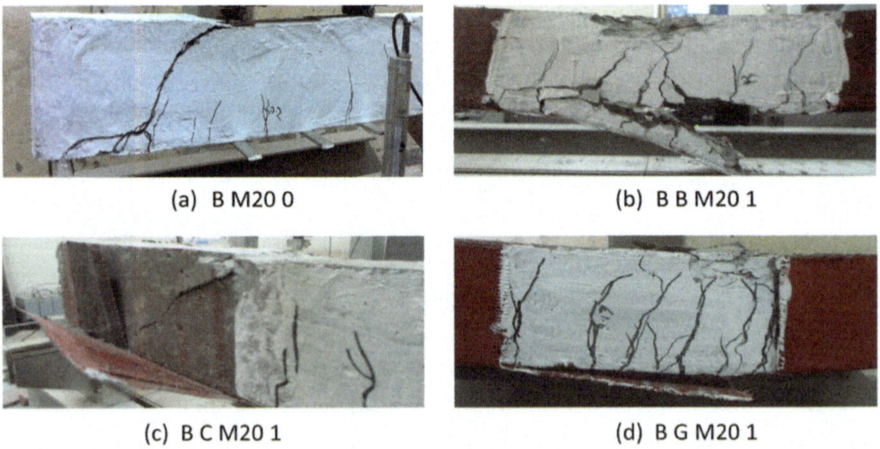

Fig. 9 Failure patterns in the tested R.C. beams

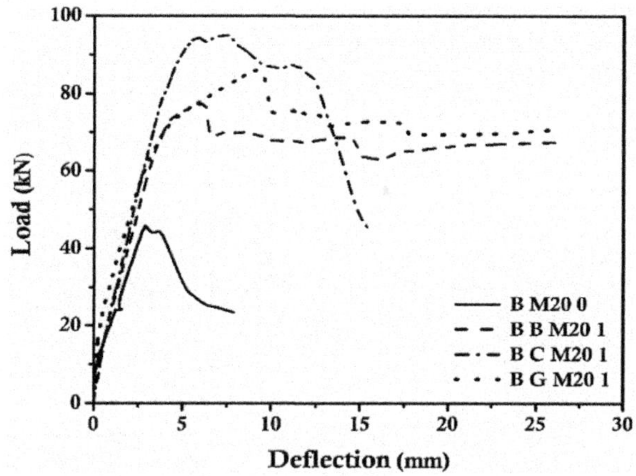

Fig. 10 Load–deflection behaviour of RC beams with full FRP wrapping

than the conventional with enhanced post-peak behaviour. The observed ultimate deflection is nearly 175% higher than the conventional specimen. This evidences the formation of flexural cracks, and the absence of shear cracks led the reinforcement to yield and finally fail in flexure. The higher modulus CFRP strengthened specimen B C M20 1 attains a peak load of 95.12 kN, which is higher than all other specimens and is 107% higher than the conventional peak load. The absence of end anchorage led the CFRP to delaminate from the strengthened zone, resulting in the shear mode of failure with decreased ultimate deflection compared to other lower modulus FRP-strengthened specimens. The average peak load of the GFRP-strengthened specimen B G M20 1 specimen is 86.52 kN, which is 88% higher than the conventional specimen. The post-yield behaviour is similar to the basalt specimen with increased deflection. The initial tangent clearly shows the effective contribution of FRP in stiffness enhancement, and the post-yield rate of loss in strength shows improved inelastic stiffness retention capacity.

R.C. Beams with External FRP Stirp U-Wrapping

Three beam specimens having a length of 800 mm and a cross-sectional size of 150 × 200 mm were strengthened for shear. Two numbers of 12 mm diameter rebars as longitudinal tension reinforcement and 6 mm diameter rebars as transverse reinforcement having a yield strength of 250 MPa at 120 mm c/c spacing have been adopted in the preparation of RC beams. Table 3 presents the details of the specimen's configuration [26].

Table 3 Details of the RC beam specimens

Specimens	Details of specimens	Width of strip (mm)	Strip spacing (mm)
B1	Conventional specimen	50	100
B2	Carbon fibre affixed specimen	50	100
B3	Basalt fibre affixed specimen	50	100

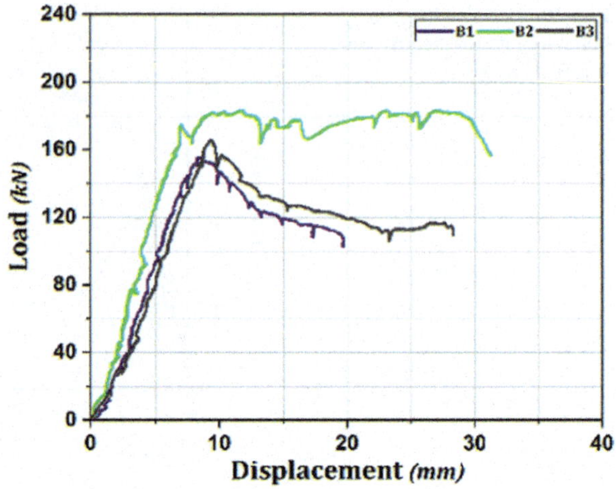

Fig. 11 Load–deflection behaviour of RC beams with strip wrapping

Out of the three specimens, one specimen is strengthened using a 50 mm wide CFRP strip (B2) and the other with a 50 mm width BFRP strip (B3) having a c/c spacing of 100mm. All specimens were tested under four-point bending having a shear–span ratio of 1.25. The load–deflection curve of tested beams shows that the CFRP strengthening enhances the stiffness and peak strength by 17% compared to the control specimen and displayed better post-yield deflection than BFRP strengthening, shown in Fig. 11. The low strength and modulus capacity of BFRP failed to resist the truss action, which led the specimen to experience diagonal cracks as shown in Fig. 12c, and has shown only 6% strength enhancement than the control specimen. CFRP strengthening in the shear span controlled the formation of diagonal cracks and led the beam to fail under flexure, as shown in Fig. 12. Flexural cracks were observed near the middle of the beam. The area under the load–displacement curve shows 140% higher energy dissipation of B2 than B1, whereas it is only 60% for specimen B3. This manifests the fact that appropriate use of FRP based on strength and demand of cross section may become an effective solution in enhancing the strength and ductility.

4 Cyclic Behaviour of FRP Strengthened Exterior Beam Column Joints

An exterior beam-column joint specimen was tested under cyclic loading in quasi-static test facility to evaluate its hysteresis behaviour under cyclic loading. Complete details of the beam-column joint specimens are given in Table 4. Figure 13 shows the beam-column joint test setup. The tested conventional beam-column joint has been retrofitted and tested again under cyclic loading. The damaged cover concrete and damaged/cracked region have been completely removed from the joint. The old

Fig. 12 Crack and failure pattern of **a** B1 **b** B2 **c** B3 specimens

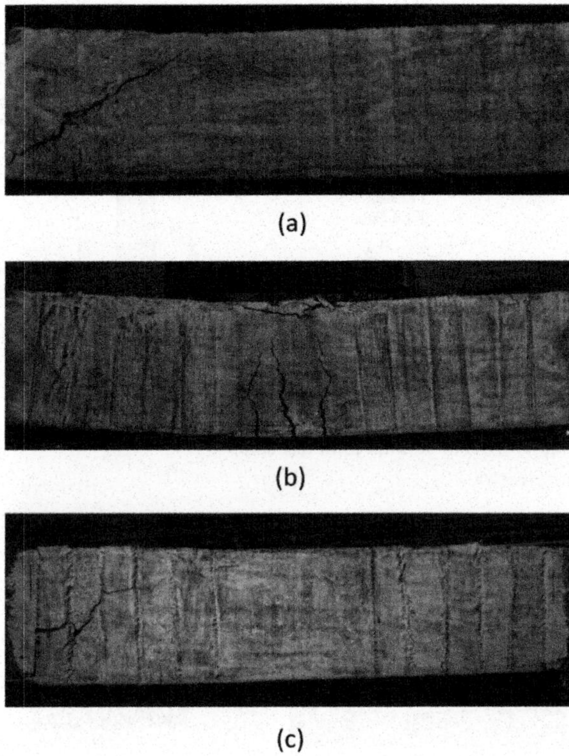

reinforcement remained the same during the retrofitting process. After removing the cover concrete, the surface was treated using latex-modified cement slurry followed by concerting. After 28 days of curing, the concrete surface was ground and cleaned for primer application. On the alternate day of primer coating, epoxy adhesives were used, and CFRP was fixed in an L-shape between the column and beam hinge region, continued by the column longitudinal directions by the GFRP confinement, as shown in Fig. 14.

The hysteresis behaviours of tested beam-column joints are shown in Fig. 15. The hysteresis behaviour of conventional specimen (J1 sudden drop in load) shows brittle failure mode since there is a sudden drop in load after peak due to joint shear failure

Table 4 Detail of the exterior beam-column joint specimen

<table>
<tr><th>ID</th><th>Transverse reinforcement</th><th>Joint configuration</th><th>Beam reinforcement</th><th>Column reinforcement</th></tr>
<tr><td>J1</td><td rowspan="2">Ø 10 mm @180 mm C/C in beam and Ø 10 mm @300 mm C/C in column</td><td>Conventional</td><td rowspan="2">All specimens are reinforced with 2 nos of 20 mm Ø at the top and bottom</td><td rowspan="2">All specimens are reinforced with 4 nos. of 20 mm Ø</td></tr>
<tr><td>J2</td><td>CFRP</td></tr>
</table>

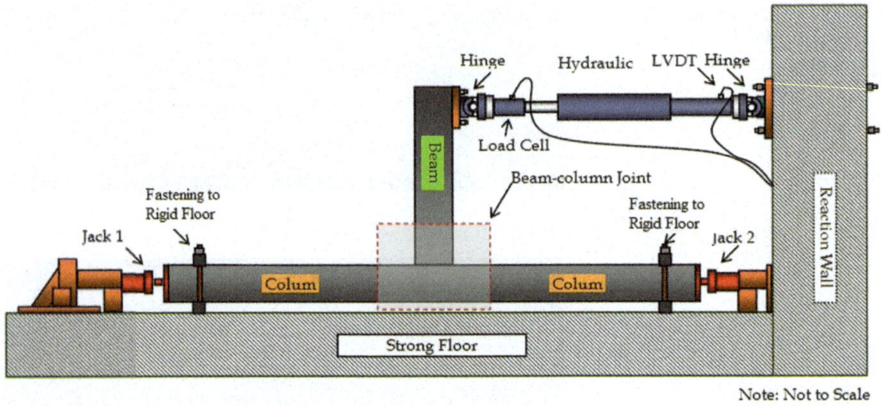

Fig. 13 Beam column joint test setup

(a) Old concrete surface treatment prior to the repair work

(b) Applying epoxy over the restored region

(c) CFRP reinforcement, along with GFRP Confinement

Fig. 14 Retrofitting sequence of CFRP strengthening

[29]. The shear cracks at the beginning stage of the test led to cover concrete spalling and allowed the longitudinal reinforcement to slip and deform. Thus, it shows a higher rate of post-peak degradation. The pinched hysteresis loop of conventional specimen indicates the absence of energy dissipation during inelastic deformation. Figure 15 a shows 80% of strength degradation at 40 mm displacement. The CFRP-strengthened specimen J2 shows enhanced load-carrying capacity with better post-peak performance compared to the specimen J1. Specimen J2 shows 50% of peak load at 40 mm displacement, which is 60% higher than the conventional specimen J1. The hysteresis curve of J2 also shows pinching in the loop, but the area of the loop shows enhanced energy dissipation capacity compared to J1.

The crack patterns noticed in tested beam-column joint specimens at failure level under cyclic testing are shown in Fig. 16. It is observed that the formation of diagonal cracks in the unconfined joint region allowed spalling of cover concrete at the early stages of loading and led to the slip of longitudinal reinforcement. It exhibits brittle failure, as shown in Fig. 16a. The joint core concrete remains intact with the top and

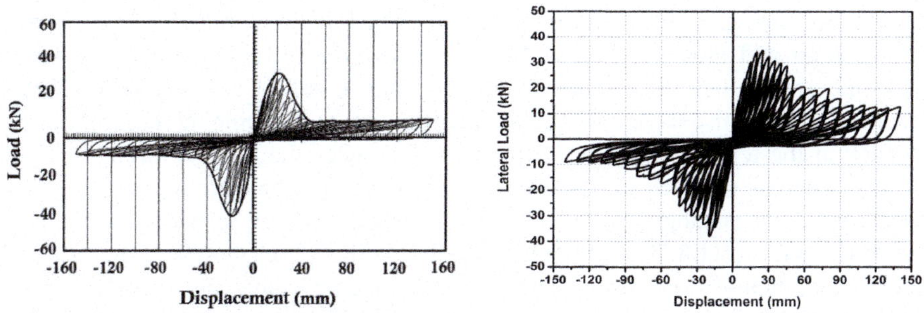

(a) J1-Conventional (b) J2 - CFRP

Fig. 15 Hysteresis behaviour of the joint specimens

(a) J1-Conventional (b) J2 - CFRP

Fig. 16 Failure pattern noticed in the joint specimens

bottom column portions, whereas the column and beam interface have lost contact. The external CFRP confinement in the joint specimen J2, showed no sign of cracks or rupture. Popup noises during higher displacement were noticed, and finally the CFRP sheet was delaminated from the joint surface, as shown in Fig. 16b. The post-failure analysis shows that the failure pattern resembles the conventional joint specimen J1. The wrapped GFRP failed to provide the required confinement when there was an increase in the drift ratio. Consequently, the GFRP ruptures followed by CFRP debonding from the column hinge region, as shown in Fig. 16b.

5 Conclusions

The chapter presents an overview of FRP strengthening for RC beams and beam–column joints. Various strengthening schemes, their associated advantages and limitations have been discussed. Following are the inferences from the literature and the experimental studies conducted.

The strength, modulus, and rupture strain of FRP plays a crucial role in strength contribution and mode of failure. In most cases, the lower modulus FRP yields and supports ductile failure rather than the higher modulus FRP. Similarly, the area fractions of FRP also have a significant contribution in deciding the failure mode.

FRP at the tension soffit face effectively enhances the flexural strength of RC beams, whereas the FRP in transverse directions enhances their shear strength. In both cases, the anchorage of FRP plays an influential role in strength enhancement and post-yield deflection capability. The end anchorage system for flexural/shear strengthening controls the early-stage debonding failure and allows for better rotations without sudden degradation in strength, which also improves ductility.

The fracture behaviour shows that the CFRP composite possesses higher fracture strength than the other FRPs. The other FRPs such as BFRP and GFRP, increase the post-peak degradation of the specimen and dissipate higher energy than CFRP. This concludes that the higher modulus FRP achieves higher strength enhancement but exhibits a brittle failure mode compared to lower modulus FRP.

All FRP-strengthened RC beam specimens show improved strength and post-peak degradation than the conventional specimen. The post-peak deflection of RC beams follows a trend similar to that of the fracture specimens. Higher modulus CFRP shows strength and stiffness enhancement rather than post-yield deflection, whereas the GFRP and BFRP strengthened specimens show contrary behaviour with higher post-yield deflections. The failure patterns also vary accordingly.

RC beam with strip U-wrapping shows that the CFRP confinement controls the formation of diagonal cracks and compels the specimen to fail in flexure rather than shear. The lower modulus BFRP could not resist the truss action and exhibited a shear failure mode. This infers that the specimen configuration and type of fibre also play a vital role in FRP strengthening. The FRP and steel reinforcement ratio also contribute to each other in deciding the performance of strengthened RC beams.

The joint shear enhancement by FRP strengthening primarily depends on the mode of strengthening and anchorage provided. Similar to the earlier observation, the strength and modulus of FRP are deciding parameters in the augmentation of strength/stiffness. There is a significant improvement in the joint strength of CFRP retrofitted specimen. Though the joint core concrete remains older, the external confinement using CFRP resisted the lateral load and exhibited better performance over conventional joint specimen. Irrespective of pinching in the hysteresis behaviour, the energy dissipation is higher than in the control specimen. The failure of GFRP confinement led to CFRP delamination from the concrete substrate and failed to resist the applied lateral forces.

References

1. Ahmed E, Rahman Sobuz H, Mohamed Sutan N (2011) Flexural performance of CFRP strengthened RC beams with different degrees of strengthening schemes. Int J Phys Sci 6(9):2229–2238
2. Dong J, Wang Q, Guan Z (2013) Structural behaviour of RC beams with external flexural and flexural–shear strengthening by FRP sheets. Compos B Eng 44(1):604–612
3. Huang L, Zhang C, Yan L, Kasal B (2018) Flexural behavior of U-shape FRP profile-RC composite beams with inner GFRP tube confinement at concrete compression zone. Compos Struct 184:674–687
4. Beber A, Campos Filho A (2005) CFRP composites on the shear strengthening of reinforced concrete beams. Rev IBRACON Estruturas
5. Triantafillou TC (1998) Shear strengthening of reinforced concrete beams using epoxy-bonded FRP composites. Struct J 95(2):107–115
6. Ary MI, Kang THK (2012) Shear-strengthening of reinforced and prestressed concrete beams using FRP: Part I—review of previous research. Int J Concr Struct Mater 6(1):41–47
7. Pellegrino C, Vasic M (2013) Assessment of design procedures for the use of externally bonded FRP composites in shear strengthening of reinforced concrete beams. Compos B Eng 45(1):727–741
8. Rousakis TC, Saridaki ME, Mavrothalassitou SA, Hui D (2016) Utilization of hybrid approach towards advanced database of concrete beams strengthened in shear with FRPs. Compos B Eng 85:315–335
9. Sas G, Täljsten B, Barros J, Lima J, Carolin A (2009) Are available models reliable for predicting the FRP contribution to the shear resistance of RC beams? J Compos Constr 13(6):514–534
10. Al-Amery R, Al-Mahaidi R (2006) Coupled flexural-shear retrofitting of RC beams using CFRP straps. Compos Struct 75(1–4):457–464
11. Sato Y, Ueda T, Kakuta Y, Tanaka T (1996) Shear reinforcing effect of carbon fiber sheet attached to side of reinforced concrete beams. In: Proceedings of the 2nd international conference on advanced composite materials in bridges and structures. ACMBS-II, Montreal
12. Hu Z, Zhou X, Guo M, Huang X, Hu B (2020) Enhancing the performance of CFRP shear-strengthened RC beams using "ductile" anchoring devices. Front Mater 7:292
13. Kalfat R, Al-Mahaidi R (2010) Investigation into bond behaviour of a new CFRP anchorage system for concrete utilising a mechanically strengthened substrate. Compos Struct 92(11):2738–2746
14. Khan A, Ayub T (2010) Effectiveness of U-shaped CFRP wraps as end anchorages in predominant flexure and shear region. In: Proceedings (CD-Rom), 5th international conference on FRP
15. Smith ST, Teng JG (2003) Shear-bending interaction in debonding failures of FRP-plated RC beams. Adv Struct Eng 6(3):183–199
16. El-Amoury T, Ghobarah A (2002) Seismic rehabilitation of beam–column joint using GFRP sheets. Eng Struct 24(11):1397–1407
17. Le-Trung K, Lee K, Lee J, Lee DH, Woo S (2010) Experimental study of RC beam–column joints strengthened using CFRP composites. Compos B Eng 41(1):76–85
18. Prota A, Nanni A, Manfredi G, Cosenza E (2001) Selective upgrade of beam-column joints with composites. In: Proceedings of the international conference on frp composites in civil engineering, vol 1, pp 919–926
19. Antonopoulos CP, Triantafillou TC (2003) Experimental investigation of FRP-strengthened RC beam-column joints. J Compos Constr 7(1):39–49
20. Lee WT, Chiou YJ, Shih MH (2010) Reinforced concrete beam–column joint strengthened with carbon fiber reinforced polymer. Compos Struct 92(1):48–60
21. Realfonzo R, Napoli A (2012) Results from cyclic tests on high aspect ratio RC columns strengthened with FRP systems. Constr Build Mater 37:606–620
22. Roy B, Laskar AI (2018) Cyclic performance of beam-column subassemblies with construction joint in column retrofitted with GFRP. Structures 14:290–300

23. ACI Committee (2008) Building code requirements for structural concrete (ACI 318–08) and commentary. American Concrete Institute
24. Teng JG, Smith ST, Yao J, Chen JF (2003) Intermediate crack-induced debonding in RC beams and slabs. Constr Build Mater 17(6):447–462
25. ACI 440.2R-17 (2017) Guide for the design and construction of externally bonded FRP systems for strengthening concrete structures. American Concrete Institute. Farmington Hills, MI, USA
26. ACI 440.2R (2017) Guide for the design and construction of externally bonded FRP systems for strengthening concrete structures. American Concrete Institute
27. Khalifa A, Gold William J, Nanni A, Aziz MIA (1998) Contribution of externally bonded FRP to shear capacity of RC flexural members. J Compos Constr 2(4):195–202
28. Satyanarayana K, Rajendran SC, Kothapalli N (2022) Influence of FRP on the shear behaviour of RC rectangular beam elements. Mater Today: Proc 64:1061–1068
29. Dadi VVSSK, Agarwal P (2013) Cyclic performance evaluation of unconfined and confined beam-column joint specimens with different type of reinforcing characteristics as per ASCE/SEI 41–06. Aust J Struct Eng 14(3):258–272

Seismic Retrofitting and Strengthening of Structures

Pranoy Debnath and Sekhar Chandra Dutta

Abstract Retrofitting the seismically damaged structures and strengthening the existing structures due to the upgradation of seismic zones continuously taking place. In fact, learning from occurring earthquakes and study of structural damages has become a very important aspect of earthquake engineering. Most of the developing countries have large numbers of habitats made of masonry, while the focus on the safety of these poor men's habitats is relatively very less. Developing countries can also afford concrete structures for dwellings, while steel structures are mostly used for industrial purposes. On the other hand, the developed countries have large numbers of steel structures as well as shear wall based structures as they can afford the cost of the same. To have a comprehensive idea about retrofitting and strengthening measures for all categorial structures, the current chapter is an attempt to bring together the existing techniques for retrofitting and strengthening all the structures under separate subheadings and also underlines the promising new techniques which are coming up for this purpose. At the end, the chapter summarizes a few important points on this aspect. The chapter may be helpful for having an overall picture of the retrofitting and strengthening of structures.

Keywords Retrofitting · Strengthening · Masonry structures · RC structures · Steel structures

1 Introduction

The massive destruction of engineered systems and facilities caused by past earthquakes has shown serious deficiencies in the current structural systems and construction standards [14, 49, 51, 74, 75, 122, 174]. These disasters have created a new

P. Debnath · S. C. Dutta (✉)
Department of Civil Engineering, Indian Institute of Technology (ISM), Dhanbad, India
e-mail: scdind2000@gmail.com; sekhar@iitism.ac.in

P. Debnath
e-mail: pranoydebnath.20dr0099@cve.iitism.ac.in

S. B. Singh and C. V. R. Murty (eds.), *RC Structures Strengthened with FRP for Earthquake Resistance*, Composites Science and Technology,
https://doi.org/10.1007/978-981-97-0102-5_6

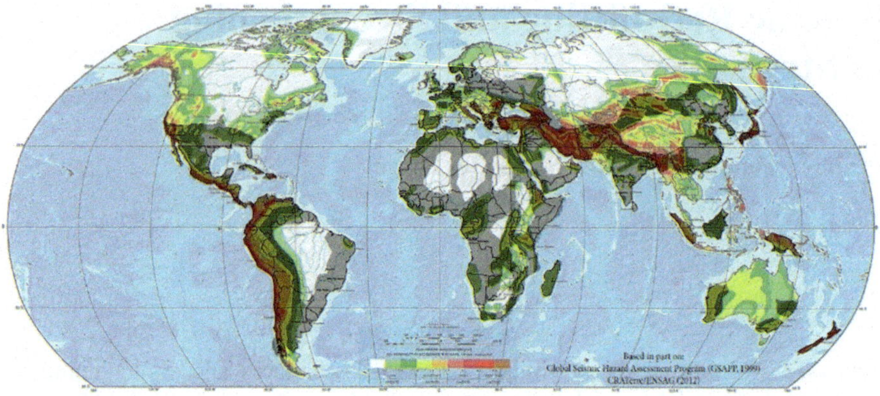

Fig. 1 World map of overlayed earthquake hazard map (red is the greatest hazard and white is the smallest) [118, 119, 121]

awareness about disaster preparedness and mitigation. Figure 1 exhibits the global distribution of earthquakes all over the world. Even though the recommended codes of practice for earthquake-resistant design do exist but those design codes throw little light on the issue of retrofitting and strengthening of the structures [62, 64, 82–86]. However, the issue of retrofitting and strengthening of damaged structures should have been of equal importance for the rebuilding of an earthquake-prone civilization. The present write-up is an effort to summarize the current state of the art on this issue and the immediate use to be catered to in the future.

1.1 Significance of Studying the Strengthening of Unreinforced Masonry (URM)

Most people in developing countries are usually poor and have little knowledge of engineering construction. In these countries, masonry structures became the primary type of habitat. It is difficult for practicing engineers to enhance the shear and flexure strength of URM structures to achieve better earthquake sustainability. In last few decades, increasing awareness about the importance of masonry structures has initiated sustainable development strategies for strengthening these masonry structures.

Traditional strengthening methods have been used for last few decades, such as (i) internal or external tensioning with steel ties; (ii) stitching of large cracks with metallic or brick elements; (iii) filling cracks by grouting; (iv) jacketing with shotcrete; and (v) using ferrocement [35, 47, 48, 111, 132]. Further, some structurally efficient methods are also suggested nowadays that are less intrusive, like fiber-reinforced polymers (FRPs) and fabric-reinforced cementitious matrix (FRCM) [68, 165]. However, despite of their advantages in structural performance when used with epoxy resins, FRP systems are considered less durable due to low resistance, sensitivity to ultraviolet rays when exposed to open air, high toxicity, and unsatisfactory long-term behaviour [163]. On the other hand, FRCMs can be combined with a fiber-reinforced grid embedded with a high-performance cementitious matrix, which can easily overcome most disadvantages of using epoxy resins as a binding agent. Furthermore, these techniques are costly to be afforded by the people who live in the masonry structures. Apart from these, some other modern materials, like organic materials and timbers, need to be highlighted, with particular importance on their eco-friendly nature and suitability characteristics.

1.2 Importance of Studying Reinforced Concrete (RC) Structures

Concrete structures undergo deterioration and degradation of strength over the course of time. For instance, the corrosive property of steel reinforcement is the primary issue causing the deterioration of concrete structures [8, 10]. Further, slow loss of strength with aging [88, 94], high-intensity loading [1, 110], freeze–thaw cycles, temperature variations [80, 108], contact with chemicals and saline water [9, 61] as well as exposure to UV radiations [92, 167] are all other factors that contribute to the deterioration. Because of this unintended deterioration of strength, the structure becomes weaker than it was designed for. Hence, such structures also need regular maintenance following the procedure of strengthening and retrofitting. Retrofitting is done to restore the strength of the weakened RC structures. It is really necessary to provide perfect and practical strategies as most old RC buildings need to be retrofitted. Once the proper retrofitting method is prepared and specified, the RC member can regain its original strength. However, it depends on the intensity of the damage that happened to the structure.

1.3 Importance of Studying Steel Structures

Although much effort has been made in recent years to strengthen URM buildings and RC frames, steel structures have not received significant attention yet. The majority of industrial buildings are examples of existing steel structures all over the world [154]. This kind of structure mainly spread during the twentieth century for its capability of easy technological solutions with affordable costs. These structures still have many drawbacks in design guidelines, like no differences were made between ductile and brittle mechanisms in designing connections. Furthermore, the deficiencies in provisions for joint classification usually led to poor design provisions (e.g., base connections) [29, 143].

Some common ideas also considered some of the technical practices that in steel structures, relatively low masses contribute to seismic action [2], like that was seen for moment-resisting frames with truss beams [146]. However, multiple seismic events proved this theorem is incorrect, as several industrial structures reported severe damages and some cases of global collapse [79, 123]. Another essential aspect that should be considered is indirect costs, like interruption expenses during productive activities for a long time which is different from residential buildings [32]. On most occasions, seismic damages occur due to local failures of different connections, claddings, or roofing [29], even though there were a few cases of global collapse [66].

Although current research has increased the variety of analysis for the seismic performance methods available in structural design firms. However, this can only be done briefly due to space constraints.

2 Common Failure Mechanisms

Construction methods are primarily determined by local patterns, engineering expertise, and various demographic factors. However, the rapid urbanization of many developing countries has significantly increased the demand for houses built without regulations or controls. Different literature in the field of structural safety classifies failures based on the following failure criteria [27, 139, 164]. Different types of structures that have been observed to exhibit common failure mechanisms in previous earthquakes are explained in this section.

- Due to different types of design errors or construction of structures.
- Due to unexpected high values of loading.
- Due to improper understanding of the structural behaviour of a system.
- Due to poor understating of the construction and using existing materials or other technologies.
- Due to improper information handling among the stakeholders.
- Due to violation of construction methods, specifications, or codes.

2.1 Failure Mechanisms of Masonry Structures

Three basic types of masonry materials are used to construct URM structures worldwide. These materials are adobe, bricks, and stones. Because of their heavy walls, adobe houses are particularly so much vulnerable to collapse failure during lower-intensity shakings. Furthermore, adobe structures are very brittle due to the lack of ductility. So, the failure of adobe structures is caused either by separating walls at the corners or cracking and failure of walls [89, 148, 155].

(a) (b)

Fig. 2 Masonry buildings failure modes **a** In-plane direction (in Mangal Bazar, Lalitpur, India) and **b** Out-of-plane direction (in Kathmandu, Nepal). Taken from Halder et al. [74, 76]

Some common failure mechanisms of brick masonry structures are corner junction failure, shear crack failure in walls, and rocking failure of the whole wall in the in-plane direction. After the walls collapse, disintegration may occur between the floors and roofing, and in most severe cases, the building may collapse completely [36, 49, 53–55, 74–76, 96, 125].

Similarly, stone masonry buildings fail during seismic activities because of their standard failure modes like delamination of the external walls from its rubble infill, overturning the long-span walls in an out-of-plane direction, and in-plane shear failure of the wall. The in-plane and out-of-plane mechanisms for masonry failures are shown in Fig. 2. Another common type of failure is out-of-plane failure due to the separation of the junction leading to collapse, as observed in many past earthquakes [6]. Mainly local failures are seen in these structures, leading to global failure later on.

2.2 *Failure Mechanisms of RC Structures*

Significant deformations in load-bearing frame elements occur across the structural elements due to the response of structures during earthquake motions. As a result of these distortions, internal forces change across the members of the building. For example, structures with lower mass and higher stiffness have smaller displacements demands. A sample failure mode of RC structures is shown in Fig. 3 during Gorkha, Nepal earthquake. On the other hand, displacement demands also have to increase. As a result, the main objective of retrofitting methods is to confirm that the displacement demand should be lower than its displacement capacity [107, 172].

Fig. 3 Reinforced concrete buildings failure in Kathmandu, Nepal [149]

This may be accomplished primarily by lowering the probable displacement demand of the building during earthquakes or by increasing the displacement capacity of the building.

2.3 Failure Mechanisms of Steel Structures

Steel structures are considered the most flexible systems among all the structures. Therefore, these structures are often prone to different types of instability because of their inherent slenderness. However, for the last five decades, it has been observed that steel structures respond relatively in an acceptable way when subjected to seismic actions, avoiding a global collapse. Failure may occur suddenly in quasi-static loadings, such as overloading, snow, earth infill in case of green roofs, or indirect actions resulting from lack of bracing or geometrical imperfections. A failure may also be initiated by any type of error in the design, construction, operation, and maintenance of the structures, lack of redundancy, robustness, and ductility. Some common failure modes in steel structures are yielding, shear fracture, brittle fracture, fatigue, corrosion, creep, etc. A typical steel roof structure failure is shown in Fig. 4.

(a) (b)

Fig. 4 Maxima shopping center steel structure roof collapsed in Riga **a** Global collapse and **b** local collapse [7, 73]

3 Strengthening Techniques

This section chronologically summarizes all the strengthening techniques available for all types of structures. The first part discusses about the strengthening techniques of URM structures, the second part is about the strengthening techniques of RC structures, and the last part is the same about steel structures.

3.1 URM Strengthening Technologies

For the reasons stated earlier, a growing amount of research has been conducted into retrofitting/strengthening existing masonry buildings to increase their probable time of collapse during earthquakes. These methods may be beneficial in reducing the huge loss of lives caused by rapid and catastrophic collapses of URM buildings. This section will compile various strategies and investigate all possible studies with their effectiveness. Different retrofitting/strengthening techniques for the URM buildings are described below.

3.1.1 Surface Treatment

Surface treatment methods consist of applying steel or polymer meshes on the exterior part of the members of the structure, which is subsequently coated with a high-strength mortar. This strategy is useful for the developing countries. It can quickly be done without any skilled labour. The system helps in the confinement of the whole masonry walls after cracking. A moderate increase in the ultimate lateral load resistance of the structures is easily possible with this method. Surface treatments can also help to improve the out-of-plane resilience of the URM buildings by increasing the height-to-thickness ratio of the walls and decreasing any arching action [4, 60].

Fig. 5 Examples of strengthening of masonry walls by shotcrete [116]

The only problem is that this method is very costly to use for the common people. For this reason, this method is not very popular.

3.1.2 Application of Shotcrete

Some researchers have tested unreinforced and retrofitted walls with shotcrete [60, 116]. During the test, applied lateral loading increased with small increments up to the ultimate failure of the wall. In the retrofitting tests, it has been noted that the lateral load resistance and the stiffness of the walls at peak loading were also increased. However, the limitations of using this method are that using shotcrete is time-consuming, creates disturbances in occupancy, reduces available spaces, and affects the aesthetics [65, 91]. A strengthening sample of URM walls by shotcrete is shown in Fig. 5.

3.1.3 Stitching, Grouting, and Epoxy Injection

These methods entail grout or epoxy injection into the URM walls to fill any cracks or voids caused by building deterioration, as shown in Fig. 6. Existing cracks on the walls can also be stitched using steel ties and mortar, as shown in Fig. 7 [131]. These methods can easily restore the wall stiffness same as in the initial condition. After epoxy injection, the increase in stiffness has been less drastic than the increase in strength [59]. Furthermore, these techniques are now very popular for the availability of materials, low cost, and ease of implementation.

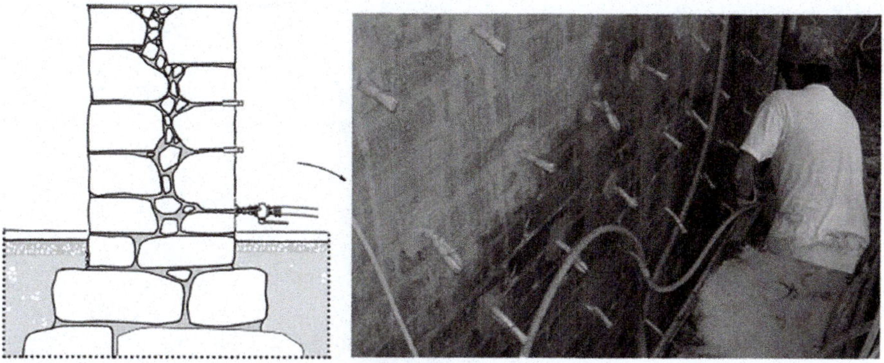

Fig. 6 Detail of the grout injection method on a masonry wall [17]

Fig. 7 Crack repair by stitching [131]

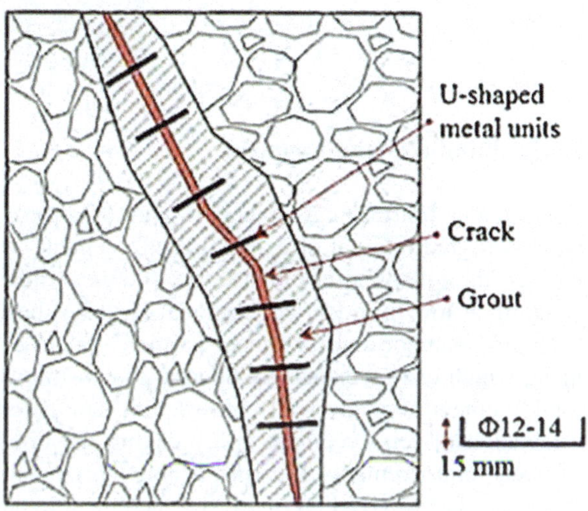

3.1.4 Repointing

When the quality of bricks is good, but the quality of mortar is poor, then the mortar may be partially replaced by a high-strength bonding material. However, adding 2% Ordinary Portland Cement (OPC) to the mortar has no such increase in lateral load resistance [155]. This allows cracks in the masonry to be stabilized by bonding. The advantages of using this method may be enlisted as minimum cost and ease of application. However, this method is not at all sustainable, and the success of this method depends on the compatibility of existing bricks with new mortar.

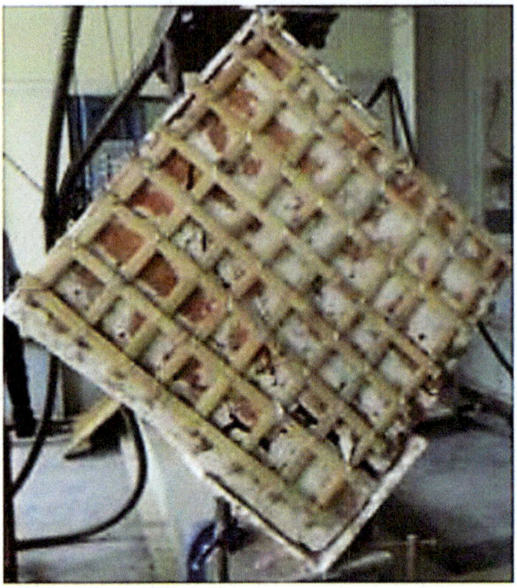

Fig. 8 Retrofitted wall panels by bamboo reinforcement at the end of the test [140]

3.1.5 Bamboo Reinforcement

The bamboo retrofitting method is part of a system that includes ring beams, buttresses, and internal or external reinforcement. This system has been shown to increase the collapse time of masonry structures while having less capacity to prevent cracking at low to moderate earthquakes. It has been proposed that a similar ring beam and bamboo reinforcement partnership can be used with vertical reinforcement, which can be externally fixed after construction [50]. Incorporating vertical reinforcement after the wall construction eliminates complications, including the orientation of reinforcement and breaking the brick. In one of the models with bamboo reinforcements and ring beams, horizontal wire mesh was used. All the structures had survived up to a 100% increment in lateral drift limits during testing. Better reinforced models withstood increases of up to 125% and one of the heavily reinforced models withstood increases of up to 400%. Pushpakumara et al. also tested the masonry wallets with bamboo retrofitting [140], which is also shown in Fig. 8.

3.1.6 Seismic Wallpaper or Fiber-Reinforced Polymer (FRP) Strengthening

This technique was tested on different URM walls made with clay bricks and low-quality mortars [57]. All these testing specimens were strengthened by using FRP strips in the vertical direction bonded with epoxy resins. Then, the walls were subjected to out-of-plane loading using an airbag system. The flexural strength of all retrofitted walls has increased significantly. Further, some other study has shown

Fig. 9 Examples of composite applications on masonry wallets by fiber-reinforced polymer (FRP) [133]

that GFRP composites are useful when arranged at 0–45° to the horizontal direction [115]. This retrofitting method has increased the tensile strength at yield up to two times and its ductility up to 7.5 times its original value in-plane diagonal shear and out-of-plane bending tests. Another one-sided strengthened masonry wallet is developed with vertical FRP strips, as shown in Fig. 9 [133]. Many researchers have studied the usefulness of various FRP rehabilitation techniques in improving the seismic resistance of URM walls. Some of the most significant contributions in this regard can be found in [44, 58, 71, 77, 99, 102, 112, 120, 138, 141, 145, 153, 168]. These studies discovered that retrofitting walls could significantly increase the ultimate load-carrying capacity, ductility, and energy absorption.

3.1.7 Strengthening by L-Shaped Reinforcement, Wire Meshes, and Polypropylene (PP) Bands

Several studies have explored the improvement in the performance of URM structures by strengthening the junctions, which are the most vulnerable part of the masonry

(a) (b) (c)

Fig. 10 Construction of URM walls using **a** horizontal L-shaped reinforcing bars [125], **b** PP bands [125], and **c** wire meshes [20, 21]

walls, using PP bands, wire meshes, and L-shaped reinforcements [18–22, 52, 124–126]. All the walls were tested on the shaking table. These experiments have shown significant strength increases due to the adopted retrofitting options (see Fig. 10).

The strengthened walls by PP bands are also tested under static loading by Macabuag [113]. This method can also improve the shear strength of the test specimens under static loading. However, there was an important issue with the mesh density at stress concentration points, such as the corners of the walls. Macabuag et al. investigated the retrofitting effect on the masonry buildings in Nepal with PP band meshing with strong ground motion shocks [114]. The study determined that using PP bands prevents the disintegration of the masonry elements. As a result, the masonry wall system can accommodate more deformation without collapse.

Tetley and Madabhushi conducted more experiments to compare several other retrofitting measures like a steel mesh cage around the wall, steel reinforcing bars, and a similar mesh made by plastic carrier bags [155]. All test specimens were built with 1:5 scale adobe blocks connected with kaolin clay and sand mortar. The results showed a significant increase in displacements before failure and an increase in resistance for acceleration by three times. A similar type of mesh using plastic carrier bags also showed increased tensile strength and ductility of the walls. The walls in the transverse direction failed at a PGA level of 0.64 g, and in-plane wall failure occurred at 1.02 g.

3.1.8 Post-tensioning Using Rubber Tires

The post-tensioning method by using scrap rubber tires with steel meshes has been proposed in some research as a potential retrofitting solution. The tires are assembled, as shown in Fig. 11. Turer et al. [93] conducted testing on a scale of 1:10 model using a simple shake table [162]. Testing was carried out using URM walls reinforced horizontally, vertically, and in both directions. After the tests, it was seen that the

Fig. 11 Strengthening or masonry walls by old tire strips [93]

cracks were spread uniformly, and the brittle failure of these structures was changed into a ductile response.

3.1.9 Mesh Reinforcement

Construction sites typically use industrial geogrids and weaker meshes as soft fences. So, two different types of polymer mesh have been used to strengthen these URM structures, as shown in Fig. 12. Blondet et al. conducted a study in which the mesh reinforcements were wrapped around the wall and coated with mud plaster [28]. Mainly three types of geogrid (50, 75, and 100% covered) and one soft mesh system (80% covered) were tested using a unidirectional shake table. Other common mesh reinforcements that are commonly used for strengthening materials are wire-reinforced mortar (WRM) [133] and (c) textile-reinforced mortar (TRM) [133]. These strengthening materials have effectively improved the load-carrying capacity and significantly increased the energy absorption capacity.

3.1.10 Steel Reinforcement for Masonry Structures

After extensive research at the Pontifical Catholic University of Peru (PUCP), the Regional Seismological Center for South America (CERESIS) decided to reinforce some adobe buildings with steel wire meshes [24]. These steel wire meshes were

Fig. 12 Examples of mesh reinforcement application on masonry wallets **a** industrial geogrid [28] **b** wire reinforced mortar (WRM) [133] and **c** textile-reinforced mortar (TRM) [133]

applied outside the adobe walls at critical points like free ends and corners. This was then enclosed with a mortar layer. A large earthquake of magnitude (M_w) 8.0 occurred in Pisco, Peru, on August 15, 2007, resulting in the collapse of over 70,000 houses and severe damage to more than 33,000 houses. However, only five houses were reinforced in 1998, and these houses survived the earthquake. Significant cracking or collapse was observed in the nearby buildings. Furthermore, another earthquake of magnitude 8.4 occurred in South Peru with similar results in 2001. Taghdi et al. conducted an experimental study using steel strips to retrofit low-rise masonry walls [150]. This study has also shown that steel strips can effectively increase the in-plane strength, energy dissipation capacity, and ductility. Another finding is that the anchor bolts running along the vertical steel strips help to keep the masonry wall from buckling.

3.2 Basic Concept of Strengthening for RC Structures

RC structures are strengthened to recover the strength of deteriorated structural concrete members and prevent further structural distress. Concrete structural members can lose strength due to design flaws in structures, poor workmanship, or deterioration caused by the aggression of harmful atmospheric agents. The main retrofitting measures of RC structures are mainly divided into two categories, global and local retrofits. The classifications are shown in Fig. 13. All the methods are briefly described below.

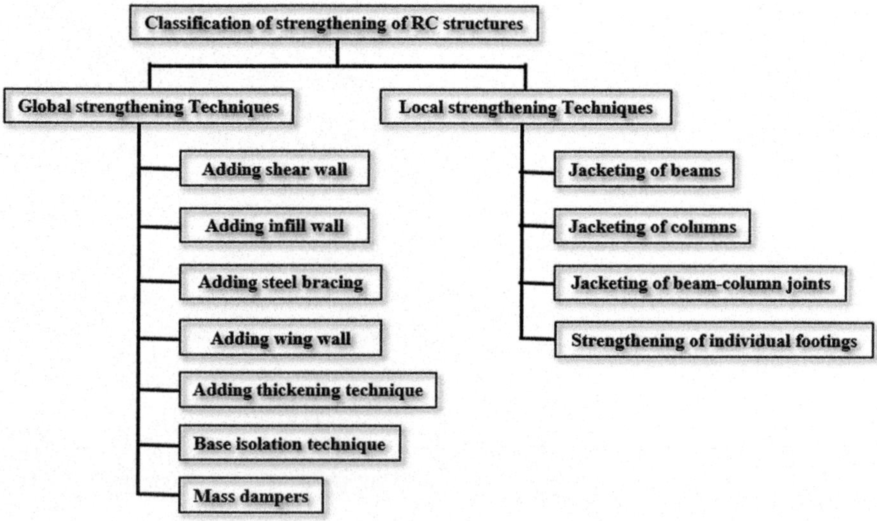

Fig. 13 Classification of the strengthening measures of RC structures

3.2.1 Adding Shear Walls/Infill Walls

Adding RC infill shear walls is the most common global strengthening method. Even when applied to damaged structures, the infill wall technique can produce adequate strength [38, 147]. The shear walls can resist most of the lateral loads and limit the displacement of the building. At the same time, the frame system can withstand high earthquake loads. Infill walls made of reinforced concrete can be used as wing walls and partial walls [34]. A sample of adding an infill shear wall is shown in Fig. 14.

3.2.2 Addition of Steel Bracings

This is a convenient solution when large openings are essential. Steel bracings are very advantageous due to their higher strength, and openings for natural light can be provided. Steel bracings can also be used for RC frames to reduce displacement demands. Steel bracings may be applied inside the system [117] or outside the frame [34]. Post-tensioning is another option that can be useful to the bracing elements [70]. In both cases, steel braces provide more pleasing solutions for various uses. However, its application inside buildings with small openings is complex. It only permits simple installation across the axes on external facades of the wall [72]. There are many literatures that contain research on various types of bracing [135].

The steel bracing members must be designed according to the specific codes and steel standards. The anchor design principles may be used to connect the existing structure and steel members. Another type of damage is to the buttress type steel walls due to the out-of-plane buckling of the compression members. Installing

Fig. 14 Infill wall application in an RC building [34]

lateral supports to the storey levels is also suggested to prevent buckling. Using such members may also economize the design of shear walls [13, 171].

3.2.3 Base Isolation Techniques

Base isolation method is defined as the separation of the superstructure from the foundation. It is the primary method for passive control of the vibration. The growing number of isolated structures shows the importance of base isolation systems that are gradually becoming accepted to mitigate earthquake hazards. This anti-seismic design technique can decrease the effect of earthquake ground motion by uncoupling the superstructure from the foundation. Significant progress has been made in isolation systems for civil engineering structures in recent years [39, 81, 87].

3.2.4 Jacketing of the RC Members

This is one of the popular methods to strengthen the RC building members. Two basic types of Jacketing are Reinforced Concrete jacketing and steel jacketing.

Jacketing the RC members may be implemented as a confinement of the columns or beams with new and higher quality RC materials, depending on the deficiencies in the structural members (Fig. 15a). Jacketing brittle columns or beams improves resistance to shear and axial loads. In that case, some variations will also arise in the capacity and stiffness of the RC member after applying the jacketing. After these modifications, the jacketed section is guaranteed to have adequate axial and shear strength. Although RC jacketing can be helpful to all types of structural members, a jacketing application is not recommended when it increases only the bending capacity

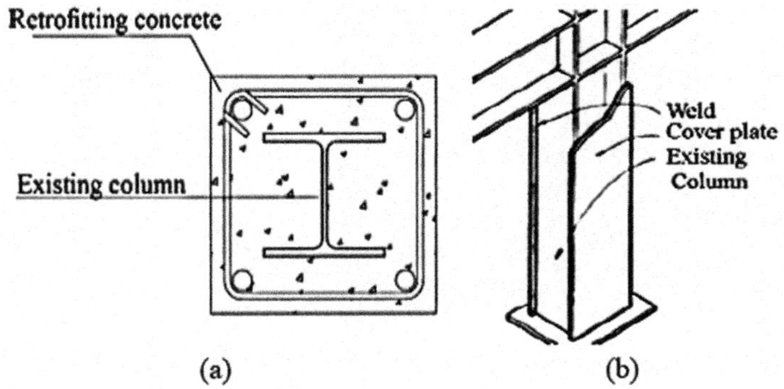

Fig. 15 Types of jacketing, **a** by concrete and **b** by steel plate [15]

of beams. It may form a strong beam-weak column, resulting in a global failure of the whole structure.

Another practical method is jacketing with steel members, which may be used efficiently for various applications. Figure 15b depicts a typical application of steel jacketing. Steel jacketing is a simple way to increase the shear strength of any RC elements. Steel jacketing may also increase the bending strength to maintain the continuity between different stories [15, 144].

In recent years, using fiber-reinforced polymer composite (FRPC) jackets have considerably increased in applications of strengthening [15]. These composites can be manufactured with various materials like glass, carbon, aramid etc. These materials can be used for various purposes, including increasing the flexural capacity of floor slabs and improving the shear capacity of columns, beams, or shear walls. FRP is used as longitudinal reinforcement. So, the additional flexural capacity can be calculated using a simple cross-section calculation. More complex calculations are required to measure changes in the axial capacity and ductility of the member, which are beyond the basic knowledge of reinforced concrete [130].

In the same manner, beam-column joints are also can be retrofitted. The retrofitting scheme is selected according to the availability of resources and seismic assessment of the building. As a result, it enables the development of plastic hinges on the beams and assures a ductile failure mode.

Retrofitting of the isolated footings can also be done by jacketing of the footings. The foundation can be retrofitted by extending the jacketing up to the level of the foundation (Fig. 16). In these cases, the foundation system can withstand a greater strain than the beams or columns. Because of the increased strength of the structure, the transmitted shear capacity between the soil and the reinforced structure may be higher [142]. Large bending moments may be generated at the base of stiff structural components, and large overturning movements may also develop large dynamic axial forces in the columns.

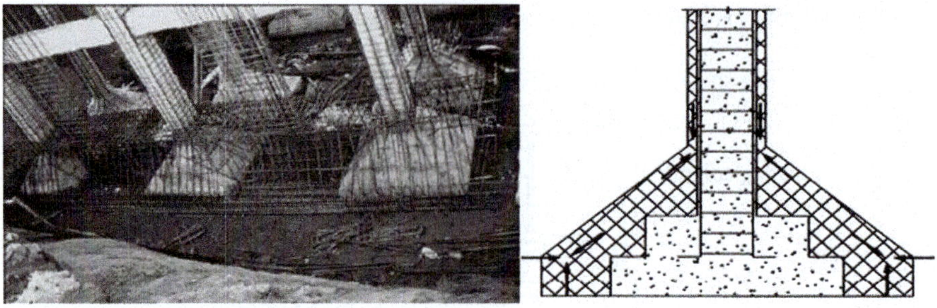

Fig. 16 Jacketing of footings [157]

(a) (b)

Fig. 17 Application of **a** epoxy resin, **b** cement grout injection in beam-column joints [157]

3.2.5 Injection in the Cracks of Concrete

Injection in cracks is a very useful, economical, and local retrofitting method for repairing RC structures [23, 90, 100, 156]. The ability of adhesive materials to enter the cracks of the damaged concrete under appropriate pressure determines the effectiveness of the repair process. Repairing the bond between steel and concrete with epoxy injection may be very difficult. This technique can be used to repair minor cracks (up to 0.1 mm), medium cracks (up to 3 mm), and large cracks (5–6 mm). Cement grout is the appropriate material for injection in larger cracks up to 20 mm wide (Fig. 17). Flexural tests on RC beams. Beam-column joints demonstrate that the repair process removes the unsightly appearance of wide cracks and restores the actual flexural strength and stiffness [95, 161]. Static and dynamic tests can show the increase in strength of joints after being repaired with epoxy resin injection.

3.2.6 Shotcrete Strengthening

Shotcrete is a strengthening method used for RC and masonry structures [26, 106]. There are two different types of shotcretes like wet-mix and dry-mix. Shotcrete

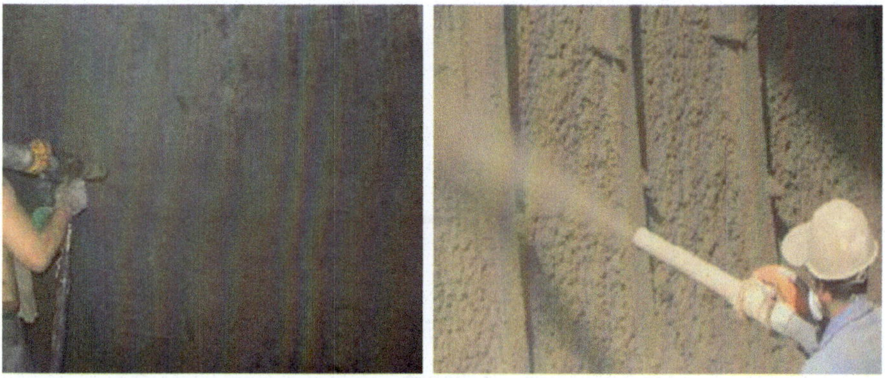

Fig. 18 Shotcrete in RC walls and tunnels [106]

may be applied to any surface and combined with other retrofit schemes [37]. It achieves excellent bonds to the majority of competent surfaces due to its high-velocity impact and low water-cement ratio. Deficiencies in shotcrete applicability are typically divided into five categories listed below [166]. These are (i) delamination of construction joints or interfaces of layers, (ii) failure of the bond of the receiving surface, (iii) insufficient material filling behind the reinforcing steel, (iv) slough due to excessive mixing water, and (v) weak interface between the concrete and steel. Figure 18 shows the injection of shotcrete in RC walls and tunnel walls.

3.2.7 Steel Plate Addition

Steel plates are mainly added with the beams. Both flexural strength and shear strength can be enhanced by this method. When thick steel plates are needed, using several thin layers of plates can reduce interfacial shear stresses. Implementing the bonding work is also very important to achieve composite action. A critical design aspect is preventing premature de-bonding of externally bonded plates [5, 11, 152]. The process is shown in Fig. 19.

3.3 Strengthening Techniques of Steel Structures

Performance of the Steel frames may be governed by the plastic hinge formation, local and global instabilities, fractures, and structural discontinuities. These behaviors or their combination are likely to influence the capacity of any structural member. Several studies have shown that capacity of steel structure is related to the response of the members, connection response, panel zone failure, and premature brittle weld failure [159]. In many cases, determining the capacity of columns is difficult. So, it

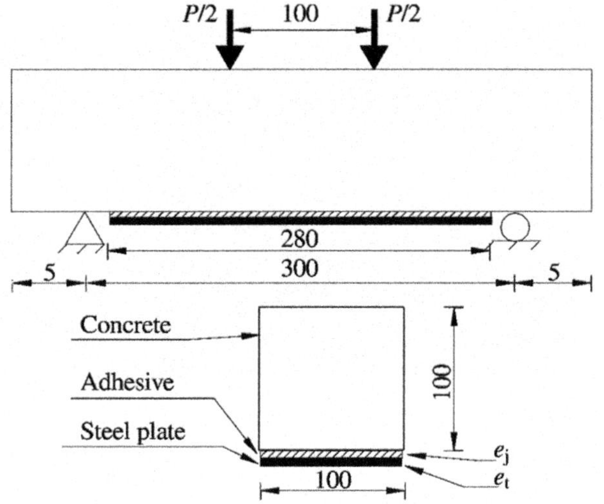

Fig. 19 Steel plate adhesion at the bottom of the beam [105, 136]

is preferable to strengthen these structures well before the next calamities occur on a large scale.

3.3.1 Side Plate Connections

A popular retrofitting method for a standard moment-resisting connection is side plate connections, as shown in Fig. 20. In this method, the side plate retrofit system is used for resisting the physical separation between the column flange and the end of the beam [46]. The steel frame is stiffened, and the zone panel deformation is removed using this type of connection due to the increased stiffness of the side plates. This connection system is entirely fillet-welded, carrying shear and moments via a combined system of transverse shear plates and fillet welds. The side plates must be strong and rigid enough to withstand any significant plastic behaviour of the connection system. FEMA 351 contains additional information [63].

3.3.2 Stiffeners at Connections

The effect of stiffeners on the welded connections has been studied for intermediate moment frames presented in AISC [109]. The behaviour of stiffeners on I-beam to hollow-column section connections is depicted in Fig. 21 [42, 98]. Connections made with triangular stiffeners exhibit the lowest rigidity, whereas those with side stiffeners exhibit higher moment capacities, ductility, and stiffness.

Furthermore, the performance of retrofitted connections was also studied, and some design guidelines were proposed [69]. The side-stiffeners on concrete-filled tabular (CFT) columns connected to I-beams have also been introduced to provide

Fig. 20 Strengthening retrofit concept by side plate retrofit connection [46]

Fig. 21 Connection of a typical I-beam to a box column [98]

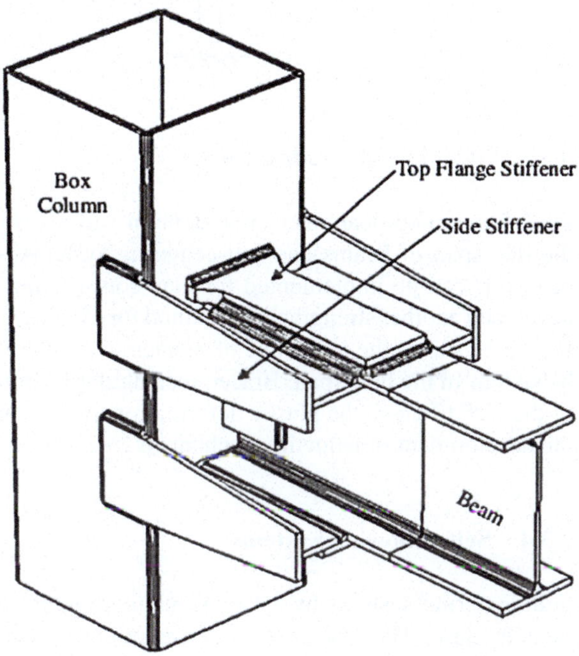

stable hysteretic behaviour and satisfactory ductility. Overall, connections with both column stiffeners and top flange stiffeners have the highest energy dissipation value, with the top flange stiffener of the beam being the most effective.

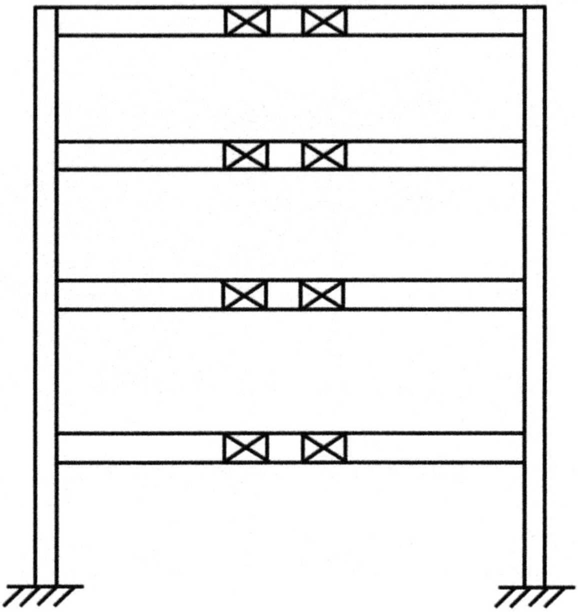

Fig. 22 Frame modified with mid-span truss opening [103]

3.3.3 Steel Frame Modifications

Steel frame modifications can be done in various ways, like frame modification at the mid-span of beams and self-centering systems by cables. Frame modification of beams can be implemented for new construction. Leelataviwat et al. have also developed another strengthening method for existing moment-resisting frames [103]. Figure 22 shows the technique of replacing specific beams by introducing a ductile behaviour in the mid-span. Braced rectangular openings are created at the mid-span of the web to move the plastic deformation away from the critical joints and ensure the development of a ductile mechanism.

3.3.4 Self-centering Systems

Self-centering systems by cables is another method that can be useful for connections to slabs. The placement methods of cables with girders or cables with beam connections are critical, particularly in tall structures. Self-centering bracings are designed using prestressed strands with friction pads or memory alloys. This system achieves energy dissipation through friction dampers, yielding seat angles, or energy-dissipating bars within tubes. Tendons can be used on multiple floors, and elastomeric spring dampers can provide an energy dissipation system. Christopoulos et al. proposed self-centering structural systems (Fig. 23) for seismic strengthening of special moment resisting frames [43].

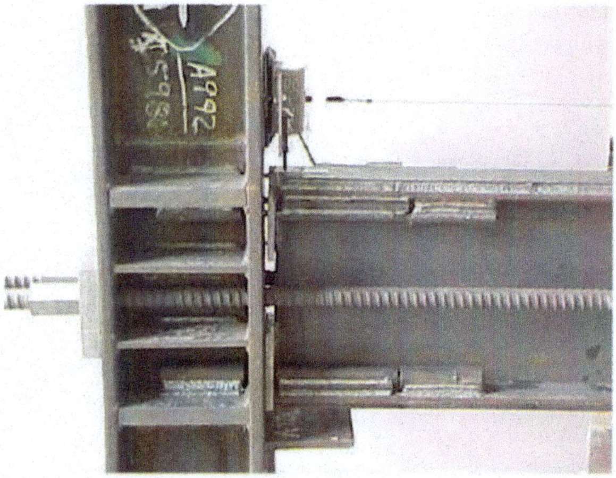

Fig. 23 PTED system for strengthening beam-column joint [43]

Later, Garlock et al. proposed a similar structural system with high-strength steel strands after installing bolted replaceable top-and-seat angles [67]. Here, the vertical shear is supported by both the angles and the friction between the beam and the column.

3.3.5 Steel Plate Shear Walls

Thorburn et al. conducted preliminary research on the design of Steel Plate Shear Walls (SPSW) and their strengthening methods [158]. SPSWs allow for shear buckling in steel buildings and can also be used as the lateral force-resisting system of the structures. Furthermore, using Reduced Beam Sections (RBS) connections, low-yield strength steel panels, and light-gauge cold-formed steel plates have been used as potential applications by Berman and Bruneau [25]. The low-yield strength steel panels demonstrate an earlier onset dissipation of energy by the panel, while perforated panel specimens (Fig. 24) can be used to control the stiffness and over-strength issues. This option is also helpful in retrofitting because it allows utilities to enter the pre-designed system.

3.3.6 Bearing Walls

Bearing walls must add stiffness to the steel frame system while tied into any framing to take the load in their weakest direction and make them equally stronger in both directions. They must also remain in place after the shock waves to retain strength for the aftershocks.

So, three approaches have been recognized to enhance building resilience by the bearing walls under progressive collapse scenarios [46]. These approaches are (i) a backup wall that builds as a secondary wall to support the existing wall, (ii) a strong

Fig. 24 SPSW specimen with cut-out corners [25]

wall that uses fabric reinforcement to control the breach area, and (iii) a ductile wall treated with polyurethane spray to prevent the punching shear. Furthermore, these walls may accommodate openings, especially in multi-story buildings, while additional improvements are required.

3.3.7 Braced Frames

These types of frames can be made with single diagonal members, x- and k-type connections, split and chevron braces, knee or lattice braces. Braced frames are commonly used in interior cores. As a result, these connections can be made quickly with wall panels inside and outside. Composite braced frames, in which concrete bracings support steel frames, are also becoming very popular.

Khatib et al. first proposed a simplified procedure for designing the suspended zipper frames [97]. This method was introduced by Leon and Yang later [104]. It consisted of inverted V-braces by adding a zipper column connecting the intersection point of the braces directly above the first floor. All intersection points of brace-to-beams are connected by zipper columns, which force compression braces to buckle the entire braced bay. As a result, the dissipated energy is distributed more evenly across the building's height [33].

Later, a system of suspension was proposed that ensured that the top storey braces remained elastic. Other braces in the compressive force direction should buckle down, and the suspended zipper struts should yield in tension [159, 170]. As a result, adequate ductility, as well as superior seismic performance, can be provided.

Eatherton et al. proposed a system for controlling the rocking phenomenon that removes residual drifts and concentrates on the major structural damages with replaceable elements using replaceable links or braced frame retrofitting concepts [56]. The system consists of three main components. Stiff steel braced frame that remains elastic but is not tied to the foundation to allow the rocking behaviour, vertical

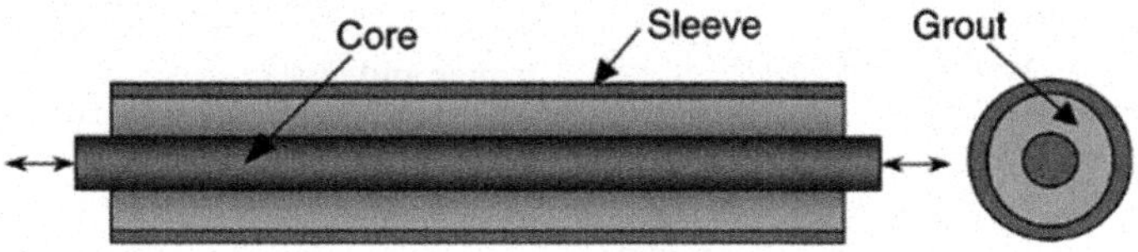

Fig. 25 Non-buckling bracing [101]

post-tensioning strands from the top of the frame to the foundation that brings the frame back to the center. The replaceable elements can absorb the energy during rocking.

3.3.8 Non-buckling Braces

Conventional non-buckling braces are likely to buckle during the tension and the tension–compression cycle during the seismic actions, resulting in less energy dissipating during compression. This may cause a chronological reduction in the hysteresis loop of the braces after some time due to forming a plastic hinge near the midspan of the structural member. Using non-buckling braces can easily bypasses this problem. These frames are also known as restrained braces [40, 101, 151]. This bracing system requires suitable strength to resist compression and rigidity to avoid buckling (Fig. 25).

Buckling Restrained Braces (BRB) have become much important in the last decade because these braces have stable hysteretic behaviour and lower fatigue life characteristics. However, in some cases, gusset plate cracking and buckling are likely the same for all braced frames.

3.3.9 Strengthening Members like Beams and Columns

Members can be reinforced by welding plates or encasing/filling the gaps with concrete. This method can be applied to any member, such as those resting on the ground floor. So, concrete encasing of columns and floor beams has been proposed [160]. This method constitutes a new strengthening method for existing steel structures. A sample of adding plates is shown in Fig. 26 [173].

4 Guidelines

The guidelines for the seismic strengthening of structures can reduce the loss of life due to the collapse of impending structures in future earthquakes. All the structures that come under a high risk of loss of life should be identified first [127]. The vision of the guidelines on seismic strengthening should be decided in such a way that

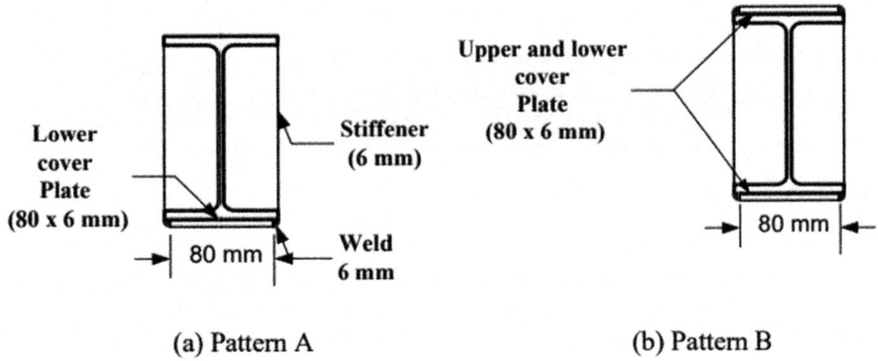

Fig. 26 Strengthening pattern of steel beam and column [173]

mandatory seismic retrofitting shall be undertaken in a prioritized manner of seismically deficient existing important structures and selected privately-owned buildings. The prioritization should be based on earthquake risk assessment of buildings and structures. The cost of strengthening shall be critically examined when deciding to undertake strengthening processes. The strengthening framework plan is shown in Fig. 27. Some crucial points for considering the proper guidelines and the extent of strengthening are discussed as follows.

Different national and international codes provide different technical criteria for assessment or strengthening solutions. However, these also include a few non-technical issues. As per [3, 16, 30, 31, 41, 64, 137], six basic non-technical issues are to be considered to take the proper decision on the strengthening techniques.

1. Overall construction cost is a critical issue for deciding whether to strengthen, according to the expected level of strengths.
2. In the case of fully or partially occupied buildings, the parameter of occupant disruption becomes a major challenge for strengthening that structure.
3. In any historic building, the conservation of historic architecture also usually has control over any particular strengthening methods.
4. The importance of a building after a disaster must be considered because it refers to whether the building has any special usage after a disaster, for instance, hospitals, airports, fire stations, etc.
5. The project's location, the availability of the required skill set of the workforce, and the material and availability of equipment may impact the choice of the strengthening method.
6. Sustainability concepts like the amount of material and machinery required, conveyance of materials, and recycled material content are the issues to be thought about for the purpose of strengthening.

The extent of strengthening is also an important point. Some agencies have been guiding in this direction. It is highly recommended that strengthening may be carried out if the existing strength of the building falls below 70% of the strength capacity

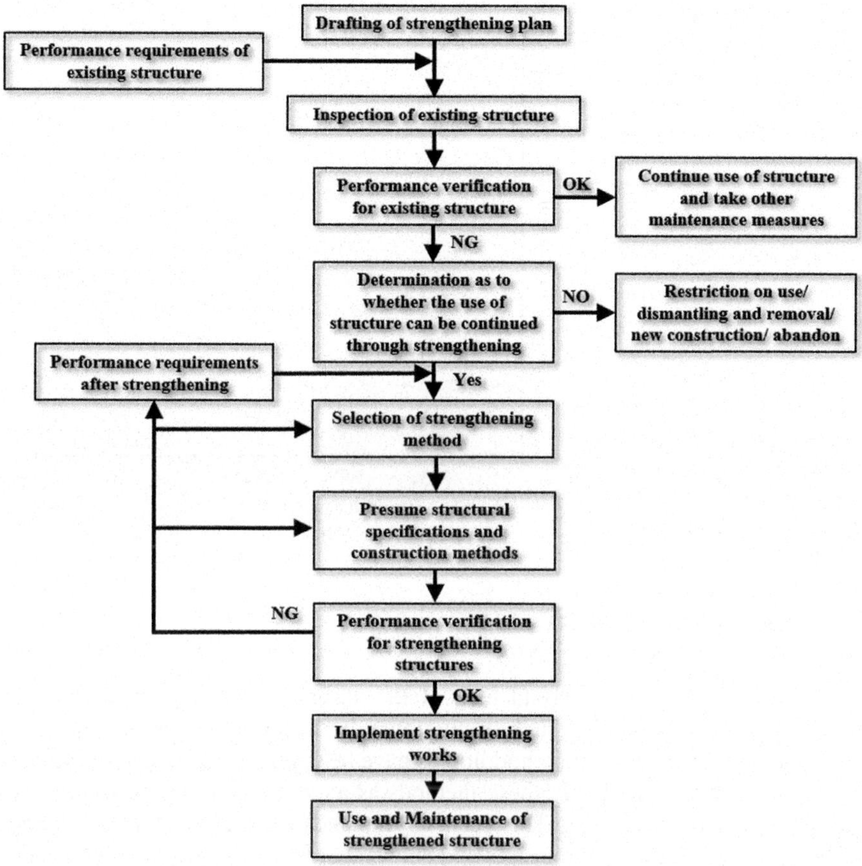

Fig. 27 Strengthening framework plan

required according to the current design standard, as suggested by [45]. When the risk is more than ten times the standard risk, building authorities should prohibit human occupancy in the interest of public safety. According to standards defined in Delhi in India, if any structure is relatively old and has lived more than half of its design life, it should be strengthened to withstand at least 70% of total design seismic loads [45], as shown in Fig. 28 [169]. These guidelines should be considered as a standard all over the world.

Some other guidelines are also available for seismic rehabilitation and strengthening existing structures [12, 37, 78, 127–129, 134]. In these guidelines, the requirements and some techniques are decided. But these guidelines are required to be revised in this current scenario of large destruction during the Turkey earthquake.

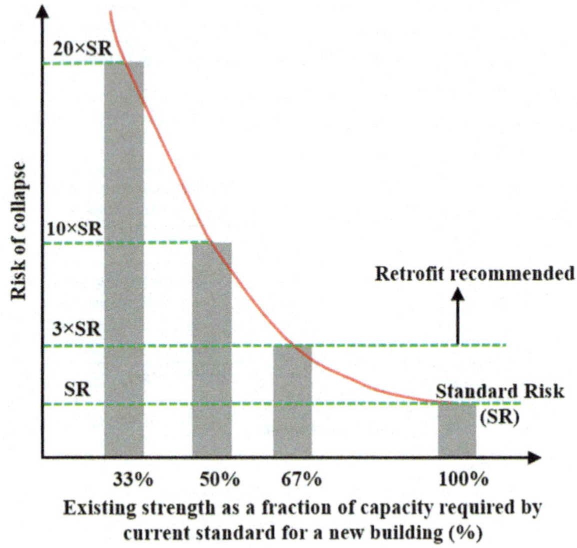

Fig. 28 Extent of retrofit [45]

5 Conclusive Remarks

The review begins with a broad idea of the casualties of previous earthquakes in last 100 years. Following such discussions, the historical evolutions of various earthquake-resisting design methodologies have been practiced for several decades. However, there is some lag in these methods as a large amount of destruction and causalities are still happening. The recent earthquake in Turkey is a burning example of the drawbacks of these processes and shows that this issue has to be considered more seriously. The failure pattern of the structures is discussed to show that the actual scenario is happening all around. Thus, such failure of structures may significantly affect human civilization. Further, existing techniques for retrofitting and strengthening URM, RC, and steel structures are elaborated. It is believed that while going through this particular chapter of the book, the reader will have at least an overall picture of the method of retrofitting and strengthening (which are existing and evolving), and one might be in a position to choose a particular one depending on the type of problem he is dealing with.

References

1. Adhikary S das, Li B, Fujikake K (2014) Effects of high loading rate on reinforced concrete beams (with Appendix). Struct J 111:651–660. https://doi.org/10.14359/51686579
2. AISC (2022) Code of standard practice for steel buildings and bridges

3. Alba-Rodríguez MD, Martínez-Rocamora A, González-Vallejo P et al (2017) Building rehabilitation versus demolition and new construction: economic and environmental assessment. Environ Impact Assess Rev 66:115–126. https://doi.org/10.1016/j.eiar.2017.06.002
4. Alcaino P, Maria HS (2008) Experimental response of externally retrofitted masonry walls subjected to shear loading. J Compos Constr 12:489–498. https://doi.org/10.1061/(ASCE)1090-0268(2008)12:5(489)
5. Ali MSM, Oehlers DJ, Bradford MA (2001) Shear peeling of steel plates bonded to tension faces of RC beams. J Struct Eng (ASCE) 127:1453–1459
6. Ali Q, Khan AN, Ashraf M et al (2013) Seismic performance of stone masonry buildings used in the Himalayan belt. Earthq Spectra 29:1159–1181. https://doi.org/10.1193/091711EQS228M
7. Anastasiadis A (2021) Failure of steel structures: rethinking some of the aftermaths. Urbanism Arhitectură Construcţii 12(2).
8. Angst UM (2018) Challenges and opportunities in corrosion of steel in concrete. Mater Struct/Materiaux et Constructions 51:1–20. https://doi.org/10.1617/S11527-017-1131-6/FIGURES/6
9. Anwar M, Roushdi M (2013) Improved concrete properties to resist the saline water using environmental by-product. Water Sci 27:30–38. https://doi.org/10.1016/J.WSJ.2013.12.003
10. Apostolopoulos C, Koulouris K (2021) Corrosion effects on durability of RC structures. Metals 11:1812. https://doi.org/10.3390/MET11111812
11. Aprile A, Spacone E, Member A, Limkatanyu S (2001) Role of bond in RC beams strengthened with steel and FRP plates. J Struct Eng (ASCE) 127:1445–1452
12. ASCE/SEI 41-06 (2007) Seismic rehabilitation of existing buildings. American Society of Civil Engineers
13. Astaneh-Asl A, Qian X, Shi Y (2019) Application of steel shear walls toward more resilient structures. In: Resilient structures and infrastructure. Springer, Singapore, pp 3–46
14. Ates S, Kahya V, Yurdakul M, Adanur S (2013) Damages on reinforced concrete buildings due to consecutive earthquakes in Van. Soil Dyn Earthq Eng 53:109–118. https://doi.org/10.1016/j.soildyn.2013.06.006
15. Aydenlou RM (2020a) Types of existing buildings: detailed introduction and seismic rehabilitation. In: Seismic rehabilitation methods for existing buildings. Elsevier, pp 193–553
16. Aydenlou RM (2020b) Seismic rehabilitation methods for existing buildings. Elsevier
17. Baltazar LG, Henriques FMA, Cidade MT (2019) Rheology of natural hydraulic lime grouts for conservation of stone masonry-influence of compositional and processing parameters. Fluids 4
18. Banerjee S, Nayak S, Das S (2018) Enhancing shear capacity of masonry wallet using PP-band and steel wire mesh. IOP Conf Ser Mater Sci Eng 431:072004. https://doi.org/10.1088/1757-899X/431/7/072004
19. Banerjee S, Nayak S, Das S (2021) Seismic performance enhancement of masonry building models strengthened with the cost-effective materials under bi directional excitation. Eng Struct 242. https://doi.org/10.1016/j.engstruct.2021.112516
20. Banerjee S, Nayak S, Das S (2020a) Shear and flexural behaviour of unreinforced masonry wallets with steel wire mesh. J Build Eng 30. https://doi.org/10.1016/j.jobe.2020.101254
21. Banerjee S, Nayak S, Das S (2020) Improving the In-plane behavior of brick masonry wallet using PP band and steel wire mesh. J Mater Civ Eng 32:04020132. https://doi.org/10.1061/(asce)mt.1943-5533.0003159
22. Banerjee S, Nayak S, Das S (2019) Enhancing the flexural behaviour of masonry wallet using PP band and steel wire mesh. Constr Build Mater 194:179–191. https://doi.org/10.1016/j.conbuildmat.2018.11.001
23. Bano S, Jaiswal G, Kumar R, Tiwari A, Karthikeyan M (2023) Experimental Study on the crack repair techniques of concrete structures: a case study. IOP Conf Ser Mater Sci Eng 1273:012006. https://doi.org/10.1088/1757-899X/1273/1/012006
24. Bartolomé AS, Quiun D, Zegarra L (2008) Performance of reinforced adobe houses in Pisco, Peru earthquake. In: Proceedings of the 14th world conference on earthquake engineering. Beijing, China

25. Bernam JW, Bruneau M (2004) Steel plate shear walls are not plate girders. AISC Eng J 95
26. Bernardo G, Guida A, Mecca I (2015) Advancements in shotcrete technology. In: Structural studies, repairs and maintenance of heritage architecture XIV. WIT Press, pp 591–602
27. Blokley DI (1980) The nature of structural design and Safety. Ellis Horwood, London
28. Blondet M, Torrealva D, Vargas J, Velasquez J, Tarque J (2006) Seismic reinforcement of adobe houses using external polymer mesh. In: First European conference on earthquake engineering and seismology, Paper no. 632
29. Bosio M, di Salvatore C, Bellotti D, Capacci L, Belleri A, Piccolo V, Cavalieri F, Lago BD, Riva P, Magliulo G, Nascimbene R, Biondini F (2022) Modelling and seismic response analysis of non-residential single-storey existing precast buildings in Italy. J Earthquake Eng. https://doi.org/10.1080/13632469.2022.2033364
30. Bostenaru Dan MD (2004) Multi-criteria decision model for retrofitting existing buildings. Nat Hazards Earth Syst Sci 4:485–499. https://doi.org/10.5194/NHESS-4-485-2004
31. Bradshaw A, Rajeev P, Tesfamariam S (2011) Multi Criteria decision making tool for the selection of seismic retrofitting techniques. In: Australian earthquake engineering society conference. Barossa Valley, South Australia
32. Braga F, Gigliotti R, Monti G, Morelli F, Nuti C, Salvatore W, Vanzi I (2014) Speedup of post earthquake community recovery: the case of precast industrial buildings after the Emilia 2012 earthquake. Bull Earthq Eng 12:2405–2418. https://doi.org/10.1007/s10518-014-9583-3
33. Bruneau M (2004) Seismic retrofit of steel structures. Sociedad Mexicana de Ingeniera Sismica, VIII SNIS, Tlaxcala
34. Bush TD, Wyllie LA, Jirsa JO (1991) Observations on two seismic strengthening schemes for concrete frames. Earthq Spectra 7:511–527. https://doi.org/10.1193/1.1585640
35. Calderini C, Cattari S, Lagomarsino S (2010) The use of the diagonal compression test to identify the shear mechanical parameters of masonry. Constr Build Mater 24:677–685. https://doi.org/10.1016/J.CONBUILDMAT.2009.11.001
36. Calderoni B, Cordasco EA, del Zoppo M, Prota A (2020) Damage assessment of modern masonry buildings after the L'Aquila earthquake. Bull Earthq Eng 18:2275–2301. https://doi.org/10.1007/s10518-020-00784-5
37. California University (1976) State-of-the-art review on shotcrete
38. Canbay E, Ersoy U, Ozcebe G (2003) Contribution of reinforced concrete infills to seismic behavior of structural systems. ACI Struct J 100:637–643. https://doi.org/10.14359/12805
39. Cancellara D, de Angelis F (2016) Nonlinear dynamic analysis for multi-storey RC structures with hybrid base isolation systems in presence of bi-directional ground motions. Compos Struct 154:464–492. https://doi.org/10.1016/j.compstruct.2016.07.030
40. Carden L, Itani A, Buckle I, Aiken I (2004) Buckling restrained braces for ductile end cross frames in steel plate girder bridges. In: 13th world conference on earthquake engineering. Vancouver, B.C., Canada
41. Caterino N, Iervolino I, Manfredi G, Cosenza E (2009) Comparative analysis of multi-criteria decision-making methods for seismic structural retrofitting. Comput-Aided Civil Infrastruct Eng 24:432–445. https://doi.org/10.1111/j.1467-8667.2009.00599.x
42. Chen SJ, Lin HY (1990) Experimental study of steel I-beam to box-column moment connection, 41–7. In: 4th international conference on steel structures and space frames
43. Christopoulos C, Filiatrault A, Uang CM (2002) Self-centering post-tensioned energy dissipating (PTED) steel frames for seismic regions. San Diego, La Jolla, CA, p 292
44. Corradi M, Borri A, Vignoli A (2002) Strengthening techniques tested on masonry structures struck by the Umbria–Marche earthquake of 1997–1998. Constr Build Mater 16:229–239. https://doi.org/10.1016/S0950-0618(02)00014-4
45. CPWD (2007) Handbook on repair and rehabilitation of RCC buildings. Delhi
46. Crawford JE (2002) Retrofit methods to mitigate progressive collapse, karagozian and case structural engineers, Glendale, CA
47. Dauda JA, Silva LC, Lourenço PB, Iuorio O (2021) Out-of-plane loaded masonry walls retrofitted with oriented strand boards: numerical analysis and influencing parameters. Eng Struct 243:112683. https://doi.org/10.1016/J.ENGSTRUCT.2021.112683

48. Debnath P, Chandra Dutta S, Mandal P (2023) Lateral behaviour of masonry walls with different types of brick bonds, aspect ratio and strengthening measures by polypropylene bands and wire mesh. Structures 49. https://doi.org/10.1016/j.istruc.2023.01.155
49. Debnath P, Halder L, Dutta SC (2022) Damage survey and seismic vulnerability assessment of unreinforced masonry structures in low-intensity Ambasa earthquake of northeast India. Structures 44:372–388. https://doi.org/10.1016/j.istruc.2022.08.005
50. Dowling D, Samali B, Li J (2005) An Improved means of reinforcing adobe walls-external vertical reinforcement. In: SismoAdobe. Lima, Peru
51. Dutta SC, Halder L, Sharma RP (2021) Seismic vulnerability assessment of low to mid-rise RC buildings addressing prevailing design and construction practices in the Northeastern region of the Indian subcontinent: a case study based approach. Structures 33:1561–1577. https://doi.org/10.1016/j.istruc.2021.05.032
52. Dutta SC, Mukhopadhyay P, Goswami K (2013) Augmenting strength of collapsed unreinforced masonry junctions: principal damage feature of walls damaged by moderate Indian earthquakes. Nat Hazards Rev 14:281–285. https://doi.org/10.1061/(ASCE)NH.1527-6996.0000096
53. Dutta SC, Mukhopadhyay PS, Saha R, Nayak S (2015) 2011 Sikkim earthquake at eastern himalayas: lessons learnt from performance of structures. Soil Dyn Earthq Eng 75:121–129. https://doi.org/10.1016/j.soildyn.2015.03.020
54. Dutta SC, Nayak S, Acharjee G, Panda SK, Das PK (2016) Gorkha (Nepal) earthquake of April 25, 2015: actual damage, retrofitting measures and prediction by RVS for a few typical structures. Soil Dyn Earthq Eng 89:171–184. https://doi.org/10.1016/j.soildyn.2016.08.010
55. Dutta SC, Nayak S, Dinakar P (2014) Lateral period and seismic vulnerability of masonry buildings. Proc Inst Civ Eng Struct Build 167:633–645. https://doi.org/10.1680/stbu.12.00043
56. Eatherton MR, Hajjar JF, Deierlein GG, Krawinkler H, Billington S, Ma X (2008) Controlled rocking of steel-framed buildings with replaceable energy-dissipating fuses. In: 14th world conference on earthquake engineering. Beijing, China
57. Ehsani MR, Saadatmanesh H, Velazquez-Dimas JI (1999) Behavior of retrofitted URM walls under simulated earthquake loading. J Compos Constr 3:134–142. https://doi.org/10.1061/(ASCE)1090-0268(1999)3:3(134)
58. El-Dakhakhni WW, Hamid AA, Hakam ZHR, Elgaaly M (2006) Hazard mitigation and strengthening of unreinforced masonry walls using composites. Compos Struct 73:458–477. https://doi.org/10.1016/J.COMPSTRUCT.2005.02.017
59. Elgawady M, Lestuzzi P, Badoux M (2004) A review of conventional seismic retrofitting techniques for URM. In: 13th international brick and block masonry conference. Amsterdam
60. Elgawady MA, Lestuzzi P, Badoux M (2006) Retrofitting of masonry walls using shotcrete. In: NZSEE conference, Paper 45
61. Fazli H, Wee T (2014) Rehabilitation of RC structures exposed to salinity and moisture using CFRP. Appl Mech Mater 567:488–493. https://doi.org/10.4028/WWW.SCIENTIFIC.NET/AMM.567.488
62. FEMA 273 (1997) NEHRP guidelines for the seismic rehabilitation of buildings
63. FEMA 351 (2000) Recommended seismic evaluation and upgrade criteria for existing welded steel moment-frame buildings.
64. FEMA 547 (2006) Techniques for the seismic rehabilitation of existing buildings. Washington DC, USA
65. Fillitsa Karantoni BV, Fardis MN (1992) Effectiveness of seismic strengthening techniques for masonry buildings. J Struct Eng 118:1884–1902. https://doi.org/10.1061/(ASCE)0733-9445(1992)118:7(1884)
66. Formisano A, Di Lorenzo G, Landolfo R (2019) Seismic retrofitting of industrial steel buildings hit by the 2012 Emilia-Romagna earthquake: a case study. In: AIP conference proceedings. American Institute of Physics Inc.
67. Garlock M, Ricles JM, Sause R (2004) Experimental studies on full-scale post-tensioned steel moment connections. In: 13th world conference on earthquake engineering. Vancouver, B.C., Canada

68. Gattulli V, Lampis G, Marcari G, Paolone A (2014) Simulations of FRP reinforcement in masonry panels and application to a historic facade. Eng Struct 75:604–618. https://doi.org/10.1016/J.ENGSTRUCT.2014.06.023
69. Ghobadi MS, Ghassemieh M, Mazroi A, Abolmaali A (2009) Seismic performance of ductile welded connections using T-stiffener. J Constr Steel Res 65:766–775. https://doi.org/10.1016/j.jcsr.2008.05.007
70. Gilmore AT, Bertero VV, Youssef NFG (1996) Seismic rehabilitation of infilled non-ductile frame buildings using post-tensioned steel braces. Earthq Spectra 12:863–882
71. Gilstrap JM, Dolan CW (1998) Out-of-plane bending of FRP-reinforced masonry walls. Compos Sci Technol 58:1277–1284. https://doi.org/10.1016/S0266-3538(98)00007-4
72. Görgülü T, Tama YS, Yilmaz S et al (2012) Strengthening of reinforced concrete structures with external steel shear walls. J Constr Steel Res 70:226–235. https://doi.org/10.1016/J.JCSR.2011.08.010
73. Gusta S (2015) Tragedy in zolitude—a lesson for contemporary society. Aleksandras Stulginskis University
74. Halder L, Dutta SC, Debnath P, Sharma RP (2021a) Seismic vulnerability assessment of low-rise unreinforced masonry buildings in Northeast India considering variability of material properties. Asian J Civil Eng 22:843–863. https://doi.org/10.1007/s42107-021-00350-7
75. Halder L, Dutta SC, Sharma RP (2020) Damage study and seismic vulnerability assessment of existing masonry buildings in Northeast India. J Build Eng 29. https://doi.org/10.1016/j.jobe.2020.101190
76. Halder L, Dutta SC, Sharma RP, Bhattacharya S (2021b) Lessons learnt from post-earthquake damage study of Northeast India and Nepal during last ten years: 2021 Assam earthquake, 2020 Mizoram earthquake, 2017 Ambasa earthquake, 2016 Manipur earthquake, 2015 Nepal earthquake, and 2011 Sikkim earthquake. Soil Dyn Earthq Eng 151. https://doi.org/10.1016/j.soildyn.2021.106990
77. Hamed E, Rabinovitch O (2010) Failure characteristics of FRP-strengthened masonry walls under out-of-plane loads. Eng Struct 32:2134–2145. https://doi.org/10.1016/J.ENGSTRUCT.2010.03.016
78. Harris JR, Price HJ, Hutchinson DA, et al (2008) NEHRP workshop on meeting the challenges of existing buildings
79. Hernández ET, Carrera JSG (2020) Damage assessment and seismic behavior of steel buildings during the Mexico earthquake of 19 September 2017. Earthq Spectra 36:250–270. https://doi.org/10.1177/8755293019878186
80. Huang KF, la Liu X (2014) Effects of temperature variations on safety of reinforced concrete structures during construction. J Shanghai Jiaotong Univ Sci 19:139–145. https://doi.org/10.1007/S12204-014-1482-X/METRICS
81. Iqbal S, Islam N (2022) Seismic base isolation of 7 storey RC structure using single friction pendulum system. Struct Int 27:200–209. https://doi.org/10.1007/978-3-031-04793-0_15/COVER
82. IS 4326 (1993) Earthquake resistant design and construction of buildings-code of practice (Second Revision). New Delhi
83. IS 13827 (1993) Improving earthquake resistance of earthen buildings—guidelines. New Delhi
84. IS 13828 (1993) Improving earthquake resistance of low strength masonry buildings—guidelines. New Delhi.
85. IS 13920 (1993) Ductile detailing of reinforced concrete structures subjected to seismic forces—code of practice. New Delhi
86. IS 13935 (2009) Seismic evaluation, repair and strengthening of masonry buildings—guidelines. New Delhi
87. Islam ABMS, Jumaat MZ, Ahmmad R, ud Darain KM (2015) Retrofitting of vulnerable RC structures by base isolation technique. Earthq Struct 9:603–623. https://doi.org/10.12989/EAS.2015.9.3.603

88. Ismail M, Muhammad B, Ismail MEG (2010) Compressive strength loss and reinforcement degradations of reinforced concrete structure due to long-term exposure. Constr Build Mater 24:898–902. https://doi.org/10.1016/J.CONBUILDMAT.2009.12.003
89. Ismail N, Khattak N (2015) Reconnaissance report on the Mw 7.5 Hindu Kush earthquake of 26th October 2015 and the subsequent aftershocks
90. Issa CA, Debs P (2007) Experimental study of epoxy repairing of cracks in concrete. Constr Build Mater 21:157–163. https://doi.org/10.1016/j.conbuildmat.2005.06.030
91. Kahn L (1984) Shotcrete retrofit for unreinforced brick masonry. In: 8th WCEE. USA, pp 583–590
92. Kamde DK, Pillai RG (2020) Effect of sunlight/ultraviolet exposure on the corrosion of Fusion-Bonded Epoxy (FBE) coated steel rebars in concrete. Corrosion 76:843–860. https://doi.org/10.5006/3588
93. Kaplan H, Yilmaz S, Nohutcu H et al (2008) Experimental study on the use of old tyres for seismic strengthening of masonry structures. In: 14th world conference on earthquake engineering. Beijing, China
94. Karapetrou ST, Fotopoulou SD, Pitilakis KD (2017) Seismic vulnerability of RC buildings under the effect of aging. Procedia Environ Sci 38:461–468. https://doi.org/10.1016/j.proenv.2017.03.137
95. Karayannis CG, Chalioris CE, Sideris KK (1998) Effectiveness of RC beam-column connection repair using epoxy resin injections. J Earthquake Eng 2:217–240. https://doi.org/10.1080/13632469809350320
96. Karic A, Atalić J, Kolbitsch A (2022) Seismic vulnerability of historic brick masonry buildings in Vienna. Bull Earthq Eng 20:4117–4145. https://doi.org/10.1007/s10518-022-01367-2
97. Khabit IF, Mahin SA, Pister KS (1998) Seismic behavior of concentrically braced steel frames. Report No UCB/EERC-88/01. Earthquake Engineering Research Center, University of California, Berkeley
98. Kiamanesh R, Abolmaali A, Ghassemieh M (2010) The effect of stiffeners on the strain patterns of the welded connection zone. J Constr Steel Res 66:19–27. https://doi.org/10.1016/j.jcsr.2009.07.010
99. Korany Y, Drysdale R (2006) Rehabilitation of masonry walls using unobtrusive FRP techniques for enhanced out-of-plane seismic resistance. J Compos Constr 10:213–222. https://doi.org/10.1061/(ASCE)1090-0268(2006)10:3(213)
100. Krauss PD, Scanlon JM, Hanson MA (1995) Evaluation of injection materials for the repair of deep cracks in concrete structures
101. Kumar GR, Satish Kumar SR, Kalyanaraman V (2007) Behaviour of frames with Non-Buckling bracings under earthquake loading. J Constr Steel Res 63:254–262. https://doi.org/10.1016/j.jcsr.2006.04.012
102. Kuzik MD, Elwi AE, Cheng JJR (2003) Cyclic flexure tests of masonry walls reinforced with glass fiber reinforced polymer sheets. J Compos Constr 7:20–30. https://doi.org/10.1061/(ASCE)1090-0268(2003)7:1(20)
103. Leelataviwat S (1998) Drift and yield mechanism based seismic design and upgrading of steel moment frames
104. Leon RT, Yang C-S (2003) Special inverted-v-braced frames with suspended zipper struts. Georgia Institute of Technology
105. Li A, Assih T, Delmas Y (2000) Influence of the adhesive thickness and steel plate thickness on the behaviour of strengthened concrete beams. J Adhes Sci Technol 14:1639–1656. https://doi.org/10.1163/156856100742456
106. Li P, Zhou Z, Chen L, et al (2019) Research on dust suppression technology of shotcrete based on new spray equipment and process optimization. Adv Civil Eng. https://doi.org/10.1155/2019/4831215
107. Lieping YE, Zhe QU (2009) Failure mechanism and its control of building structures under earthquakes based on structural system concept
108. Liu H, Wang X, Jiao Y (2016) Effect of temperature variation on modal frequency of reinforced concrete slab and beam in cold regions. Shock Vib. https://doi.org/10.1155/2016/4792786

109. LRFD (1999) Specification for structural steel buildings
110. Lu Y (2019) Impact on reinforced concrete structures. Encycl Continuum Mech 1–24. https://doi.org/10.1007/978-3-662-53605-6_227-1
111. Ma P, Xin R, Yao J (2021) Assessment of failure mode and seismic performance of damaged masonry structures retrofitted with grout-injected ferrocement overlay reinforcement (GFOR). Constr Build Mater 305:124778. https://doi.org/10.1016/J.CONBUILDMAT.2021.124778
112. Maalej M, Lin VWJ, Nguyen MP, Quek ST (2010) Engineered cementitious composites for effective strengthening of unreinforced masonry walls. Eng Struct 32:2432–2439. https://doi.org/10.1016/J.ENGSTRUCT.2010.04.017
113. Macabuag J (2007) An introduction to modelling and retrofitting of non-engineered masonry buildings under seismic loading
114. Macabuag J, Guragain R, Bhattacharya S (2012) Seismic retrofitting of non-engineered masonry in rural Nepal. Proc Inst Civ Eng Struct Build 165:273–286. https://doi.org/10.1680/stbu.10.00015
115. Man L (2008) Improving the earthquake resistance of non-engineered adobe masonry buildings. Oxford University
116. Mardani A, Dehkordi MR, Moghadam AS, Yekrangnia M (2018) Example of a two-story unreinforced masonry building retrofitted by shotcrete. In: Advanced design examples of seismic retrofit of structures. Elsevier, pp 13–118
117. Masri AC, Goel SC (1996) Seismic design and testing of an RC slab-column frame strengthened by steel bracing. Earthq Spectra 12:645–666
118. Michiels T, Adriaenssens S, Dejong M (2019) Form finding of corrugated shell structures for seismic design and validation using non-linear pushover analysis. Eng Struct 181:362–373. https://doi.org/10.1016/J.ENGSTRUCT.2018.12.043
119. Michiels T, Adriaenssens S, Jorquera-Lucerga JJ (2017) Parametric study of masonry shells form-found for seismic loading. J Int Assoc Shell Spatial Struct 58:267–275. https://doi.org/10.20898/J.IASS.2017.194.892
120. Mosallam AS (2007) Out-of-plane flexural behavior of unreinforced red brick walls strengthened with FRP composites. Compos B Eng 38:559–574. https://doi.org/10.1016/J.COMPOSITESB.2006.07.019
121. Moudi M, Yan S, Bahador B et al (2019) Statistical model for earthquake economic loss estimation using GDP and DPI: a case study from Iran. Qual Quant 53:583–598. https://doi.org/10.1007/s11135-018-0776-8
122. Mulas MG, Perotti F, Coronelli D et al (2013) The partial collapse of "Casa dello Studente" during L'Aquila 2009 earthquake. Eng Fail Anal 34:566–584. https://doi.org/10.1016/j.engfailanal.2013.02.031
123. Narayan S, Shrimali MK, Bharti SD, Datta TK (2018) Collapse of damaged steel building frames because of earthquakes. J Perform Constructed Facil 32. https://doi.org/10.1061/(asce)cf.1943-5509.0001125
124. Nayak S, Banerjee S, Das S (2018) Augmenting out-of-plane behaviour of masonry wallet using PP-band and steel wire mesh. In: IOP conference series: materials science and engineering. Institute of Physics Publishing
125. Nayak S, Dutta SC (2016) Improving seismic performance of masonry structures with openings by polypropylene bands and L-shaped reinforcing bars. J Perform Constr Facil 30:04015003. https://doi.org/10.1061/(asce)cf.1943-5509.0000733
126. Nayak S, Dutta SC (2016) Failure of masonry structures in earthquake: a few simple cost effective techniques as possible solutions. Eng Struct 106:53–67. https://doi.org/10.1016/j.engstruct.2015.10.014
127. NDMA (2014) Seismic retrofitting of deficient buildings and structures. India
128. NIST (2012) program plan for the development of seismic design guidelines for port container, wharf, and cargo systems
129. NIST (2013) Review of past performance and further development of modeling techniques for collapse assessment of existing reinforced concrete buildings.

130. Oyguc R (2016) Seismic performance of RC school buildings after 2011 Van earthquakes. Bull Earthq Eng 14:821–847. https://doi.org/10.1007/s10518-015-9857-4
131. Oyguc R, Oyguc E (2017) 2011 Van earthquakes: lessons from damaged masonry structures. J Perform Constructed Facil 31. https://doi.org/10.1061/(asce)cf.1943-5509.0001057
132. Padalu PKVR, Singh Y, Das S (2020) Cyclic two-way out-of-plane testing of unreinforced masonry walls retrofitted using composite materials. Constr Build Mater 238:117784. https://doi.org/10.1016/J.CONBUILDMAT.2019.117784
133. Padalu PKVR, Singh Y, Das S (2019) Out-of-plane flexural behaviour of masonry wallettes strengthened using FRP composites and externally bonded grids: Comparative study. Composites Part B 176. https://doi.org/10.1016/J.COMPOSITESB.2019.107302
134. Pekelnicky R, Poland C (2012) seismic evaluation and retrofit rehabilitation of existing buildings. In: SEAOC 2012 convention proceedings
135. Perera R, Gómez S, Alarcón E (2004) Experimental and analytical study of masonry infill reinforced concrete frames retrofitted with steel braces. J Struct Eng 130:2032–2039. https://doi.org/10.1061/(asce)0733-9445(2004)130:12(2032)
136. Piekarczyk M, Grec R (2012) Application of adhesive bonding in steel and aluminium structures. Arch Civ Eng 58:309–329. https://doi.org/10.2478/v.10169-012-0018-8
137. Power A (2008) Does demolition or refurbishment of old and inefficient homes help to increase our environmental, social and economic viability? Energy Policy 36:4487–4501. https://doi.org/10.1016/j.enpol.2008.09.022
138. Prakash SS, Alagusundaramoorthy P (2008) Load resistance of masonry wallettes and shear triplets retrofitted with GFRP composites. Cem Concr Compos 30:745–761. https://doi.org/10.1016/J.CEMCONCOMP.2007.11.005
139. Pugsley AG (1951) Concepts of safety in structural Engineering, Pugsley on concepts of safety in structural engineering.
140. Pushpakumara B, Mendis W, de Silva S (2016) Bamboo as a new retrofitting material for existing masonry walls with improving the walls aesthetic view. Built Environ Spatial Sci
141. Roca P, Araiza G (2010) Shear response of brick masonry small assemblages strengthened with bonded FRP laminates for in-plane reinforcement. Constr Build Mater 24:1372–1384. https://doi.org/10.1016/J.CONBUILDMAT.2010.01.005
142. Roeder CW, Banerjee S, Jung DR, Smith SK (1996) The role of building foundations in seismic retrofit. Earthq Spectra (EERI) 12:925–942
143. Scozzese F, Terracciano G, Zona A, Corte GD, Dall'Asta A, Landolfo R (2018) Modeling and seismic response analysis of italian code-conforming single-storey steel buildings. J Earthq Eng 22:2104–2133. https://doi.org/10.1080/13632469.2018.1528913
144. Sen D (2017) Behaviour and strength of RC columns retrofitted with steel angles and strips under eccentric axial loads. Bangladesh University of Engineering and Technology
145. Shrive NG (2006) The use of fibre reinforced polymers to improve seismic resistance of masonry. Constr Build Mater 20:269–277. https://doi.org/10.1016/J.CONBUILDMAT.2005.08.030
146. da Silva LS, Silva LC, Tankova T, Craveiro HD, Simões R, Costa R, D'Aniello M, Landolfo R (2021) Performance of modular hybrid cold-formed/tubular structural system. Structures 30:1006–1019. https://doi.org/10.1016/j.istruc.2021.01.066
147. Sonuvar M, Ozcebe G, Ersoy U (2004) Rehabilitation of reinforced concrete frames with reinforced concrete infills. ACI Struct J 101. https://doi.org/10.14359/13335
148. Sumerente G, Lovon H, Tarque N, Chácara C (2020) Assessment of combined in-plane and out-of-plane fragility functions for adobe masonry buildings in the peruvian andes. Front Built Environ 6:52. https://doi.org/10.3389/FBUIL.2020.00052/BIBTEX
149. Suwal R (2018) Failure study of reinforced concrete buildings of kathmandu valley in gorkha earthquake 2015. Int J Modern Res Eng Manag (IJMREM) 2581–4540
150. Taghdi M, Bruneau M, Saatcioglu M (2000) Seismic retrofitting of low-rise masonry and concrete walls using steel strips. J Struct Eng 126:1017–1025. https://doi.org/10.1061/(ASCE)0733-9445(2000)126:9(1017)

151. Takeuchi T, Wada A (2018) Review of buckling-restrained brace design and application to tall buildings. Int J High-Rise Build 7:187–195
152. Taljsten B (1997) Strengthening of beams by plate bonding. J Mater Civil Eng 9
153. Tan KH, Patoary MKH (2004) Strengthening of masonry walls against out-of-plane loads using fiber-reinforced polymer reinforcement. J Compos Constr 8:79–87. https://doi.org/10.1061/(ASCE)1090-0268(2004)8:1(79)
154. Tartaglia R, Milone A, Prota A, Landolfo R (2022) seismic retrofitting of existing industrial steel buildings: a case-study. Materials 15. https://doi.org/10.3390/ma15093276
155. Tetley R, Madabhushi G (2007) Vulnerability of adobe buildings under earthquake loading. In: Proceedings of the international conference on earthquake geotechnical engineering. Greece
156. Tetteh AR, Liang Y (2020) Repair of cracks on concrete structures. Int J Sci Res (IJSR) 9:82–92. https://doi.org/10.21275/sr20324212517
157. Thermou GE, Elnashai AS (2006) Seismic retrofit schemes for RC structures and local-global consequences. Prog Struct Mat Eng 8:1–15
158. Thorburn LJ, Kulak GL, Montgomery CJ (1983) Analysis of steel plate shear walls. Edmonton, Alberta, Canada
159. Tsavdaridis KD (2014) Strengthening techniques: code-deficient steel buildings. Encyclopedia of earthquake engineering. Springer, Berlin, pp 1–26
160. Tsavdaridis KD, D'Mello C, Huo BY (2013) Experimental and computational study of the vertical shear behaviour of partially encased perforated steel beams. Eng Struct 56:805–822. https://doi.org/10.1016/j.engstruct.2013.04.025
161. Tsonos AG (2002) Seismic repair of reinforced concrete beam-column subassemblages of modern structures by epoxy injection technique. Struct Eng Mech 14:543–563. https://doi.org/10.12989/SEM.2002.14.5.543
162. Turer A, Korkmaz SZ, Korkmaz HH (2007) Performance improvement studies of masonry houses using elastic post-tensioning straps. Earthq Eng Struct Dyn 36:683–705. https://doi.org/10.1002/EQE.649
163. Turk AM (2013) Seismic response analysis of masonry minaret and possible strengthening by fiber reinforced cementitious matrix (FRCM) materials. Adv Mater Sci Eng. https://doi.org/10.1155/2013/952497
164. Turner BA (1978) Man-made disasters. Wykeham Publications, London
165. Valluzzi MR, Tinazzi D, Modena C (2002) Shear behavior of masonry panels strengthened by FRP laminates. Constr Build Mater 16:409–416. https://doi.org/10.1016/S0950-0618(02)00043-0
166. Warner J (1995) Understanding shotcrete—the fundamentals. Concr Int 17(5)
167. Wasserman I (2019) Behaviour of concrete structures upon the impact of UV and rain of natural acidity. J Int Sci Publ: Mater Methods Technol (Online) 13:210–229
168. Willis CR, Yang Q, Seracino R, Griffith MC (2009) Damaged masonry walls in two-way bending retrofitted with vertical FRP strips. Constr Build Mater 23:1591–1604. https://doi.org/10.1016/J.CONBUILDMAT.2007.09.007
169. Yadav P, Khursheed S, Solanki SK, Kumar Paul V (2022) A review of retrofitting techniques of masonry structures.
170. Yang CS, Leon RT, DesRoches R (2008) Design and behavior of zipper-braced frames. Eng Struct 30:1092–1100. https://doi.org/10.1016/j.engstruct.2007.06.010
171. Ying B (2021) Investigating the behavior of boundary elements in steel shear walls with different connections. Stavební Obzor—Civil Eng J 30. https://doi.org/10.14311/cej.2021.04.0071
172. Yılmaz S, Tama YS, Kaplan H, Caliskan O, Solak A (2011) Experimental program on design and application of external retrofit walls for low ductility RC frames. In: 4th international conference on advances in experimental structural engineering. Isprira, Italy
173. Yossef NM (2015) Strengthening steel I-beams by welding steel plates before or while loading. Int J Eng Res Technol (IJERT) 4
174. Zhao B, Taucer F, Rossetto T (2009) Field investigation on the performance of building structures during the 12 May 2008 Wenchuan earthquake in China. Eng Struct 31:1707–1723. https://doi.org/10.1016/j.engstruct.2009.02.039

Seismic Retrofit of Reinforced Concrete Structures Using Fibre Reinforced Polymer

Aparna (Dey) Ghosh, Chaitali Ray, and Dhiraj Biswas

Abstract Aging and deteriorating infrastructure, and more strict seismic design standards, call for the need of seismic retrofitting. Amongst various seismic retrofit strategies, use of fibre reinforced polymer (FRP) is both effective and advantageous over many other techniques, chiefly due to its high strength-to-weight ratio and good fatigue strength. Further, the speed of installation of FRP being high, it results in reduced downtime, that is extremely beneficial in case of seismic retrofitting of crucial infrastructure. The retrofit techniques with FRP are designed to add ductility, confinement, moment and shear capacity to existing structural members. Thus, local strengthening of members is achieved and alterations to overall structural dynamic properties are minimal. Amongst the commonly used fibres, glass fibres are suitable in low-cost seismic retrofit applications in comparison to carbon. However, caution needs to be exercised over the performance of FRP under elevated temperatures, as fire is a common hazard associate with earthquakes. This chapter discusses the various aspects of FRP retrofitting of reinforced concrete structural elements, as well as of masonry infill walls, for enhanced seismic performance.

Keywords Fibre Reinforced Polymer (FRP) · Seismic retrofit · Ductility · Confinement · Lateral strength · Strength-to-weight ratio

1 Introduction

There are several structures that do not meet the seismic requirements as specified by the prevailing codal provisions, and are hence termed 'deficient'. Needless to say, such structures pose a very serious threat to the safety of the population, not only under strong shaking but even under mild tremors. The deficiencies could be inherent in the structure from the time it was built or it could have developed due to the aging

A. Ghosh (✉) · C. Ray · D. Biswas
Department of Civil Engineering, Indian Institute of Engineering Science and Technology, Shibpur 711103, Howrah, India
e-mail: aparna@civil.iiests.ac.in

S. B. Singh and C. V. R. Murty (eds.), *RC Structures Strengthened with FRP for Earthquake Resistance*, Composites Science and Technology,
https://doi.org/10.1007/978-981-97-0102-5_7

of the building over time, or they could also be caused due to unplanned modifications imposed on the structure. According to IS 13935: 2009 [1], upgrading the earthquake resistance of the deficient structure to the level of the current codes through the adoption of appropriate seismic strengthening techniques is defined as the seismic retrofitting of the structure. Construction of a new structure in place of the deficient structure is always an option, however, that is generally avoided, chiefly because that is usually more expensive than seismic retrofitting and may lead to greater disruption of normal activities. Moreover, retrofitting is faster than re-construction and also does not cause an unwanted change in the look of the structure.

The chief aim of seismic retrofitting is to improve the lateral strength of the structure. Moreover, it is also very important to enhance the structural integrity, so as to ensure a proper transfer of load from the relatively weaker to the stronger elements. Further, the ductility of the structure is to be improved by avoiding the brittle modes of failure. These aims are attempted to be met through two broad categories of methods of seismic retrofitting. The first attempts to reduce the seismic demand on the structure, while the second focuses on upgrading the seismic capacity of the structure. The retrofit techniques can also be classified into global and local techniques, depending on whether the technique is applied to the structure as a whole or to some particular member(s). Various retrofit techniques include incorporation of additional infill walls, shear walls, insertion of steel braces, addition of frames [2], all of which aid in enhancing the lateral strength and stiffness of the structure. While these methods are very effective in improving the seismic performance of the structure, they may lead to loss of existing space and functionality. Moreover, being construction-intensive, they can significantly disrupt or lead to a suspension of the daily activities within the structure. In this scenario, fibre reinforced plastic or polymer (FRP) composites provide an alternate solution for strengthening of existing structural sections, without the afore-mentioned problems.

The civil construction industry depends heavily on reinforced concrete (RC) material and therefore a major chunk of civil infrastructure, such as buildings, tunnels, bridges, airports, drainage and hydraulic projects, is made of reinforced concrete. An engineer often experiences a challenge while working in the concrete industry, which is to increase the strength and performance of the concrete as well as to enhance its durability throughout its service life. With an ever-intensifying competition to construct tall buildings worldwide, there is a trend to use modern and newer materials, and construction techniques. FRP composites have gained remarkable acceptance worldwide, especially in the aircraft and marine industries since several decades, due to their various favourable properties, and most importantly, being lightweight. Now, these materials are increasingly being adopted in civil engineering construction, especially in bridges and buildings [3]. The FRP material started its journey in civil engineering in the 1990s. Along with its lightweight property, FRP composites possess many other advantageous properties, such as, increased speed of construction, easy and rapid installation, reduced environmental impact, less thermal conductivity that aids to maintain ideal internal temperatures in buildings subjected to severe weather conditions. They also have the capability of providing a better service life. The composite materials are commonly used in civil engineering in two ways, namely, (i) in new construction, and (ii) in retrofitting of existing structures. Composites have been more popular in repairing and retrofitting of damaged structures than in new construction.

This chapter first discusses the properties of FRP owing to which it plays such an important role in seismic retrofitting. Some of the relevant codes are then mentioned, followed by a description of the different forms in which FRP are used and their application procedures. Next, the retrofitting of RC columns, beams and masonry infill wall members and subsequently, the seismic performance of RC framed buildings retrofitted with FRP are discussed. Lastly, a brief commentary on the problems associated with FRP retrofitted structural members under fire is made, followed by the concluding remarks.

2 Advantages of FRP in Retrofitting

Retrofitting of existing civil engineering structures using FRP composites possesses the following advantages over conventional repair and strengthening systems.

1. Fast and clean to install
2. Reach desired strength faster
3. Lightweight and easy to transport material
4. Material is flexible and can conform to any geometry
5. High strength-to-weight ratio
6. Excellent resistance to corrosion
7. Resins used in the composite are non-hazardous and odourless making the system ideal for work in occupied buildings
8. Low construction cost.

In addition, FRP also serves as a waterproofing membrane and does not require a large work/setup area. With the above considerations, FRP has been established as more economical and durable than conventional repairing materials. Under seismic conditions, the most significant advantage of FRP is its high strength-to-weight ratio, as it will limit the increase in the seismic base shear that might occur due to the added retrofit material.

2.1 Design Specifications

Some standard design codes are available that offer appropriate guidelines for the retrofitting and repairing of RC structures using FRP.

(a) Japan Society of Civil Engineers (JSCE) [4]

The recommendations include (i) Design and construction standards for upgrading the existing concrete structures using FRP sheets to improve their strength and durability, (ii) Evaluation of the effect of FRP sheet on flexural capacity, shear capacity and ultimate deformation of columns, and (iii) Upgradation of the process and material, detailed inspection of existing concrete structures and maintenance of upgraded concrete structures.

(b) Canadian Standards [5, 6]

This Standard covers the manufacturing process requirements of FRP bars that are part of a grid for use in non-prestressed internal reinforcement of concrete components of structures (e.g., bridges, buildings, and marine structures).

(c) ACI Committee 440 Codes [7–10]

These have published various design and construction guidelines for the use of FRP in RC structures.

(d) European Standard [11]

This provides design guidelines on the externally bonded FRP reinforcement for repair and strengthening of concrete structures, practical execution and quality control.

3 Types of FRP Composites and Installation

FRP composites for strengthening of civil engineering structures are available today mainly in the form of: (a) thin unidirectional strips (with thickness in the order of 1 mm) fabricated by pultrusion; (b) flexible sheets or fabrics or textiles made of fibres in one, two or more directions. The different types of strengthening systems using FRP, available in the present day, are summarised below.

1. Pre-cured systems: manufactured in various shapes by pultrusion or lamination. Pre-cured systems are directly bonded to the structural member.
2. Wet lay-up systems: manufactured with fibres lying in one or more directions, e.g. FRP sheets or fabrics, and impregnated with resin at the construction site.
3. Prepreg (pre-impregnated) systems: manufactured with unidirectional or multi-directional fibre sheets or fabrics pre-impregnated at the manufacturing plant with partially polymerized resin. They may be bonded to the member to be strengthened with (or without) the use of additional resins.

These systems correspond to several manufacturers and suppliers, and are based on different configurations, types of fibres, adhesives, etc. Furthermore, the suitability of each system depends on the type of structure that is to be strengthened. The overall strengthening techniques are taken up in two stages, namely, (i) Treatment and surface preparation, and (ii) Adhesive curing. The essentials of each step are depicted in the following.

(i) ***Treatment and Surface Preparation***

The overall success and performance of patching materials applied to the concrete surface are highly dependent on the quality of the cleaning and surface preparation performed. The surfaces should be free from dust, dirt, oil, previously applied coatings, etc., on which the FRP laminates are to be applied. Any depression or concavity in the surface should be eliminated by filling with putty as a stretched FRP sheet will lose contact with the surface at such locations. The FRP sheets are also prone to tearing at corners and sharp edges due to stress concentrations, so corners should be rounded to a minimum radius of 25 mm. If the enhancement of the structural performance of the repaired member relies on the bond to the laminate, then the concrete surface preparation is a very important factor concerning the application of FRP composites to concrete structures.

Most of the methods for concrete surface preparation are related to mechanical abrasion, by grinding, sandblasting, water-jetting and scraping as shown in Fig. 1. The effects of these surface preparation methods strictly depend on material properties, concrete strength, aggregate size, type of epoxy and ACM fibreglass mat used, worker's operation skill, construction environment, etc. ACI 515.1R-79 [10] has provided a general procedure for roughening, cleaning, and checking of the concrete surface.

(ii) ***Adhesive curing***

An adhesive is applied to the RC structure surface to attach the FRP composites. The goal of this approach is to provide the mechanical bond between two elements so that they stick together, as shown in Fig. 2. The application of adhesives between the substrate and compound substance is vital in the process of preparation of surfaces before the application of FRP. The adhesives should not be applied in the wet condition of the surface as it may weaken the bond or cause negative effects. Again, the environmental conditions like temperature, direct contact with rain or dust may accelerate or retard the resin curing time. The following features of adhesives must be maintained to achieve superior strengthening.

1. It must be easy to handle the process of mixing, applying and curing.
2. It should not be susceptible to ordinary alteration in the moisture content of the surface.

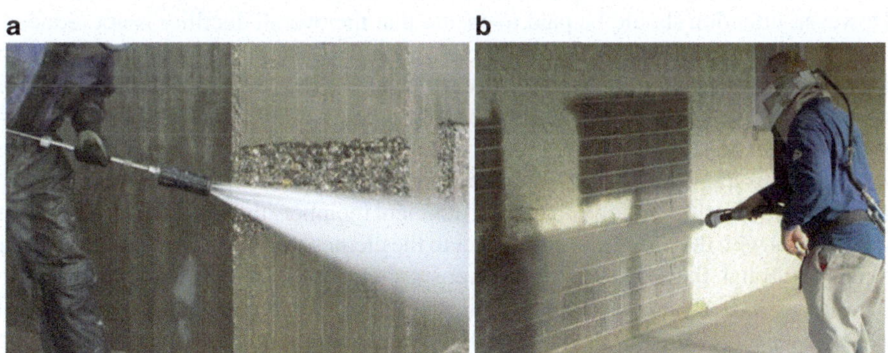

Fig. 1 Treatment and surface preparation; **a** water jetting, **b** sandblasting

Fig. 2 Cross sectional configuration of concrete, adhesive and FRP composite

3. It should maintain compatible thermal characteristic with both concrete and FRP.
4. It must provide a low value of creep.

The appropriate curing is required to prevent transfer of moisture from the surface of concrete to the outermost layer of FRP sheet as bonding may be affected by the presence of moisture.

4 Retrofitting of RC Structures

Composite materials can be used effectively to retrofit reinforced concrete structures subjected to seismic load. The objective is to strengthen buildings that do not meet the safety requirements defined by the current seismic codes under the design seismic action, with respect to one or more limit states. Once a preliminary seismic assessment is performed on the existing structure, the strengthening intervention is designed based on its outcome. The entire process goes through the following steps: (a) identification of safety requirements, (b) definition of protection levels (which yields the intensity of the seismic action), (c) choice of analysis methods, (d) choice of verification criteria, (e) assessment of seismic safety, (f) choice of material properties. Regarding the criteria for selection of the FRP strengthening method, it is widely recognized that irregularities in stiffness cannot be solved by applying FRPs. An intervention performed with FRP is classified as a selective technique, since irregularities in strength can be adjusted by strengthening a selected number of elements. However, attention should be paid to ensure that the overall ductility is not reduced.

The design of a strengthening intervention with FRP should include the following activities, (a) justification of the intervention type, (b) selection of techniques and/or materials, (c) preliminary design of the strengthening intervention, (d) structural analysis of the upgraded structure. As mentioned above, FRP strengthening is regarded as a selective intervention technique from the seismic standpoint, aiming at, (a) increasing the flexural capacity of deficient members through the application of composites with the fibres placed parallel to the element axis, (b) increasing the shear strength through the application of composites with the fibres placed transversely to the element axis, (c) increasing the ductility (or the chord rotation capacity) of critical zones of beams and columns through FRP wrapping (confinement), (d) improving the efficiency of lap splice zones through FRP wrapping, (e) preventing buckling of longitudinal rebars under compression through FRP wrapping, (f) increasing the tensile strength of the panels of partially confined beam-column joints through the application of composites with the fibres placed along the principal tensile stress direction.

During retrofitting of RC structures, each type of member is treated on the basis of its specific requirement and procedure.

5 Retrofitting of Axial Members

Strengthening of reinforced concrete axial members (columns) subjected to high axial load levels is a challenge often faced by a structural engineer. Low concrete strength combined with poor site curing conditions may result in axial load ratios in individual columns of multi-storey buildings that are much higher than the values anticipated during design. There are several ways to enhance the axial load capacity of such members with deficiencies in strength. One of them is through passive confinement of the concrete. As the ultimate load is approached, the concrete in the column dilates and exerts pressure on the confining element. This interaction leads to a triaxial state of stress in the concrete core of the axial member that results in an increase in the ultimate compressive strength of the concrete. Thus, the axial load carrying capacity of the members is also increased. Confinement of concrete also substantially increases the deformation capacity of reinforced concrete columns, changing their mode of failure from brittle to ductile. FRP composite jackets applied around the perimeter of reinforced concrete columns confine the concrete member. External jacketing of columns with FRP is a promising retrofitting technique for enhancing seismic performance of sub-standard or damaged reinforced concrete (RC) buildings. The enhancement in deformation capacity and shear strength of jacketed members helps to avoid the brittle collapse mechanism of buildings with inadequate ductility. Further, the increase in ductility due to the concrete confinement is highly beneficial in the seismic retrofit scheme.

The enhancement in strength and ductility is proportional to the thickness of the FRP wrapping, but is limited by the ultimate strain of the FRP. FRP composite jackets may also provide restraint against buckling of the longitudinal column bars, which may be significantly advantageous in columns of old buildings located in seismic regions because, very often, the hoops are widely spaced and are unable to provide an effective restraint. It is to be noted that the effects of FRP confinement on a column are most successful for a circular column.

With the assumption of 45° cracking, the thickness of the FRP jacket, t_j, for shear retrofitting of columns under seismic loads may be determined from the following (2).

$$t_j = \frac{V_f}{2\varepsilon_{fe} E_j D} \tag{1}$$

where,

V_f $V_u - (V_c + V_s)$
V_u shear demand based on the flexural capacity in the potential hinge locations
V_s shear capacity from steel ties
V_c shear capacity of concrete (IS 456:2000, Clause 40.2)
ε_{fe} effective usable strain in the FRP wrap
D the column dimension in the direction of loading
E_j elastic modulus of composite

6 Retrofitting of Flexural Members

6.1 Flexural Strengthening

Flexural strengthening with FRP materials may be carried out by applying one or more laminates/sheets to the tension side of the element as shown in Fig. 3. The flexural capacity is attained if either the compressive strain in concrete or the tensile strain in FRP reaches its ultimate value. However, the capacity after strengthening cannot be greater than twice the initial capacity. Furthermore, according to the capacity design approach, flexural strengthening should be designed to avoid the activation of shear failure mechanisms. As a member strengthened with FRP is generally loaded at the time of retrofitting application, the existing state of strain in the structure should be considered.

The flexural capacity of a RC beam retrofitted with FRP laminate may be evaluated from the relevant constitutive relationships, equilibrium of forces and strain compatibility conditions. The ultimate flexural capacity, M_{uR}, which should be greater than the ultimate moment demand, is obtained from the following Eq. (2).

$$M_{uR} = 0.87 f_y A_s (d - 0.42 x_u) + \psi_f A_f f_f (h - 0.42 x_u) \tag{2}$$

a b

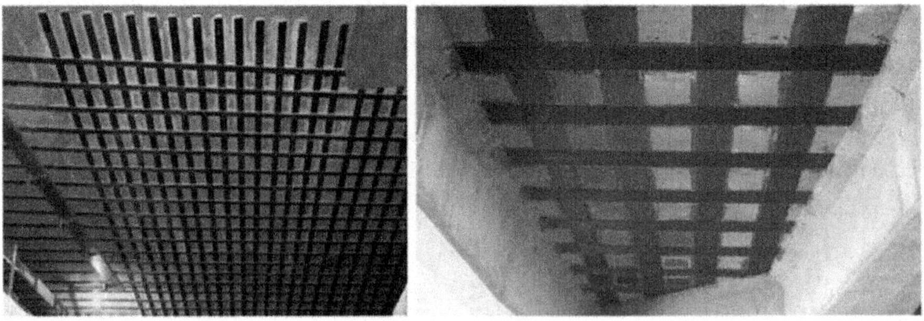

Fig. 3 Retrofitting of **a** Bridge deck panel, **b** RC slab

The definition of the symbols used in the above equation is given below.

f_y yield stress of the steel reinforcement
A_s area of the steel reinforcement
d effective depth of the steel reinforcement
x_u depth of the neutral axis
ψ_f reduction factor equal to 0.85 is used for the strength from the FRP
A_f area of the FRP composite
f_f stress in the FRP composite
h depth of the beam

Several researchers [12–17] have carried out investigations on the performance of FRP retrofitted beams under seismic effect. Experimental studies on the confinement of M30 RC beams using GFRP (Glass fiber reinforced polymer), CFRP (Carbon fiber reinforced polymer), BFRP (Basalt fiber reinforced polymer), with double and four layers of wrapping using bi-directional fibre mat [17] indicated the superior performance of the four layer CFRP both in terms of increase in strength and reduction in deflection. A comprehensive literature review was presented [18] on strengthening using prestressed CFRP laminates in conventionally and internally prestressed concrete structures (PCSs). The flexural behavior, failure modes, and serviceability performance of the strengthened concrete members were discussed. The effects of the prestressing level on ductility and energy absorption and a summary of observations on the recommended optimum prestressing level was presented in the review [18].

6.2 Shear Strengthening

Shear strengthening in a flexural member is necessary when the shear demand is greater than the shear capacity of the flexural member, evaluated on the basis of the contributions of both concrete and shear reinforcing steel. It may also be necessary after designing flexural strengthening, in order to re-establish the strength hierarchy between bending and shear failure mechanisms. Shear strengthening is usually obtained by applying one or more layers of FRP, externally bonded to the surface of the structural member to be strengthened. External FRP reinforcement can be applied in a discontinuous manner, with gaps between strips or continuously, with strips next to each other. Figure 4 shows U-wrapped FRP strengthening configurations. For U-wrap strengthening of rectangular or T-sections, delamination of the end portions of FRP reinforcement can be avoided by using laminates/sheets and/or bars installed in the direction of the longitudinal axis of the member. There can also be shear strengthening using inclined wraps.

The shear capacity of a FRP strengthened RC member is the summation of the shear capacities of the steel reinforcement, the concrete and that of the FRP. The latter is the shear strength of the FRP reduced by a factor, the value of which is

Fig. 4 U-wrap shear strengthening of rectangular beam

limited to 0.85. The shear strength that can be derived from the FRP is dependent on the orientation of the fibres and on the considered crack pattern. In case of discrete strips of FRP and cracks with an angle of inclination of 45°, the shear strength of the FRP is evaluated from the following expressions (2).

$$V_f = \frac{A_{fv} f_f (\sin\alpha + \cos\alpha) d_f}{S_f}. \tag{3}$$

with, $A_{fv} = 2nt_f w_f$

and, $f_f = \varepsilon_{fe} E_f$

Here,

- A_{fv} total vertical area of the FRP wrap
- f_f stress in the FRP wrap
- α inclination of the FRP wrap with respect to the beam axis
- d_f effective depth of the FRP wrap
- E_f elastic modulus of FRP
- n number of wraps
- S_f spacing of the FRP wrap
- w_f width of the FRP wrap
- t_f thickness of the FRP wrap
- ε_{fe} effective usable strain in the FRP wrap

The above expression is modified in case of continuous FRP sheets, the change being based on the length of the FRP sheet intercepted by an inclined crack.

Fig. 5 Retrofitted specimen at ultimate load

7 Retrofitting of Beam Column Joints

The beam, and the connection between beam and column, are highly vulnerable components of a building structure during the occurrence of an earthquake. FRP wrapping helps to improve the performance of these critical elements under seismic conditions. The increased confinement of the joints due to various types of wrapping [19] helps to overcome inadequacies related to ties and anchorage at the joints. Mosallam et al. [20] investigated the structural behaviour of retrofitted beam column joints by using different FRP composites and hybrid connector through numerical and experimental modelling. Allam et al. [21] conducted experimental investigations to evaluate the seismic performance of FRP retrofitting in beam column joint (Fig. 5). The results indicate that the FRP composite systems proposed in their study have succeeded in enhancing the strength, stiffness and ductility of the seismically deficient reinforced concrete beam-column joint. The test setup was capable of capturing the shear deficiency behaviour of the beam-column joint and the enhancement in its behaviour after retrofit. They used high-strength carbon fibre reinforced polymer composites, E-glass fibre reinforced polymer composites and high-modulus carbon fibre reinforced polymer composites for the study. Using the proposed retrofitting systems, a significant enhancement in the joint shear strength was achieved. A few other researchers [22, 23] have presented the techniques of FRP retrofitting of beam column joints.

8 Retrofitting of Masonry Walls

RC structures have masonry infill walls that are highly susceptible to failure under earthquake excitation due to their low tensile and shear resistance. Local crushing at the corners, shear cracking along the bed joints and diagonal compression failure due to buckling are some of the common modes of failure of masonry infill walls. FRP composites can be used to provide an economic and workable solution for the seismic retrofit of masonry infill walls in RC structures [24]. Different configurations of FRP laminates, such as full surface bonding, X-frame, H-frame, Picture frame, etc. can be adopted for the purpose. The shear strength of the retrofitted masonry wall is augmented by the shear strength of the FRP laminate, which is obtained in case of vertical FRP laminates as a product of the elastic modulus of the FRP, the area of the vertical FRP laminate and the effective usable strain in the FRP laminate. The effective usable strain in the FRP is evaluated from half the ultimate tensile strain in the FRP laminate, multiplied with an environmental reduction factor.

9 Performance of Seismic Retrofitting by FRP in RC Structures

Despite the large amount of research work carried out on improving the load carrying capacity and ductility of RC members using FRP composite materials, only a few studies have focused on the behaviour of FRP retrofitted RC structural systems, and in particular the structures that are code-compliant. A case study to evaluate the efficiency of glass fibre reinforced polymers (GFRPs) in improving the seismic performance of an 8-storey moment resisting framed RC building was reported in literature [25]. A numerical investigation on the seismic enhancement of code-compliant and poorly-confined RC buildings was presented by them. Nonlinear static analysis of the frames was carried out in order to simulate the performance of the buildings under ground motion. Plastic behaviour of the elements was characterised using two nonlinear hinges at the end sections of the members based on the lumped plasticity concept. The nonlinear properties of plastic hinges were calculated based on the well-established models for concrete and steel reinforcement. After verification of the adopted assumptions for nonlinear characterisation of members, the results of pushover analysis were implemented in the seismic assessment of the frames. The following retrofitting strategies were adopted in the investigation to increase the displacement capacity of the structure and provide more energy dissipation under seismic loads.

(i) Fully exploiting the rotational capacity of beams and columns through GFRP wrapping of potential plastic hinge regions of each column in both code-compliant and poorly-confined frames, as shown in Fig. 6.
(ii) Increasing the ductility of beams in the code-compliant structure in addition to column confinement. This could be achieved using web bonded FRP with fibres oriented in the direction of transverse reinforcement in regions prone to inelastic behaviour. The stress–strain model proposed by Lam and Teng [26, 27] was selected for FRP-confined concrete as shown in Fig. 7.

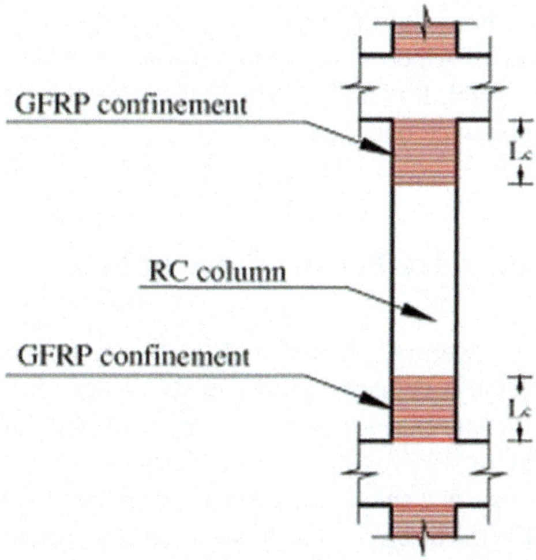

Fig. 6 FRP jacketed regions of columns for seismic retrofitting

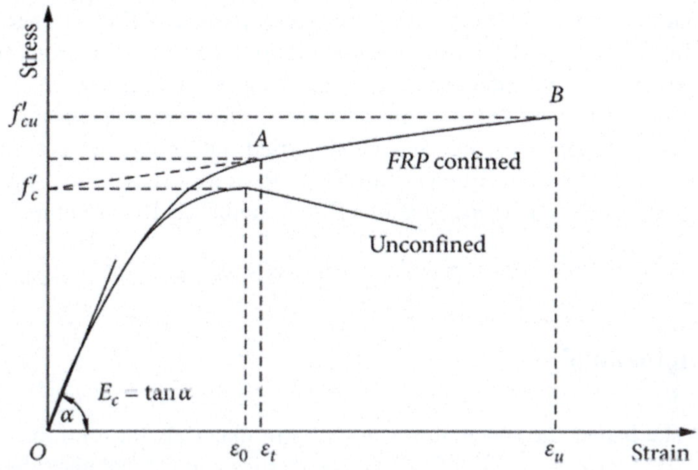

Fig. 7 The stress–strain model proposed by Lam and Teng [26, 27]

The poorly confined structure could resist much higher ground motion (PGA = 0.52 g) by adopting FRP wrap, which increased the ductility and energy dissipation capacity. However, further studies are required to be conducted on the FRP-retrofitting of pre-existing RC buildings in order to not only study their seismic response but also provide a more economical retrofitting scheme for these types of buildings.

Details of a real-life implementation of FRP for seismic retrofitting of a ten storied residential RC building of the 1970s may be found in [28]. The strategy of FRP retrofitting was selected amongst other options as it was more economical and had minimum interference with the living spaces. It provided the requisite strength, without significantly impacting the stiffness, which would have caused a change in the distribution of the lateral forces and a consequent need to retrofit other elements as well. It is noteworthy that during the entire operation of retrofitting, the building remained occupied.

10 Fire Performance of FRP

It is well recognized that fire is one of the major hazards following an earthquake. Though the performance of RC structures under elevated temperatures is quite good, RC structures that have been retrofitted with FRP need to be specially dealt with, as there are problems with the performance of FRP under fire. Beyond about 200 °C, there is a substantial loss in both the tensile strength and Young's modulus of FRP [29]. Moreover, the bond strength of near-surface mounted FRP system degrades significantly with increase in temperature, and only 20% strength is retained at 200 °C. Hence, FRP-retrofitted RC members should be properly insulated for fire protection. The various factors that influence the fire performance of FRP-strengthened RC structures are the steel-to-FRP reinforcement ratio, load level, fire scenario, insulation scheme, etc. Simple alterations, such as changing the reinforcement ratio, or location of the FRP bars, can help in achieving higher fire resistance of FRP-RC members. Though some research has been carried out on the fire performance of RC structural members retrofitted with FRP [30], more detailed investigations are required to enhance the performance of FRP-strengthened RC structures subjected to fire.

11 Conclusions

FRPs are established and effective seismic retrofit materials for reinforced concrete structures. They possess several favorable properties and can enhance ductility and strength, without significantly affecting the dynamic property of the structure. Several retrofit projects that have utilized FRP strengthening techniques have been very successful, with the retrofitted structures performing as designed under strong excitation. Moreover, the retrofit project with FRP allows the functionality of the project to be maintained, which is a substantial advantage of this retrofit strategy over other options, especially in the post-earthquake scenario. Detailed experimental studies and investigations with actual earthquake scenarios have led to the development of reliable guidelines for design methodologies, testing methods and construction practices for FRP to be used by the engineering community. The performance of RC structures strengthened with FRP under elevated temperatures, as in the outbreak of a fire in the aftermath of an earthquake, is likely to deteriorate, and requires further study.

References

1. IS 13935: 2009 Seismic evaluation, repair and strengthening of masonry buildings–Guidelines
2. Seismic retrofit of buildings, CPWD, IBC in association with IIT Madras. Narosa Publishing House
3. Nanni A, De Luca A, Zadeh HJ, Reinforced concrete with FRP bars: mechanics and design. CRC Press
4. Japan Society of Civil Engineers (JSCE) (1999) Recommendation for design and construction of concrete structures using continuous fibre reinforcing materials. Concr Eng Ser
5. Canadian Standards Association (CSA) (2018) Specification for fibre reinforced polymers. CAN/CSA S807–10, 2nd edn. Toronto
6. ISIS Canada Design Manual-4-Strengthening Reinforced Concrete Structures with Externally Bonded Fibre Reinforced Polymer (2001) The Canadian Network of Centres of Excellence on Intelligent Sensing of Innovative Structures
7. ACI 440.2–17 (2017) Guide for the design and construction of externally bonded FRP systems for strengthening concrete structures
8. ACI 440.3R-12 (2012) Guide Test Methods for Fiber Reinforced Polymer (FRP) composites for reinforcing or strengthening concrete and masonry structures
9. ACI 440.1R-15 (2015) Guide for the design and construction of concrete reinforced with FRP bars
10. ACI 440.11–22 (2023) Building code requirements for structural concrete reinforced with Glass Fiber-Reinforced Polymer (GFRP) bars—code and commentary
11. European Commission (2012) Mandate for amending existing eurocodes and extending the scope of structural eurocodes. Brussels.
12. Pham H, Al-Mahaidi R (2004) Assessment of available prediction models for the strength of FRP retrofitted RC beams. Compos Struct 66(1–4):601–610
13. Buyukozturk O, Hearing B (1998) Failure behaviour of precracked concrete beams retrofitted with FRP. J Compos Constr 2(3), ascelibrary.org
14. Amini Pishro A, Zhang S, Zhang Z, Zhao Y, Pishro MA, Zhang L, Yang Q, Postel V (2022) Structural behavior of FRP-retrofitted RC beams under combined torsion and bending. Materials 15(9):1–17
15. Danraka MN, Mahmod HM, Oluwatosin OJ (2017) Strengthening of reinforced concrete beams using FRP technique: a review. IJESC 7(6):13199–13213
16. Pham H, Al-Mahaidi R (2004) Experimental investigation into flexural retrofitting of reinforced concrete bridge beams using FRP composites. Compos Struct 66(1–4):617–625
17. Durgadevi S, Karthikeyan S, Lavanya N, Kavitha C (2021) A review on retrofitting of reinforced concrete elements using FRP panel. Mater Today Proc, Part 2 45:1050–1054
18. Havez AA, Al-Mayah A (2023) Flexural strengthening of concrete structures using externally bonded and unbonded prestressed CFRP laminates—a literature review. J Compos Constr 27(4). https://doi.org/10.1061/JCCOF2.CCENG-4053
19. Mukherjee A, Joshi MV (2002) Seismic retrofitting techniques using fibre composites. Indian Concr J 76(8):496–502. The Associated Cement Companies Ltd.
20. Mosallam A, Allam K, Salam M (2019) Analytical and numerical modeling of RC beam-column joints retrofitted with FRP laminates and hybrid composite connectors. Compos Struct 214:486–503
21. Allam K, Mosallam AS, Salam MA (2019) Experimental evaluation of seismic performance of interior RC beam-column joints strengthened with FRP composites panel. Eng Struct 196:1093–1108
22. Pohoryles DA, Melo J, Rossetto T (2019) Seismic retrofit schemes with FRP for deficient RC beam-column joints: state-of-the-art review. J Compos Constr 23(4):2–18
23. Bindurani P, Ganesan N, Indira PV (2023) Strengthening of exterior RC moment-resisting beam-column connections using fibre-reinforced polymers and ferrocement. IJST. https://doi.org/10.1007/s40996-023-01078-x

24. Surya Prakash S, Alagusundaramoorthy P (2007) Experimental study on masonry wallettes and shear triplets externally bonded with glass fibre fabric. J Inst Eng (India) 88:24–31
25. Eslami A, Ronagh HR (2013) Effect of FRP wrapping in seismic performance of RC buildings with and without special detailing–a case study. Compos Part B: Eng 45:1265–1274
26. Lam L, Teng JG (2003) Design-oriented stress–strain model for FRP-confined concrete. Constr Build Mater 17:471–489
27. Lam L, Teng JG (2003) Design-oriented stress–strain model for FRP-confined concrete in rectangular columns. J Reinf Plast Compos 22(2003):1149–1186
28. Witt S, Gilda G (2016) Seismic retrofit with fibre reinforced polymers-using ASCE-41 to retrofit a multi-story concrete building. Struct Mag
29. Kodur VKR, Naser MZ (2020) Structural fire engineering. McGraw Hill Professional, New York, p 480
30. Kodur VKR, Bisby LA, Green MF (2007) Preliminary guidance for the design of FRP-strengthened concrete members exposed to fire. J Fire Prot Eng 17(1):5–26

Conventional and Emerging Materials Used in FRP-Concrete Composites for Earthquake Resistance

Sanchit Gupta and Sandeep Chaudhary

1 Introduction

Fibre-reinforced polymer (FRP) has gained significant applications in concrete composites. A FRP consists of tendons or fibres encased in a polymer matrix or bonding agent [1]. Different FRPs can be produced by selecting fibres and polymers of suitable materials. A typical FRP consists of non-metallic fibres, like glass and carbon fibres. The use of non-metallic fibres makes FRPs non-corrosive, chemically resistant and magnetically inert [1]. Additionally, non-metallic fibres have a lower weight than steel, improving FRP's strength-to-weight ratio [1]. The high strength-to-weight ratio of FRP is one of the essential properties of earthquake-resistant structures. Literature shows that frames designed using FRP, due to their lightweight, have 40 and 88.5% lower shear force at the base due to earthquakes than steel and concrete frames, respectively [2]. The fibres in an FRP are aligned to provide tensile strength along the fibres and flexibility along other directions; therefore, FRPs are much easier to install than other steel-based composites. The high corrosion resistance, strength-to-weight ratio and low maintenance costs of FRP have made them a fitting solution to design concrete composites, especially for earthquake resistance [3].

FRP concrete composites are structural elements in which FRP act as the reinforcing material and transfers part of the stress of concrete. FRP concrete composites are typically prepared using FRP sheets, FRP reinforcements, FRP tubes or a combination of them [1, 4–6]. FRP sheets are typically used for strengthening, i.e., improving the properties by applying FRP sheets after the construction of structures [4]. FRP reinforcements are also used as a substitute for conventionally used

S. Gupta · S. Chaudhary (✉)
Department of Civil Engineering, Indian Institute of Technology Indore, Simrol, Indore 453552, India
e-mail: schaudhary@iiti.ac.in

S. B. Singh and C. V. R. Murty (eds.), *RC Structures Strengthened with FRP for Earthquake Resistance*, Composites Science and Technology,
https://doi.org/10.1007/978-981-97-0102-5_8

steel reinforcement [1]. Literature shows that FRP can be designed per the desired geometry to substitute fabricated structural elements [3].

The FRPs can be engineered to have properties as per the desired applications. These properties are based on the materials used for the development of FRP. For example, glass fibre-reinforced polymers (GFRP) have higher creep and lower alkaline resistance than other FRPs [3]. Therefore, GFRP is typically not recommended for prestressed concrete [3]. A suitable selection of FRP depends on the selection of materials used to engineer the same. Therefore, the present chapter focuses on the various conventional and emerging materials used in FRP concrete composites. The materials used in FRP have been studied in three categories: fibres, polymer matrix and special materials used for improving earthquake resistance. The overall discussion will present a better understanding of the various materials used in FRP and their viability towards earthquake-resistant structures.

2 Fibres Used in FRP

Fibres provide the primary tensile characteristics associated with FRP. Some of the commonly used materials in FRP are glass fibres (GFRP), carbon fibres (CFRP), basalt fibres (BFRP) and aramid fibres (AFRP) [3]. In recent years, new fibres have been identified for their potential in FRP, like polyethylene terephthalate (PET) fibres and polyethylene naphthalate (PEN) fibres. However, their investigations are limited in the literature.

2.1 Glass Fibres

Glass fibres are tendons produced with glass as the raw material. Several types of glass fibres can be used to produce glass fibre-reinforced polymers (GFRP). The most commonly used type of glass used in GFRP is E-glass [1]. GFRP have good thermal stability, making them suitable for composite applications. The tensile strength, compressive strength and modulus of elasticity for E-glass fibres can be as high as 3400 MPa, 340 MPa and 76 GPa, respectively [1, 7]. The rupture strain of GFRP (2.97%) is low, as compared to steel (38.0%) [7]. In terms of durability, GFRP offers better resistance against corrosion [1]. The good mechanical characteristics, low density (~2.54 g/cm^3) and improved durability make GFRP a viable substitute for steel reinforcements [1].

E-glass fibres are commonly referred to as glass fibres only [8]. Other notable types of glass used in GFRP are known by different notations, i.e., A, AE, C, D, Advantex, ECR, AR, R, S, and T [8–10]. The properties and advantages of using some of the other glass fibres can be:

- C-glass fibres provide resistance to both acidic and alkaline corrosive environments [11]. However, their application is limited due to low-strength properties [11].
- ECR-glass, AR-glass and Advantex are specially designed to have good mechanical characteristics like E-glass fibre but with higher acidic and alkali corrosion resistance [9]. These fibres can have structural applications with better durability characteristics. AR-glass and Adnatex are highly alkali-resistant and commonly used for direct addition in fibre-reinforced concrete [9]. ECR-glass shows better corrosion resistance in acidic environments [8].
- S-glass has better mechanical properties than E-glass fibres [8]. Their tensile strength and modulus of elasticity can be as high as 4900 MPa and 87 GPa, respectively [8]. They are designed by glass with higher silica content [8].

Among the various glass fibres, S-glass can be used for earthquake engineering applications requiring high mechanical characteristics (or better strength-to-weight ratio). And other glass types can be used where durability parameters, like corrosion and alkali-silica reaction, also act on the structure. For example, ECR-glass can be used in GFRP concrete structures exposed to seawater [9].

2.2 Carbon Fibres

Carbon fibres are fibres prepared with carbon polymer as the primary material. Mainly, two types of carbon fibres are used based on the raw materials: PAN type and pitch type [12]. PAN or polyacrylonitrile type carbon fibre have high tensile strength and elastic modulus, making them highly suitable for carbon fibre-reinforced polymers (CFRP) [12]. Pitch type carbon fibres are typically cheaper but have lower mechanical properties [12]. Carbon fibres are generally recognized by their elastic modulus, which can range between 200 and 700 GPa [13]. The tensile strength of carbon fibres can be more than 3500 MPa (as high as 7000 MPa) for intermediate elastic modulus (250–350 GPa) [13].

Carbon fibres have high strength, low density and corrosion resistance, making them widely applicable in CFRP concrete structures [1]. Carbon fibres have high fatigue resistance and can be used for cyclic loadings, like foundations of heavy machines and bridges [1]. Although the initial costs for CFRP are higher than steel, the lower maintenance cost makes up for the difference [1]. Carbon fibres exhibit the highest tensile strength and best durability characteristics among all conventional fibres; however, the associated costs are also high. Therefore, it is recommended to use CRFP in corrosion-resistant structures which require FRP strengthening.

2.3 Basalt Fibres

Basalt fibres are prepared by melting and drawing basalt rocks. Basalt fibres are cost-effective with high durability, making them a good solution for basalt fibre-reinforced polymer (BFRP) concrete composites [1]. Compared to GFRP and CFRP, BFRP shows higher thermal stability and can sustain exposure temperatures of 600 °C [14]. It should be noted that at high-temperature exposures (above 70 °C), the failure of FRP depends on the glass transition temperature of the adhesive and polymer matrix. Therefore, BFRP has no unique application in fire-resistant construction despite having one of the highest thermal stabilities among various fibres. In terms of alkali and UV exposure, BFRP performs better than GFRP but shows lower durability than CFRP [14]. BFRP shows an ultimate tensile strength of up to 1500 MPa, lower than other conventionally used fibres [14, 15]. Due to this, BFRP has been recommended for moderate structural strengthening [14]. Some recent studies show that the tensile strength of basalt fibres can be as high as 5000 MPa, but the application of high-strength basalt fibre was not observed in the case of FRPs [16, 17].

The primary advantage of basalt fibre is the low cost, carbon emissions and embodied energy associated with the production of fibres [17]. Considering the production costs and associated emissions, basalt fibres have the highest cost-to-benefit ratio among all conventional fibres [1, 18]. Therefore, basalt fibres can be considered a low-cost and low-carbon option for earthquake-resistant structures requiring moderate strengthening.

2.4 Aramid Fibres

Aramids are synthetic fibres produced by creating polymer chains of aromatic amide [19]. Unlike other synthetic fibres, aramid has a higher melting point and tensile strength [1, 19]. As a result, aramid fibres have substituted most natural and synthetic fibres in protective clothing [1]. Aramid fibres are also used to produce aramid fibre-reinforced polymers (AFRP) [19]. AFRP have high abrasion tolerance, low thermal conductivity and high tensile strength, making them a good solution for surface applications [1, 19]. However, AFRPs are sensitive to salts, acids and UV light, limiting their application to low-durability structures [1, 19]. In case of earthquake resistance, AFRP can be limited to indoor structures with less severe environmental exposures.

2.5 PET and PEN Fibres

PET fibres are produced by recycling PET bottles [20]. Similarly, PEN fibres are produced by recycling PEN waste. PET and PEN fibres are more environmentally friendly than conventional fibres as they promote waste upcycling [20]. Among the two, PET fibres are more abundantly available and cost-effective, while PEN fibres have better resistance to chemical attacks [21]. Higher chemical resistance also helps in improving the durability of FRP concrete composites [21].

PET and PEN fibres have a large rupture strain (more than 5%) and show ductile failure. Furthermore, PET FRP and PEN FRP show a bi-linear stress–strain behaviour with a yield point of around 1% strain value [20]. The conventional fibres, i.e., glass, carbon, aramid, and basalt fibres, show a brittle failure with a small rupture strain [20]. As a result, the previously discussed FRPs show brittle failure with small rupture strains (up to 3%) [20]. In contrast, PET FRP and PEN FRP show large rupture strains (LRS) and are also termed LRS FRP [20]. The drawback to this large rupture strain is the relatively low elastic modulus and low ultimate strength [20]. Dai et al. [20] observed that the ultimate strain for PET FRP, PEN FRP and AFRP were 8.7, 6.3 and 3.2%, respectively. At the same time, the ultimate strength for PET FRP, PEN FRP and AFRP were 751, 768 and 3732 MPa, respectively [20]. PET FRP and PEN FRP show higher axial and lateral strain in FRP composite columns [20]. Results reported in the literature show that LRS FRP prepared from PET fibres and PEN fibres can prevent sudden failure of FRP concrete composites [20]. On account of ductile failure, the PET fibres and PEN fibres are recommended for FRP with application in seismic retrofitting of concrete structures [20].

2.6 PA Fibres

PA or polyacetal fibres are prepared by drawing polyoxymethylene or polyacetal materials [12]. PA fibres can be prepared from virgin and recycled materials and may also be considered a waste recycling-based solution. PA fibres, similar to PET fibre and PEN fibre, provide a solution for developing LRS FRP [12]. Compared with PET fibres and PEN fibres, the PA fibres have higher rupture strain and tensile strength [12, 22]. Despite the higher toughness of PA fibres, limited experimental investigations have been carried out on their application as FRP [23]. Iihoshi et al. [23] demonstrated the application of PA fibres on post-strengthening of concrete columns under lateral seismic loading. The study shows that FRP prepared with PA fibres provides a technically superior and cost-effective seismic retrofitting solution than conventional fibres [23].

2.7 Flax Fibres

Flax is a naturally occurring fibre with relatively high tensile strength compared to steel [24]. Flax fibre has a higher cellulose content, fibre length and mechanical properties as compared to various natural fibres [25]. Flax fibre has been demonstrated for developing a flax FRP [25]. Flax, being a naturally available resource, can further lower the environmental cost of FRP [24]. The major drawback of flax fibre is its hydrophilic nature, while polymer matrix is hydrophobic [24]. On exposure to moisture, additional stress is created between the fibre and polymer matrix, causing accelerated degradation of FRP [24]. Furthermore, flax fibres are also weak against exposure to heat, UV rays and chemicals [24]. In its present state, flax FRP shows poor durability characteristics, which limits its field application [24].

Like flax fibres, other natural fibres, like jute, have also been explored for FRP development [26]. However, such investigations are limited, and their performance is weaker than flax fibres [26]. Therefore, other natural fibres have not been discussed separately in the chapter.

2.8 Overview

A comparison between the properties of different fibres has been shown in Table 1. It can be observed from the table that conventional fibres, i.e., glass fibre, carbon fibre, basalt fibre and aramid fibre, have high tensile strength. High tensile strength helps in improving the strength of FRP concrete composites. The primary drawback of conventional fibres is the low rupture strain and associated brittle failure, which can be overcome through emerging fibres [27]. It can be observed that PET fibre, PEN fibre and PA fibre have higher rupture strains than other fibres. Due to higher rupture strains, PET, PEN, and PA fibre can be used to make LRS FRP concrete composites. LRS FRP concrete composites have better ductility and are more suitable for earthquake resistance [27]. Among the three, PA fibres also show high tensile strength, making them suitable for strengthening and ductility improvement.

It should be kept in mind that low elasticity and high rupture strain improve earthquake resistance for FRP concrete composites where bonding is ensured. Ko

Table 1 Properties of various fibres used in FRP [12, 22, 24]

Fibre	Sub-type	Tensile strength (MPa)	Modulus of elasticity (GPa)	Rupture strain (%)	Density (g/cm^3)
Glass	E-glass	1700–3400	72	2.4	2.6
	S-glass	2500–4600	87	2.9	2.5
Carbon	PAN	2300–3790	242–370	0.6–1.6	1.8–1.9
	Pitch	1100–3210	41–940	0.4–2.8	1.6–2.2
Basalt	–	1000–1560	50	3.0–1.0	2.2
Aramid	–	2060–2350	79–118	1.8–3.0	1.4–1.5
PET	–	740	10	7.0–8.7	1.4
PEN	–	790	15	5.0–6.3	1.4
PA	–	1760	40	6.0–9.0	1.4
Flax	–	300–1000	20–100	–	1.2–1.3

and Sato observed the issue of debonding in FRP concrete composites where FRP sheets are transferring the loads [28]. The study showed that FRP sheets with higher elastic modulus could resist more loads before debonding [28]. Therefore, in FRP concrete structures where adequate bonding is not possible, conventional FRP with a high modulus of elasticity should be used for earthquake resistance.

3 Matrix Used in FRP

Matrix holds the fibres together and can bond FRP to the concrete surface. The mechanical properties of FRP are less affected by the matrix than the fibres. Matrix plays a more important role during the load transferring across the fibres.

Resins are primarily used as polymer matrix in the FRPs [12]. Resins can be made from either thermosetting or thermoplastic materials [12]. In the case of FRP, thermosetting resins are used for better mechanical properties and thermal stability [12]. Thermosetting resins form irreversible cross-linking polymer chains after curing and cannot be recycled or remoulded [12]. Literature shows that three types of thermosetting resins are commonly used as polymer matrix in FRPs, i.e., epoxy, vinyl ester and polyester [12]. At the same time, resins like phenolic resin and polyurethane (PU) resin have started gaining attention. Table 2 shows the properties of different thermosetting resins used in FRP. Other than the resins listed in Table 2, bismaleimide (BMI) resin and cyanoacrylate resin can also be used for developing FRPs; however, their application has not been explored in the case of concrete.

3.1 Epoxy Resin

Epoxy is one of the most commonly used polymer matrix in FRPs [18]. Epoxy matrix is prepared by mixing epoxy resin and hardener in desired proportions [18]. Epoxy resins show relatively higher strength, rupture strain, elasticity and thermal stability

Table 2 Properties of various resins used in FRP [12, 29, 30]

Matrix	Density (g/cm^3)	Glass transition temperature (°C)	Tensile strength (MPa)	Modulus of elasticity (GPa)	Rupture strain (%)
Epoxy	1.2–1.3	100–270	60–80	2.0–4.0	1.0–8.0
Polyester	1.2–1.3	70–120	20–70	2.0–3.0	1.0–5.0
Vinyl ester	1.1–1.2	102–105	68–82	3.5	3.0–4.0
Phenolic	1.0–1.3	260	30–50	3.6	1.8–2.5
Polyurethane	1.1	77–116	11–18	0.3–1.1	6.0–10.0

[12]. The primary drawback of using an epoxy matrix is the high material cost [12]. The properties of the epoxy resin can be engineered (by adjusting the proportion of epoxy resin) to provide a matrix with desired characteristics. It should be noted that epoxy resin can be designed for large rupture strains, making them highly suitable for LRS FRP [12].

3.2 Polyester Resin

Polyester resins are ambient cured unsaturated polyesters [12]. They have a relatively lower cost and are easy to manufacture among all thermosetting resins [12]. Their properties are slightly inferior to epoxy resins in terms of mechanical characteristics. The major drawback of polyester resins is the low glass transition temperature (70–120 °C) [12]. In the given temperature range, polyester resins can be damaged by steam. The low glass transition temperature lowers the thermal stability and mechanical performance of FRP concrete composites [31]. As a result, it is recommended to limit the application of polyester resins in FRP composites without exposure to elevation in temperature. If polyester resins are used, adequate thermal insulation should be provided.

3.3 Vinyl Ester Resin

Vinyl ester resins are a mid-way solution of epoxy and polyester resins [12]. Vinyl ester consists of unsaturated esters of epoxy resins, providing the resin with good mechanical properties and ease of processing [12, 32]. Vinyl ester resins have better thermal stability than polyester resins [12]. The addition of flame retarders can further improve the thermal stability of vinyl ester resin [32]. Vinyl ester resins also show better resistance against alkaline attack than polyester resin and can help protect encased fibres [9]. The major drawback of vinyl ester resin is the low rupture strain, making it ineffective in LRS FRP. In earthquake-resistant structures, other than LRS FRP, vinyl ester resin provides a better solution due to relatively lower cost, good mechanical properties and satisfactory thermal stability.

3.4 Phenolic Resin

Phenolic resins show the best thermal stability among the discussed resins [33]. They have high glass transition temperature and low flammability [12, 33]. The major drawback associated with phenolic resin is the high viscosity of the material, making it challenging to produce FRP [12]. Alternatively, phenolic resins can be used as a thermal insulation material on the external surface of FRPs. Functionally layered FRP of such kind still requires further investigation.

3.5 Polyurethane Resin

Polyurethane resin provides an alternative to polyester resin. Polyurethane resin is a fast-curing, low-cost, low shrinkage and easy-to-apply polymer [34, 35]. Polyurethane resin shows low glass transition temperature, making it susceptible to damage from exposure temperatures [34]. The key difference of polyurethane resin is its large rupture strain and relatively lower costs [35]. In the future, polyurethane resin can be applied in LRS FRP, but the same was not observed in the available literature.

4 Other Materials Used in FRP Concrete Composites

4.1 Shape Memory Alloys

Shape memory alloy, or SMA, has a high potential in earthquake-resistant FRP composites, as it can provide active confinement during earthquakes and help post-earthquake damage repairs [18, 36, 37]. SMA is a special material class that regains its original shape on the application of heat by phase transformations [18, 37]. Han et al. [18] demonstrated the application of iron-based SMA coupled with FRP on concrete columns. The SMA was pre-stretched and anchored around concrete columns [18]. On application of heat, the SMA regains its original shape and generates an active confining stress on the columns [18]. Active confinement acts as a prestressing force and improves load-bearing capacity at smaller strain values [18, 37]. A jacket of FRP coupled with SMA provides passive confining stress [18, 37]. Passive confinement improves load-bearing capacity at higher strain values [18, 37]. The combination of active and passive confinement provided improved performance of FRP concrete composite columns [18, 37]. Literature shows that SMA can be an emergency repair solution for damaged concrete [18, 37].

Among SMA, a special material class exists called superelastic alloys [38]. Superelast alloys, or SA, show a characteristic flag-like stress–strain curve [36, 38]. During the application of strain, SA initially shows elastic deformation followed by plastic deformation [36, 38]. On removing strain, SA shows both elastic and plastic recovery without requiring the application of heat [36, 38]. SA are highly suitable for applications which require large recoverable deformations [38]. Zafar and Andrawes [36] used the SA as fibres for developing FRP reinforcing bars and explored their potential for earthquake resistance. As compared to conventional steel bars, SA-based FRP bars can reduce the inter-story drift and improve the earthquake resistance of structures, especially in case of aftershocks [36].

4.2 FRP Anchors

Anchors are used to join FRP sheets with concrete structures during jacketing [39, 40]. Anchors are typically employed when sufficient concrete surface is not available to bond FRP sheets using adhesives [40, 41]. Unlike conventional anchors, FRP anchors are prepared using FRP of the same material type [39, 41]. They are prepared by bundling fibres in the form of a long rope of cylinder, which is then partly cured using bonding agents [40, 41]. The bonded fibres are then inserted in concrete and used as anchor dowels [40, 41]. The uncured part of the fibres is spread in a fan-like formation and bonded on the FRP sheets or concrete surface using the curing agent [40, 41]. FRP anchors can prevent debonding failure in FRP concrete composites [39–41].

4.3 Thermal Insulation

FRP sheets, while strengthening most mechanical properties, have little impact on the fire performance of concrete structures [31, 42]. The lack of impact on fire performance results from the decomposition of adhesive between FRP and concrete [31, 42]. On exposure to temperatures above the glass transition temperature (T_g) the adhesive loses its bond properties [31]. The loss in bond properties results in delamination of FRP from concrete, making them ineffective at higher temperatures [31]. The glass transition temperature for adhesives can be as low as 70 °C, which limits their application in high-temperature environments like kitchens. As the probability of fire increases during an earthquake, FRP concrete composites can become ineffective during structural fires. This challenge can be resolved through thermal insulation [31, 42].

Thermal insulation can be applied on the exposed surface of FRP sheets to delay the temperature rise. The delay in temperature rise increases the time after which adhesives attain the glass transition temperature. The delay in failure of FRP improves the fire protection of the FRP concrete composites [31]. The time of attaining glass transition temperature depends on several factors, including exposure temperature, insulation material type and thickness [31, 42]. Thermal insulation can be designed to ensure a minimum time duration before the adhesives start to fail [31, 42]. Literature shows that thermal insulation can improve the mechanical properties of FRP against exposure temperatures as high as 400 °C [42]. Thermal insulation can also protect against post-earthquake fire and provide added safety [43].

5 Summary

FRP concrete is a better substitute for steel-reinforced concrete or steel–concrete composite in earthquake-resistant structures due to advantages like higher strength-to-weight ratio and corrosion resistance. FRP concrete composites primarily consist of two essential materials, i.e., fibres and polymer matrix. Fibres govern the mechanical behaviour of FRP and have been extensively investigated in the literature. Conventional fibres, i.e., glass fibre, carbon fibre, basalt fibre and aramid fibre, have high tensile strength and modulus of elasticity. Higher mechanical strength favours the application of conventional fibres to improve the strength of concrete, which helps improve earthquake resistance. The emerging fibres, i.e., PET fibres, PEN fibres and PA fibres, have large rupture strain and low modulus of elasticity. The emerging fibres can be used for developing LRS FRP, which improves the ductility of concrete and is highly effective in earthquake resistance.

Polymer matrix have a more dominant role in the stability of FRP and load transfer across fibres. Epoxy resin provides large rupture strain, making it highly suitable for LRS FRP. On the other hand, vinyl ester resin can provide a viable solution for most engineering applications. In addition to the two primary materials, several new materials have emerged for improving the overall performance of FRP concrete composites, like SMA, FRP anchors and thermal insulators. These materials compensate for the drawbacks of FRP concrete composites and improve their performance as earthquake-resistant structures. Understanding various materials will help suitably identify the materials and develop the desired FRP concrete composite for earthquake-resistant structures.

References

1. Rafieizonooz M, Kim J-HJ, Varaee H, Nam Y, Khankhaje E (2023) Testing methods and design specifications for FRP-prestressed concrete members: a review of current practices and case studies. J Build Eng 73:106723. https://doi.org/10.1016/j.jobe.2023.106723
2. Mincigrucci L, Civera M, Lenticchia E, Ceravolo R, Rosano M, Russo S (2023) Comparative structural analysis of GFRP, reinforced concrete, and steel frames under seismic loads. Materials (Basel) 16. https://doi.org/10.3390/MA16144908
3. Zou Y, Yu K, Heng J, Zhang Z, Peng H, Wu C, Wang X (2023) Feasibility study of new GFRP grid web—concrete composite beam. Compos Struct 305:116527. https://doi.org/10.1016/J.COMPSTRUCT.2022.116527
4. Turkowski P, Łukomski M, Sulik P, Roszkowski P (2017) Fire resistance of CFRP-strengthened reinforced concrete beams under various load levels. Procedia Eng 172:1176–1183. https://doi.org/10.1016/J.PROENG.2017.02.137
5. Shi X, Guo T, Song L, Yang J (2023) Cyclic load tests and finite element modeling of self-centering hollow-core FRP-concrete-steel bridge columns. Alexandria Eng J 70:301–314. https://doi.org/10.1016/J.AEJ.2023.03.001
6. Xiao Y (2004) Strengthening of concrete columns applications of Frp composites. 7:335–344
7. Boru E (2023) Torsional strengthening of steel I beams with different GFRP configurations. Structures 56:104859. https://doi.org/10.1016/j.istruc.2023.07.049

8. Dwight DW, Begum S (2017) Glass fiber reinforcements. Compr Compos Mater II(1):243–268. https://doi.org/10.1016/B978-0-12-803581-8.03812-1
9. Benmokrane B, Wang P, Ton-That TM, Rahman H, Robert J-F (2002) Durability of glass fiber-reinforced polymer reinforcing bars in concrete environment. J Compos Constr 6:143–153. https://doi.org/10.1061/(asce)1090-0268(2002)6:3(143)
10. Rubino F, Nisticò A. Tucci F, Carlone P (2020) Marine application of fiber reinforced composites: a review. J Mar Sci Eng 8:26. https://doi.org/10.3390/JMSE8010026
11. Safwat EM, Khater AGA, Abd-Elsatar AG, Khater GA (2021) Glass fiber-reinforced composites in dentistry. Bull Natl Res Cent 45. https://doi.org/10.1186/s42269-021-00650-7
12. Ye YY, Da Liang S, Feng P, Zeng JJ (2021) Recyclable LRS FRP composites for engineering structures: current status and future opportunities. Compos Part B Eng 212. https://doi.org/10.1016/j.compositesb.2021.108689
13. Mirdehghan SA (2021) Fibrous polymeric composites. Eng Polym Fibrous Mater 1–58. https://doi.org/10.1016/B978-0-12-824381-7.00012-3
14. Sim J, Park C, Moon DY (2005) Characteristics of basalt fiber as a strengthening material for concrete structures. Compos Part B Eng 36:504–512. https://doi.org/10.1016/j.compositesb.2005.02.002
15. Mohamed OA, Al Hawat W, Keshawarz M (2021) Durability and mechanical properties of concrete reinforced with basalt fiber-reinforced polymer (Bfrp) bars: towards sustainable infrastructure. Polymers (Basel) 13. https://doi.org/10.3390/polym13091402
16. AL-Kharabsheh BN, Arbili MM, Majdi A, Alogla SM, Hakamy A, Ahmad J, Deifalla AF (2022) Basalt fibers reinforced concrete: strength and failure modes. Materials (Basel) 15. https://doi.org/10.3390/ma15207350
17. Al-Rousan ET, Khalid HR, Rahman MK (2023) Fresh, mechanical, and durability properties of basalt fiber-reinforced concrete (BFRC): a review. Dev Built Environ 14:100155. https://doi.org/10.1016/j.dibe.2023.100155
18. Han T, Dong Z, Zhu H, Wu G, Zhao X (2023) Compression behavior of concrete columns combinedly confined by FRP externally wrapped Fe-SMA strips. Eng Struct 294:116754. https://doi.org/10.1016/J.ENGSTRUCT.2023.116754
19. Dharmavarapu P, Sreekara SR (2022) Aramid fibre as potential reinforcement for polymer matrix composites: a review. Emergent Mater 5:1561–1578. https://doi.org/10.1007/s42247-021-00246-x
20. Dai J-G, Bai Y-L, Teng JG (2011) Behavior and modeling of concrete confined with FRP composites of large deformability. J Compos Constr 15:963–973. https://doi.org/10.1061/(asce)cc.1943-5614.0000230
21. Zeng JJ, Zhuge Y, Da Liang S, Bai YL, Liao JJ, Zhang L (2022) Durability assessment of PEN/PET FRP composites based on accelerated aging in alkaline solution/seawater with different temperatures. Constr Build Mater 327:126992. https://doi.org/10.1016/j.conbuildmat.2022.126992
22. Ye YY, Zeng JJ, Li PL (2022) A state-of-the-art review of FRP-Confined Steel-Reinforced Concrete (FCSRC) structural members. Polymers (Basel) 14. https://doi.org/10.3390/POLYM14040677
23. Iihoshi C, Fukuyama H, Matsumoto Y, Abe S (1999) Strengthening effect of reinforced concrete elements with polyacetal fiber sheets (1999)
24. Chen C, Li X, Li C, Zhou Y, Sui L (2022) Optimized flax FRP stirrup in reinforced concrete beam: material property and shear performance. Compos Struct 302:116219. https://doi.org/10.1016/j.compstruct.2022.116219
25. Liu Z, Wang H, Yang L, Du J (2022) Research on mechanical properties and durability of flax/glass fiber bio-hybrid FRP composites laminates. Compos Struct 290:115566. https://doi.org/10.1016/j.compstruct.2022.115566
26. Chen C, Yang Y, Yu J, Yu J, Tan H, Sui L, Zhou Y (2020) Eco-friendly and mechanically reliable alternative to synthetic FRP in externally bonded strengthening of RC beams: natural FRP. Compos Struct 241:112081. https://doi.org/10.1016/j.compstruct.2020.112081

27. Liu X, Li Y (2018) Experimental study of seismic behavior of partially corrosion-damaged reinforced concrete columns strengthened with FRP composites with large deformability. Constr Build Mater 191:1071–1081. https://doi.org/10.1016/j.conbuildmat.2018.10.072
28. Ko H, Sato Y (2008) Bond stress—slip relationship between FRP sheet. J Compos Constr 11:419–426. https://doi.org/10.1061/(ASCE)1090-0268(2007)11
29. Zamri FA, Primus WC, Shaari AH, Sinin AE (2019) Technical note: Effects of polyurethane resin on the physical and mechanical properties of wood fiber/Palm kernel shell composite boards. Wood Fiber Sci 51:448–454. https://doi.org/10.22382/wfs-2019-043
30. Sales FCP, Ariati RM, Noronha VT, da Costa RRC, Ribeiro JE (2021) PU tensile tests: conventional and digital image correlation analysis. Procedia Struct Integr 37:389–396. https://doi.org/10.1016/j.prostr.2022.01.100
31. Dong K, Hu K, Gao W (2016) Fire behavior of full-scale CFRP-strengthened RC beams protected with different insulation systems. J Asian Archit Build Eng 15:581–588. https://doi.org/10.3130/jaabe.15.581
32. Zhang X, Wu J, Geng Z, Qin Z, Pan YT, Zhang W, Yang R (2023) Effect of diluent on the properties of intrinsically flame-retardant vinyl ester resins and their fiberglass-reinforced composites. Compos Commun 37:101441. https://doi.org/10.1016/J.COCO.2022.101441
33. Liu TQ, Wang R, Zhen S, Feng P (2023) A binary resin system of epoxy and phenol-formaldehyde for improving the thermo-mechanical behavior of FRP composites. Constr Build Mater 389:131790. https://doi.org/10.1016/J.CONBUILDMAT.2023.131790
34. Al-Lebban Y (2017) Polyurethane fiber reinforced polymer strengthening of shear deficient reinforced concrete beams. College Eng Comput Sci. https://stars.library.ucf.edu/etd/5670. Accessed 13 Sept 2023
35. Huang S, Bachtiar EV, Yan L, Kasal B (2022) Bond behaviour and thermal stability of flax/glass hybrid fibre reinforced polymer–timber structures connected by polyurethane. Constr Build Mater 322:126456. https://doi.org/10.1016/J.CONBUILDMAT.2022.126456
36. Zafar A, Andrawes B (2015) Seismic behavior of SMA–FRP reinforced concrete frames under sequential seismic hazard. Eng Struct 98:163–173. https://doi.org/10.1016/J.ENGSTRUCT.2015.03.045
37. Suhail R, Amato G, McCrum DP (2020) Active and passive confinement of shape modified low strength concrete columns using SMA and FRP systems. Compos Struct 251:112649. https://doi.org/10.1016/J.COMPSTRUCT.2020.112649
38. Tsuchiya K (2011) Mechanisms and properties of shape memory effect and superelasticity in alloys and other materials: a practical guide. Woodhead Publishing Limited. https://doi.org/10.1533/9780857092625.1.3
39. Dong K, Gao Y, Yang S, Yang Z, Jiang J (2023) Experimental investigation and analytical prediction on bond behaviour of CFRP-to-concrete interface with FRP anchors. Case Stud Constr Mater 19:e02510. https://doi.org/10.1016/J.CSCM.2023.E02510
40. Khorasani M, Muciaccia G, Consiglio AN, Mostofinejad D (2023) Evaluating the behavior and bond properties of FRP spike anchors under confined conditions and elevated temperature. Compos Struct 322:117407. https://doi.org/10.1016/J.COMPSTRUCT.2023.117407
41. Li Z, del Rey Castillo E, Henry RS, Thompson A (2023) Axial compression testing of concrete prisms confined by FRP spike anchors and estimation of failure modes. Compos Struct 322:117403. https://doi.org/10.1016/J.COMPSTRUCT.2023.117403
42. Altunişik AC, Akbulut YE, Adanur S, Kaya A, Günaydin M, Mostofi S, Mosallam A (2023) Evaluating the high-temperature endurance of FRP-strengthened concrete using an innovative insulation system: experimental investigation. J Build Eng 73:106444. https://doi.org/10.1016/J.JOBE.2023.106444
43. Risco GV, Zania V, Giuliani L (2023) Numerical assessment of post-earthquake fire response of steel buildings. Saf Sci 157:105921. https://doi.org/10.1016/J.SSCI.2022.105921

Retrofitting of RC Structures Using FRP Techniques—Case Studies

Mangesh V. Joshi and S. V. Vivek

Abstract Fiber Reinforced Polymer (FRP) based strengthening and retrofitting methods have gained importance even in developing countries such as India, with the advent of professionally managed and technically sound contracting companies. Numerous advantages of FRP, such as fast and clean application, exceptional tensile strength, free of corrosion, negligible deadload addition, etc. are being capitalized while staying within the limits of safety by understanding the limitations of the same. The author has completed over 2,500 retrofitting projects on reinforced cement concrete (RCC) structures in professional capacity in India and abroad, from industrial and commercial structures to bridges, dams and heritage buildings. Most of these projects required major utilization of FRP, in the form of FRP rods, laminates, wraps, anchors, NSM applications, crack-stitch and fiber reinforced cementitious matrix (FRCM) hot-wrap. In this chapter we shall discuss some of these projects as case studies and get an introduction into some of its design concepts and practical aspects.

Keywords Fiber Reinforced Polymer (FRP) · Retrofitting of RC structures · Bridge repair · Prestressed FRP laminates · Carbon laminates · CFRP wrapping

1 Introduction

FIB Bulletin 14 by European Committee for Concrete published in 2001 stated that the issue of upgrading the existing civil engineering infrastructure has been of great importance [1] for over the previous 10 years, i.e., from at-least the early

M. V. Joshi (✉)
Sanrachana Structural Strengthening Pvt. Ltd., Harinivas Circle, Thane West, Thane 400602, Maharashtra, India
e-mail: mangesh@sanrachana.in

S. V. Vivek
Structural Design Team, Sanrachana Structural Strengthening Pvt. Ltd., Harinivas Circle, Thane West, Thane 400602, Maharashtra, India

S. B. Singh and C. V. R. Murty (eds.), *RC Structures Strengthened with FRP for Earthquake Resistance*, Composites Science and Technology,
https://doi.org/10.1007/978-981-97-0102-5_9

1990s. Several reasons such as change in use of the existing structures, building additional floors on them, increased axle loads of design vehicles of bridges, changes in the design guidelines such as ductile detailing for earthquake resistance, mistakes committed in design and in construction, etc. have been identified as the common causes for this surge in demand for structural retrofitting [2]. Nearly thirty years later, among many techniques in practice in India and abroad, FRP based solutions appears to have gained popularity due to their versatility, ease of application and instant strength contribution as opposed to traditional concrete based techniques. The American Concrete Institute ACI has recognized this in their 2017 publication of FRP retrofitting guidelines [3] by naming FRP a viable option for repair and retrofitting, highlighting advantages such as light-weight, non-corroding nature and high tensile strength. Even in areas where it is hard to implement traditional strengthening measures, thin profiles of FRP may help achieve desired results.

Due to the extent of distress or design deficiency, several RCC bridges in India which were constructed decades ago have been demolished. The author has come across road bridges in dry areas of Gujarat and Rajasthan which have lost significant flexural and shear strength in less that 5 years of construction, due to corrosion and poor construction practices. It is worthwhile noting that as of 2009, roughly 15% of all the 1,47,523 bridges on Indian rail network were 80–100 years old [4]. While addition of deadload and dynamic loads due to placing of concrete for jacketing may be risky for aged and dilapidated structures, the light-weight FRP based solutions seem appropriate, after performing structural repairs and satisfying the minimum prerequisites set forth by design guidelines such as ACI-440-2R-17. Century old bridges such as the Chitpore bridge in Kolkata, India has been retrofitted using FRP based techniques in similar manner (Fig. 1).

It is well known that FRP could be used to impart strength and ductility to a column through lateral confinement. As opposed to RCC jacketing, FRP confinement does not increase the stiffness of the column. Through parametric modelling of a symmetric RC framed multi-storey building with identical columns supporting uniform symmetric static loading with 8 different cases of RC jacketed columns (Fig. 2), it was found that RC jacketed columns attract larger loads through it even when self-weight of the jacket is ignored (Table 1), because of their increased stiffness [5]. Thereby the other structural members connected to the RCC jacketed column such as beams, slabs and foundation, may be subjected to increased shear forces, bending moment and axial loads.

This effect may be even more significant when considering rapidly varying and stress-reversing loads such as seismic forces. Hence even for piers of bridges in seismic zone 5 of India (zone with the highest risk of intense seismic activity) where insufficient ties were provided, FRP based solutions have been used to compensate the design deficiency in place of RC jacketing and thereby enabling the structure to achieve the standards laid down by IRC and IS codes without impacting the behaviour of the superstructure or foundation. A case study of such a bridge in Assam has been discussed later in this chapter.

Fig. 1 Before and after structural repairs and FRP based retrofitting measures were performed on 100-year-old Chitpore bridge in Kolkata

2 Current Scenario of Retrofitting in India

The trend in the industry for retrofitting (especially seismic retrofitting) in India as of early 2020s is as follows. The asset owner or the asset maintenance head, seeks advice from a structural consultant upon observation of visible distresses, perceivable serviceability concerns or when revisions occur in compliance standards. This may also be the case when permissions are granted for increasing the floor area of an existing building by construction of additional storeys. In either case, consultant proceeds to analyse the structural model using commercially available analysis software with or without FEM. As per latest strength parameters obtained from non-destructive testing (NDT), the model is modified and accordingly it is determined whether the structure needs any structural intervention or not.

The existing and required cross-sectional properties of each member can be tabulated through this analysis, in terms of increased tension steel area, stirrup area, stirrup spacing, total sectional area and grade of concrete. Otherwise, the factored bending moment, shear force and axial force that the existing member shall be made capable

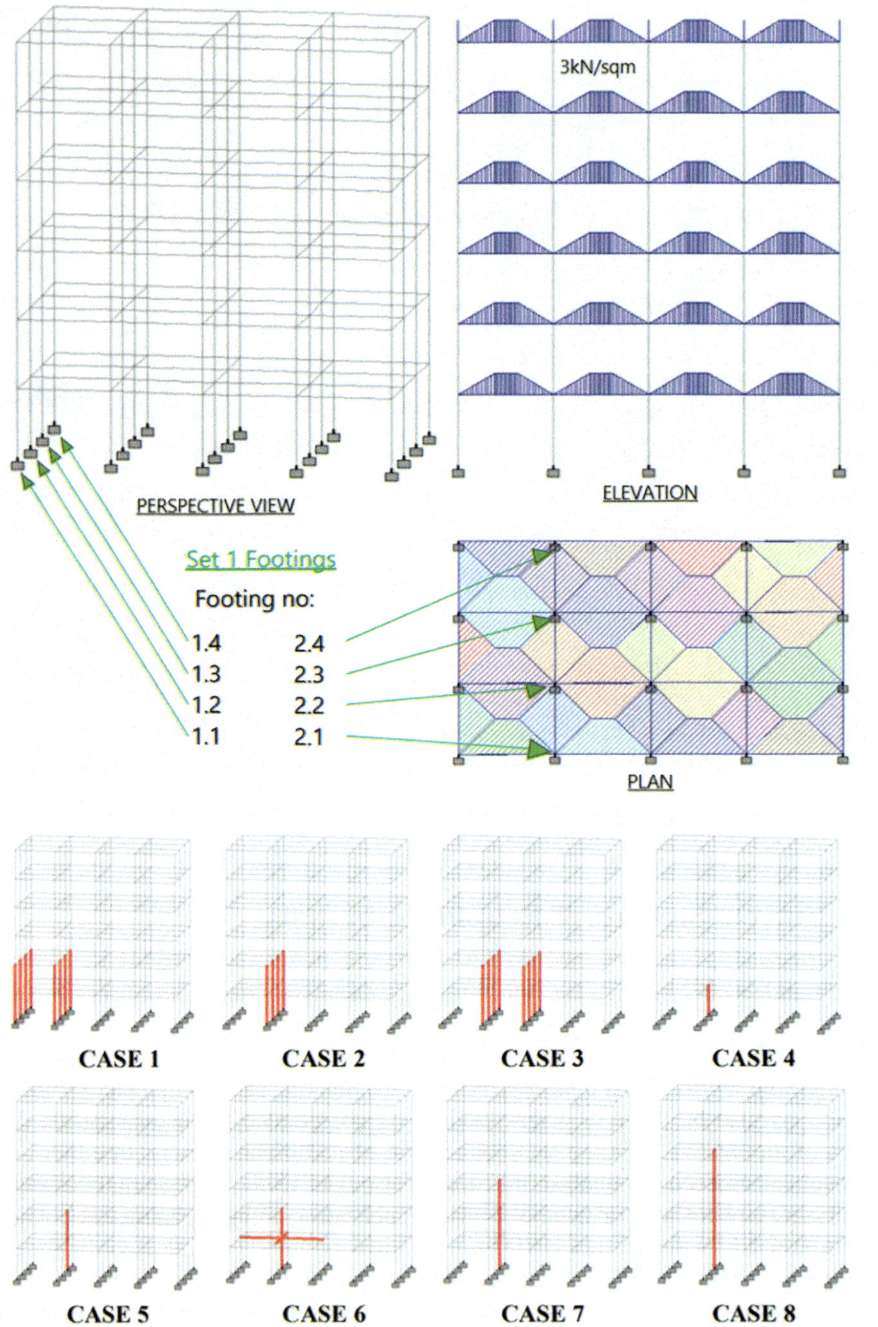

Fig. 2 Structural model of multi-storey portal framed RCC structure, with loading condition. The 8 different cases of RC jacketing of columns are shown, jacketed columns in red

Table 1 Reaction forces obtained at each foundation in kN for different jacketing cases and control (no jacketing) when all self-weights including that of RC jacket were ignored. Reaction load at footing no. 23 is highlighted and colour contoured, indicating highest (yellow) and lowest (green) reactions

Set 1 Footings

Case No:	Original	CASE 1	CASE 2	CASE 3	CASE 4	CASE 5	CASE 6	CASE 7	CASE 8
Footing No:									
11	59.6	56.9	56.8	57.0	59.7	59.7	59.7	59.7	59.8
12	107.1	106.1	102.6	102.9	106.7	106.2	106.2	105.9	105.6
13	107.1	106.1	102.6	102.9	105.4	103.7	103.0	102.6	102.4
14	59.6	56.9	56.8	57.0	59.2	58.8	58.8	58.5	58.3

Set 2 Footings

Case No:	Original	CASE 1	CASE 2	CASE 3	CASE 4	CASE 5	CASE 6	CASE 7	CASE 8
Footing No:									
21	113.8	115.5	117.4	114.2	114.1	114.3	114.5	114.4	114.5
22	204.8	213.3	216.4	216.4	200.2	197.1	196.0	195.0	193.7
23	204.8	213.3	216.4	211.3	218.9	229.1	232.0	235.6	239.4
24	113.8	115.5	117.4	114.2	109.8	107.1	106.5	105.5	104.7

Set 3 Footings

Case No:	Original	CASE 1	CASE 2	CASE 3	CASE 4	CASE 5	CASE 6	CASE 7	CASE 8
Footing No:									
31	116.3	113.3	113.1	116.9	116.3	116.3	116.4	116.4	116.4
32	208.9	204.2	203.9	215.8	208.4	207.9	207.9	207.6	207.4
33	208.9	204.2	203.9	215.8	206.7	205.2	204.4	204.2	203.6
34	116.3	113.3	113.1	116.9	115.4	115.4	115.3	115.1	114.9

of handling by retrofitting, is tabulated. This data is sent to retrofitting consultants, who design the optimal solution based on the nature and condition of the structure, nature of loads, percentage increase in strength required, desired extension in life expectancy of the structure, environmental exposure conditions, type of retrofit materials to be used, etc. Although this process exempts the retrofitting consultant from performing structural analysis and bearing the associated liability, the retrofitting consultant must make sure that their optimised solution equates to the requirements adjudged by the structural consultant, and also make sure that it is practically (and financially) feasible to implement the same. At times, the retrofitting consultant by themselves develop the structural model through site visit, reinforcement scan data, NDT data and assumptions.

The calculated strength and serviceability parameters of a retrofitted section may be equal to or better than that of the consultant's recommendation. However, the performance of a retrofitted section in reality is highly reliant on the workmanship of the retrofitting applicator. The novelty of structural retrofitting and its specialised nature implies that the applicator skills are a very critical criteria in achieving desired results. A wide variety of retrofitting products are also getting applied in practice. Accordingly, the true margin of safety in a retrofit design is heavily dependent on the exposure that the applicator has, in installation of the given product. It becomes imperative that retrofitting consultants have their own well-trained execution team, whose skills are constantly updated as and when new products and application methods are launched in the industry.

3 Prerequisites for Using FRP Based Solution

Several structural consultants prescribe FRP-wrapping as a blanket solution for structural strengthening in RCC. In Rashtrapati bhavan, which is a heritage structure that serves as the office and residence of the President of India, a dilapidated embellished lime-concrete chajja (sun-shade) of the length of 1,200 m constructed about 100 years ago needed repair and strengthening. By conducting a detailed condition assessment, the most appropriate solution was identified as recasting with structural lightweight concrete with cathodic protection.

This project was later recognized by American Concrete Institute (ACI) India Chapter as the best retrofitting project in 2022 at national level. In this project, an FRP based solution could not be modelled or proposed as the pre-requisites laid down by ACI were not met. ACI-440-2R-17, one of the most commonly used FRP retrofitting code of practice in India, states that the substrate, including all bond surfaces between repaired areas and the original concrete, should have sufficient direct tensile and shear strength to transfer force to the FRP system. For bond-critical applications, the tensile strength should be at least 200 psi (1.4 MPa), determined by using pull-off adhesion test. Bond critical and contact critical applications are discussed later in this chapter. The code goes on to say that "FRP systems should not be used when the concrete substrate has a compressive strength f_c' less than 2500 psi (17 MPa)". It must be noted that the ACI code was meant for practice in USA, where strength of concrete is measured in cylinder strength. An equivalent cube strength (f_{ck} value) is estimated by increasing the cylinder strength by a factor of 0.8, i.e., minimum cube compressive strength shall be 21 MPa.

In addition to the minimum concrete grade, there are some prerequisites laid down by ACI-440-2R-17 which are called the strengthening limits of the RC members. These measures tend to preserve the structure against the possibility of a collapse, in case the FRP incorporated into an RC structure for strengthening purpose gets damaged by vandalism, fire or debonding activities such as nailing, hammering, etc. Equation 9.2 in ACI-440-2R-17 stipulates that, the existing strength of the member, when reduced by the factor ϕ shall be more than the sum of 1.1 times new deadload and 0.75 times new live load. Since deadload can be relatively accurately assessed, the factor of 1.1 is used. The statistical mean of yearly average live load factor as per ASCE is 0.5, thereby to exceed this, 0.75 is used by ACI. However, in case the design live load acting on the member has a high likelihood of being present for a sustained period of time, then instead of 0.75, the live load factor used shall be 1.0 in the equation. Φ can be obtained from ACI 318, based on strain in tensile steel. This factor ensures that brittle sections have higher factor of safety than ductile sections.

Strengthening solution using FRP may affect the fire-rating requirement of the structure since at the glass-transition temperature T_g, the mechanical and bond properties get depleted. The resin used in the FRP thereby becomes the weakest link in case of fire. The code says that the T_g of commercially available ambient temperature-cured FRP systems typically ranges from 60 to 82 °C. In practice, to comply with the fire-rating norms of RCC structures, the FRP is either protected with a layer of fire-resistant mortar (typically in buildings) or shall be encased within micro-concrete jacketing (typically in bridges, especially in areas where impacts are anticipated, such as in the submerged piers in Assam bridge project shown in Fig. 3).

In RCC structures where constant exposure to sun-light is expected such as in exposed faces of road bridges, the applied FRP shall be protected from UV rays of the sunlight using specialized UV-protection coatings. The love-grove flyover in Worli, Mumbai is a good example (Fig. 4). In areas where extreme fire-protection is required, the chance of fire in itself can be eliminated by totally submerging the FRP-applied portion in sand, wherever possible such as in the case of structural retrofitting of fire-damaged hydrocarbon storage tanks in Dhamra, India (discussed as case study later). More than 180 circular c/s columns supporting a large hydrocarbon tank were repaired at first and then strengthened using CFRP wrapping, followed by cement plastering over the FRP layer. The entire space between the ground level and soffit of the tank was then filled with sand so that in no case shall a fire event in its close vicinity affect the polymers in the FRP.

ACI has put in place a minimum strength criterion based on strength reduction of the structure during fire, as exposure to fire reduces the strength of both steel and concrete. It says, that the resistance of a structural member to load, with the factored concrete and steel strengths and without FRP contribution, can be compared to the

Fig. 3 Water from the melted snow and the rain, combines to reach high flood water levels, causing complete submergence of piers, exposing them to impact from floating bodies such as uprooted trees

Fig. 4 Love Grove flyover in Mumbai, India showing carbon laminates exposed to direct sunlight, protected by UV resistant coating

load demand on the member in the event of a fire, to ensure that a strengthened member passes the fire-rating load. ACI 440-2R-17 can be used to check that this condition is met. Fire would majorly affect the epoxy-based compounds that form the matrix phase of an FRP or the saturant/adhesive used to bond the FRP to the substrate. However, since the FRP is of small thickness and is bonded closely to the concrete, the effects may be different from that of exposing a piece of CFRP laminate to fire. Section 9.2.1 of ACI 440-2R-17 states: "Although the FRP system itself is significantly affected by exposure to elevated temperature, a combination of the FRP system with an existing concrete structure may still have an adequate fire resistance". Also, when fire-protection systems are applied on the installed FRP, they are applied so that they shield most of the member and not just the FRP. This may help reduce the extent of loss of strength of the embedded rebar and concrete due to fire as well. The equation 9.2.1a in ACI 440-2R-17 essentially says that the reduced capacity of the member (reduced due to fire) shall be sufficient to support 1.0 X new dead load +1.0 X new live load.

In light of these pre-requisites of ACI, it is generally recommended to use FRP as a secondary reinforcement in strengthening schemes. The code itself says "FRP strengthening systems use FRP composite materials as supplemental externally-bonded or near-surface mounted reinforcement". If the factor of safety incorporated in original RCC design has been depleted by flaws in design/detailing/execution or by external factors such as floods, fire or earthquake, it is wiser to use traditional retrofit materials as primary strength providers and use FRP to supplement them. The case study of Ambet river bridge discussed later, is a good example.

The extent of deterioration of the hammerhead portion (the 12–16 m long RCC box sections that transfer the load of the two spans onto the pier) in Ambet river bridge was too high due to corrosion, carbonation and excessive reduction of concrete quality.

After improving the microstructure of the concrete using low-viscosity monomer application through pressure grouting followed by brush application of migratory corrosion inhibitor, additional reinforcing steel was provided and section was rebuilt using M50 grade micro-concrete. Once the structure is made sound through this jacketing mechanism, FRP based strengthening in terms of prestressed and non-prestressed carbon laminates were applied, over which shear strengthening was performed using carbon FRP wraps.

One other pre-requisite of application of FRP is training of the workforce. For all products used in construction and specifically in retrofitting, it is mandatory to follow the manufacturer's instructions and the material safety datasheet (MSDS) as most materials used in retrofitting are outcomes of recent research and hence these are not familiar to most engineers and workmen. Newer and more effective products constantly replace the existing. Ignorance leads to mistakes, which may not only affect the structural performance, but also the health and safety of the personnel involved. FRP based strengthening is often performed after structural repairs, which involve activities like application of anti-corrosive chemicals, pressure grouting, crack filling, primer application, etc. Most of these applications require mixing a base with a hardener, initiator, catalyst, solvent, or so on. Some of these components could be highly volatile, inflammable, corrosive, initiator of undesired reactions or a health hazard. FRP based materials tend to cause itching or irritation of the skin, eyes and may cause breathing difficulties as well, since minute fibres could break off from the material easily. Storage and handling of these materials must therefore be performed only under strict supervision.

4 Points of Focus in FRP Application

For the FRP based retrofitting solution to take effect in a desired manner, the following points shall be taken care of.

4.1 Substrate Strength and Quality

The structures that require strengthening, are often found to be left with significantly low of concrete strength and quality. Poor construction practices such as inadequate compaction and high water-cement ratio, leaves the concrete with a less dense matrix, honeycombed areas and porous substrate. This in-turn leads to early onset of corrosion and carbonation, leading to even faster deterioration in the future. Ultimately, the basic pre-requisites for performing FRP based strengthening may not be met. By performing NDT tests such as rebound hammer test and ultrasonic pulse velocity test, the quality of concrete can be ascertained. Carbonation test and half-cell potential test can give an initial understanding of extent of carbonation and probability of active corrosion respectively. Breakout windows can be opened in concrete to directly observe the state of corrosion in embedded steel. Advanced techniques such as galva-pulse and i-corr may also be performed if detailed corrosion study is required. Surface pull-off strength can be used to ensure that the minimum tensile strength criteria is met. Core strength is generally considered a reliable measure of compressive strength of concrete. Various chemical tests such as pH, chloride profile, chloride content, sulphate content, etc. can further back-up the inferences made from visual observation and the NDTs.

All these tests help identifying the characteristics of the concrete in the spot where the test has been conducted. The number of test locations and the locations of those tests shall be selected such that they become a statistically representative sample of the bridge. Test readings shall be reported in a segregated manner, such that say the results from the girder and the deck slab shall not be mixed. Often times, the cast-in-situ deck may be in a state of severe distress whereas the girders may have relatively better concrete quality, such as in the case of Baghajatin flyover (Fig. 5). This is especially true if the girders are post tensioned, as precast PT structures undergo rigorous quality control during casting while the decks are cast in situ, whose quality varies based on the quality of site supervision and workmanship. In Baghajatin flyover, certain parts of the deck slab were to be demolished and recast, whereas the girders only had a few locations to be rectified, mostly due to honeycombing.

Fig. 5 A portion of Baghajatin flyover showing the old and new bridges side by side

4.2 Surface Preparation

FRP must be applied only on structurally sound and smooth concrete substrate. This means that the surface cracks, pitting, undulation, moisture, laitance, etc. must be removed and must have sufficient shear and tensile strength. These are gauged indirectly using core test, CAPO test and surface pull-off strength test. The CAPO (cut and pull off) and surface pull-off strength tests are good indicators of tensile strength requirement of concrete for FRP application. However, the concrete on which FRP is to be applied, usually needs repairs. Hence it is seldom included in the initial NDT. Typical repairs may include rectification of microstructure of concrete by pressurized injection or flooding of low-viscous monomers or other suitable compounds. These penetrate the substrate through the interconnected pores and voids in it. Upon solidification of these compounds, the permeability of the concrete is significantly reduced, thereby improving its microstructure. After such treatments, the concrete usually passes for the tensile strength criteria.

Concrete must however be checked for cracks that may be structural or corrosion related. Unless the root cause is removed, the cracks may keep on widening, exerting un-desirable local stresses on the FRP that may be applied across it. Hence any case of hollowness must be checked by delamination survey, so that the loose concrete shall be removed, corrosion shall be addressed and sound concrete can be packed in its place. Several types of repair mortars and pre-mix concretes are commercially available for such activities. Non-corrosion related cracks shall be opened up by grooving and sealed appropriately, so that they may be packed while pressure grouting. Structural cracks can only be addressed permanently if the root cause is identified and appropriate solution is worked out. Crack stitching may be required to counteract the widening forces acting across the crack. In areas of active water/moisture ingress, the source needs to be identified and curtailed to prevent further seepage or dripping, such that after sufficient surface dryness is attained, surface preparation can be performed.

The surface on which FRP is to be applied, shall ultimately be made as good as new. Hence, surface grinding (Fig. 6) must be performed whether the concrete is old or freshly cast/repaired. This removes any material (paint, oil or laitance) adhering to the surface and exposes the original concrete for retrofitting applications. Whether it is beam or column, FRP wraps cannot be applied over sharp edges, since these become sites of stress concentration, leading to premature failure of FRP. Hence while grinding, the edges of the beams and columns are ground and rounded to a minimum radius of 25 mm. After grinding, pressurised air blowing is performed so that the concrete dust is removed from the surface. The prepared surface typically needs to be primed before FRP application.

Fig. 6 Concrete surface grinding on piers

4.3 Workmanship in FRP Application

Factory-cured FRP laminates and on-site applied FRP wraps are the most frequently used forms of FRP for structural retrofitting in India as of now. Both are installed in slightly different manner, but the surface preparation requirements remain the same. FRP wraps are flexible fabrics thinner than 1 mm, made of the continuous structural fibers only. These could either be carbon, glass, aramid, or a combination of these fibers (Fig. 7). The weight class (in GSM) and number of layers to be used in structural application are determined by the retrofitting design engineer. A 600 GSM UD (unidirectional) CFRP has thrice as much thread-count (carbon fibers) as in 200 GSM. Wraps can be unidirectional (UD) or bi-directional (BD). FRP laminates on the other hand, are unidirectional and are manufactured in factories with precise proportion of fiber and matrix (Fig. 8). This ensures effective quality control, such that thicknesses of up-to 3 mm can be achieved while ensuring monolithic behaviour.

Before applying FRP, the surface undulations shall be removed by physically grinding the surface to level and applying structurally sound thixotropic putty. These fill the surface voids and also prevent bubbles from forming during curing of the saturating resins. Sometimes it may be needed to apply a layer of polymer modified mortar (PMM) or even micro-concrete before applying thixotropic putty, based on the extent of undulation. The surface profiling has to be carried out in regions of sharp edges and corners. Primer application follows as the next step. Selection of primer depends of FRP manufacturer's recommendation. This is to prevent excessive absorption of saturant into the substrate, leading to unsaturated FRP. Primers also

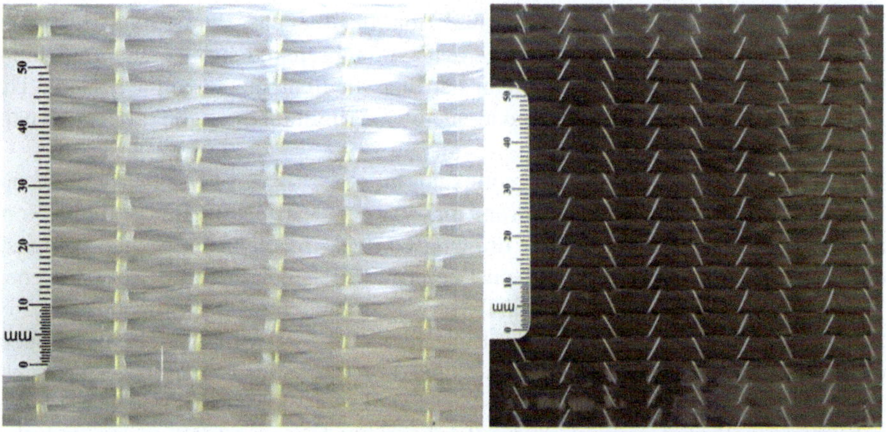

Fig. 7 **Left**—Glass aramid UD fabric (930 GSM) and **Right**—Carbon UD fabric (900 GSM)

Fig. 8 100 mm wide carbon laminate

provide a good bonding surface for the saturant (in case of wraps) and structural adhesive (in case of laminates).

On this treated surface, FRP wraps may be installed using wet-layup or dry-layup process. Dry-layup process requires application of saturant on the substrate, once the layer becomes tacky, fixing the wrap by saturant-roller application and if required, application of another coat of saturant. This method may be good for fabrics of lighter weight-class. Those above 600 GSM shall be installed only using wet-layup method. In wet-layup, the fibers are pre-saturated by passing them through a pool of saturant, such that a single layer of fabric emerges from a dry roll of FRP wrap, enters the pool of saturant and passes through rollers to be spooled into a saturated roll of FRP wrap.

This saturated roll is then taken for installation. In either case, while fixing the wrap, the FRP fabric shall be gently pressed against the prepared surface with a medium nap roller to remove any entrapped air between the fiber sheet and the surface. The fiber sheet shall be applied stretched along the longitudinal direction (along the main fiber direction).

Insufficient saturation of FRP with saturant causes part of the fabric to be not bonded to the substrate. Such errors may cause early debonding of FRP. Non-adherence to the detailing as per drawing can also result in the same. Sufficient development length and overlaps must be provided as detailed by retrofit designer. Detailing of laps and splices is usually provided in accordance with clause 14.2 of ACI-440-2R-17. In cases where multiple layers of FRP wraps are to be applied, designers often include fiber anchors, which are short and stiff FRP rods of length 75–200 mm, with a tuft of loose fibers on the other end. These are drilled onto the substrate usually before application of subsequent layers of FRP wraps, such that the tuft of loose fibers can then be flayed and fixed on to the previous FRP layer in a spread manner using the same saturant. It is difficult to quantify the exact increase of bonding force obtained by fiber anchors but it is a good detailing practice. Typically, the design engineers limit the number of layers of FRP to three or four.

For FRP laminate application, the primed surface and underside of the laminate are coated with slight surplus of structural adhesive. These are then adhered together carefully, in such a manner that air-voids do not form between the laminate and the substrate. The ends of the laminates are fixed onto the parent concrete using plate-bolt systems, usually made of aluminium alloys, to reduce the chances of delamination (end peeling). Laminates could also be post tensioned, for which specially designed (usually proprietary) machinery is required. For these, specialised end plates are used. Multiple layers of laminates are not usually applied in practice. FRP wrapping at the curtailed ends of the laminates, suffices in place of the bolted end-plates.

Contact critical applications need the entire substrate to be structurally sound and in-compressible. Loose, hollow, honeycombed or porous concrete may affect the effectiveness of contact critical applications significantly. In bond-critical applications (eg. Strengthening for flexure and shear), the only way in which load transfer happens between concrete and FRP, is through the bond between them, whose quality as long as the recommended chemicals are used, purely depends on the workmanship quality. Also concrete of poor strength and quality could fail while transferring loads to the FRP. In general, the FRP systems work by capturing minute strains in concrete and converting them into huge resistive forces by taking advantage of their large Young's modulus and tensile load carrying capacity. Any slack in an applied FRP resulting from air entrapment, dry patches, etc. due to incompetency or negligence in workmanship, can render the entire application ineffective, as the minute strains generated in RCC would then be insufficient to stress the FRP.

Upon curing, FRP exposed to the environment is sprayed on with UV-resistant coating for the longevity of the resins used. If fire-resistant mortars, regular cement plaster or micro-concrete is to be applied, a layer of coarse-grained washed silica sand is pasted on the FRP surface over a layer of saturant in tacky condition. This provides mechanical keying between the FRP and the upcoming cementitious layer.

4.4 Bond Critical and Contact Critical Applications

4.4.1 Bond Critical

The applications where a strong bond between FRP and the RCC substrate is required for load transfer are called bond critical applications. In such cases, no load is transferred to FRP in the absence of bond. When FRP is applied as a supplementary tensile reinforcement or as shear reinforcement, the application is typically bond critical (Fig. 9). For instance, for installing an FRP laminate on the bottom sagging face of a simply supported beam, high strength structural adhesive is applied in the interface between the laminate and the beam bottom.

As design load is applied on a regular simply supported beam, the concrete tends to crack in the tension face as expected, since it is weak in resisting tension. Hence these sections are designed to have sufficient main reinforcement embedded in the bottom face to handle 100% of all tensile forces generated. In practice, FRP application is usually required on structures that have previously been subjected to design loads, thereby the beams are usually cracked sections. Before application of FRP, some superimposed dead loads and almost all live loads are removed from the area of application and from nearby locations. Also propping is done from lower side so that existing strain level at the soffit of the beam from time of application of FRP till time of curing shall be minimal and unaffected by any impact loading.

After curing of FRP and on application of design loads, the previously formed cracks in the tension face tend to open up. However, since the FRP is strongly bonded to the tension face, the opening up of those cracks will happen only if a corresponding strain is developed in the FRP, i.e. the FRP needs to get stretched as well. Since the FRPs have high Young's modulus in tension, this stretching requires tremendous

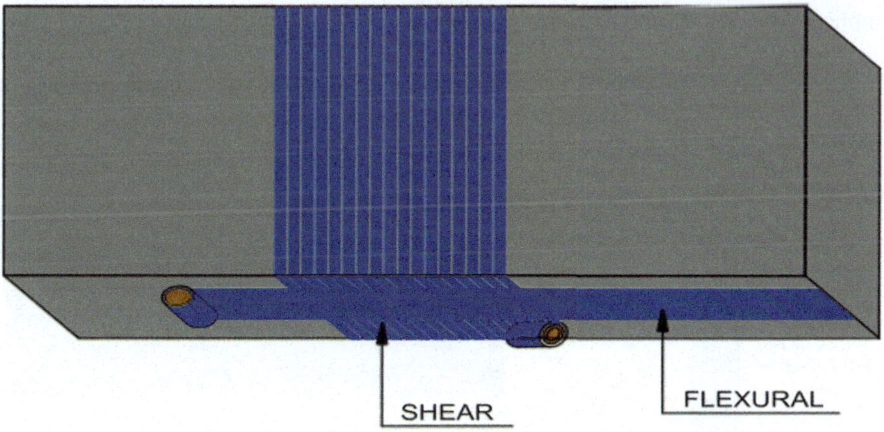

Fig. 9 Typical direction of unidirectional fiber in FRP wrapping for shear and flexure. Wraps for shear shall be analogous to stirrups and that for flexure be analogous to main reinforcement

stress generation in FRP, for which large amount of force needs to get transferred from the concrete to the FRP. This is the mechanism by which bond-critical application of FRP increases the load-carrying capacity of a structural member. Evidently, the load transfer happens through the bond, and therefore the name "bond critical". Thus, the mechanical properties of the bonding material act as the limiting factor in the design of FRP based retrofit solution.

Failure of FRP through rupture is sudden and hence it is not preferred. The preferred failure mode of FRP is through debonding, as it may be less drastic in comparison. However, when the substrate quality and strength are lower than that of the bond, the cover concrete fails instead of the bond, which is called cover delamination. If the bonding between concrete/FRP is strong enough, then the next weak link in the concrete/FRP interface is the concrete. The minimum tensile and compressive strength pre-requisites set by ACI, aims to avoid such incidents.

4.4.2 Contact Critical

When the tensile force required to be generated in the FRP can be transferred to it without the necessity of an adhesive bond between the FRP and substrate, it is called a contact critical application. Here the bond is not necessary as long as the contact between the two is intact. Increasing the compressive load carrying capacity of an axially loaded concrete member, by applying FRP along the Hoop direction, orienting the unidirectional fibres transverse to the longitudinal axis of the member, may be considered as a contact critical application. Clause 12.1 in the code ACI-440-2R-17 says that the transverse Hoop fibers act similar to helical or tie reinforcements. This contributes to passive confinement of the compression member, which until dilation and cracking of the concrete, remains unstressed. Thereby, intimate contact between FRP jacket and concrete member is critical." Hence the term contact-critical application (Fig. 10).

When a compressive axial load is applied on a non-slender column of uniform cross-section, its longitudinal dimension decreases and the lateral dimensions

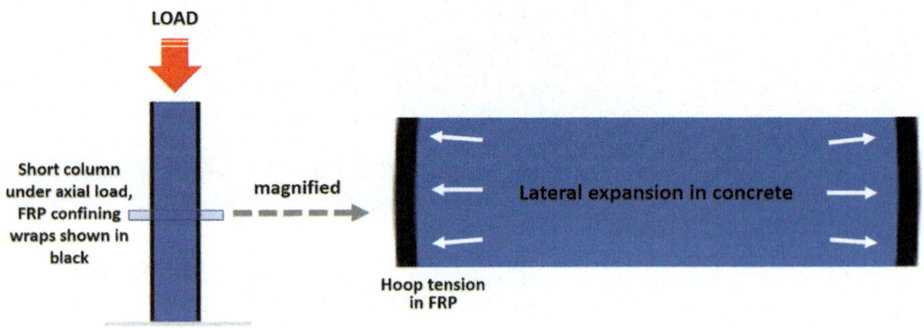

Fig. 10 Confining FRP wraps on short column. Dilation of concrete generates Hoop tension in FRP

increase based on Poisson's ratio. This leads to decrease in height of column and increase in the perimeter of transverse sections. When load is increased, the column could get compressed further and the radially outward acting pressure could keep on increasing, causing increase in the perimeter. If a material sufficiently strong in tension, is wrapped around the perimeter of the column so as to form a confining ring of constant diameter, then increase in perimeter of the column can be restrained. Confinement of concrete enhances it strength and ductility by restraining lateral dilation [6]. To achieve this, typically uni-directional FRP are chosen for wrapping around the column, with the fiber direction along the perimeter, in single/multiple layers as required by design, with lengthwise overlap. While application of FRP on the column, the saturant used will create a bond between FRP and concrete, but it is not necessary as long as the ring diameter is maintained by adhesion of the overlapped FRP with itself.

In a test conducted for the author as part of the Assam bridge project, four commercially available brands of uni-directional FRP wraps were tested, to compare their ability in enhancing the compressive strength of standard concrete test cylinders of same grade by contact critical application. 3 specimens of FRP wrapped cylinders were prepared per brand. There was one sample set of 900 GSM, three sample sets of 600 GSM and one sample set of control experiment, making a total of 15 specimens. The 900 GSM specimens were given 3 wraps each and the 600 GSM specimens given 4 wraps each. By performing uni-axial strain-controlled compression testing, (Fig. 11) the stress–strain curves of the wrapped samples and bare sample (control) were obtained in terms of cylinder strength (MPa) versus axial strain. In the resulting stress–strain curve, it was found that the concrete samples after FRP wrapping were able to resist high stresses even as the compressive strain exceeded 0.0100 value, as opposed to the control specimens that failed at around stain of 0.0032 as expected. ACI permits maximum compressive strain of 0.0100 in FRP wrapped concrete, hence the stress–strain curves were manually limited at that point.

The test report concludes that all the CFRP wrapped specimens showed significant improvement in performance in terms of ultimate load as well as ultimate strain compared to the control specimens made of M40 grade bare concrete. Ultimate load carrying capacity of wrapped specimens are observed to be more than 250% compared to control specimens. Ultimate strain developed in wrapped specimens are observed to be more than 300% compared to control specimens. Ultimate strain of 0.32% with ultimate stress of 37 MPa (equivalent cube compressive strength = 46 MPa) was recorded for control specimens. Ultimate strain of 1% with average ultimate stress of 122 MPa (minimum 97 MPa and maximum 152 MPa) was recorded for strengthened specimens (Fig. 12). This implies significant enhancement in ductile behaviour as well as improved load carrying capacity.

In case of earthquake resistance, it is important that the structural members accommodate the large strains which are generated by the ground movements, without themselves failing in such a way as to totally lose their structural integrity and thereby leading to collapse of the structure partially or globally. In the lab experiment, FRP wrapping was found to make the specimens capable of handling much higher strains than bare concrete. No apparent signs of distress were observed on

Fig. 11 Experimental setup for testing the FRP wrapped specimens in axial compression. Strain gauges attached

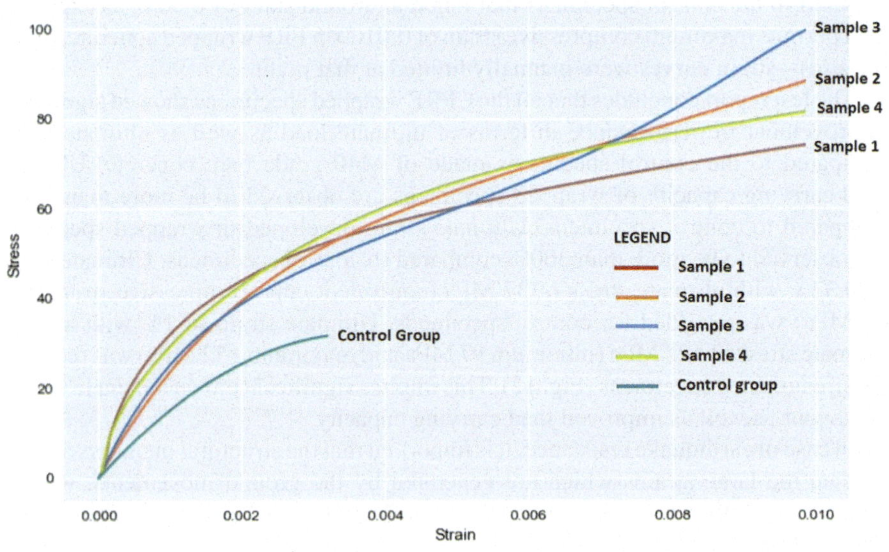

Fig. 12 Stress strain behaviour of different samples of FRP wrapped concrete in axial compression

CFRP wrapped specimens even when loaded up-to a load equal to two times of ultimate load of control specimens (bare concrete specimens). The ultimate load-carrying capacity of the wrapped samples were found to be more than 250% of the control (bare specimens). Ultimate strains developed were more than 300% higher as well. Average ultimate stress was found to be 122 MPa, where the highest and least values obtained were 152 and 97 MPa respectively. Area under the stress–strain curve for CFRP specimens was more than 8 times corresponding to area under the stress–strain curve for control specimens. This implies significant improvement in energy absorption capacity as compared to bare specimens, which is desirable for structures that are expected to resist earthquake and impact loading. All the CFRP wrap systems tested showed similar stress–strain behaviour, very close to the theorised bi-linear curve with high initial slope and lesser slope afterwards.

5 Case Studies

5.1 Case study 1: Assam bridge project

In addition to other factors, every now and then, structural deficiencies occur because of design issues such as usage of outdated design codes, inadequate detailing, incorrect consideration of design loads, etc. These may either be the designer's mistake, or could be lapse in communication, where site execution gets done based on older versions of drawings. Ultimately, when the structure is built and commissioned, service loads are subjected on it and the users experience serviceability issues in the form of vibrations, excessive deflections, or cracks may start appearing unexpectedly, even before full design loads are applied. Sometimes, cracks may appear during construction phase itself, due to exertion of loads arising from say shifting of precast sections of a segmental PT girder, stacking of construction materials on slabs, vibrations/movements of construction equipment, etc.

In case of this project, the new asset owner performed a structural due diligence and found that the piers of this bridge did not comply with the latest seismic standards. The structural consultant found that the vertical c/c spacing of lateral ties in the piers is too high and thereby to comply with the latest code, the spacing must be lesser. This can either be performed by anchoring additional ties and jacketing the pier with RCC or by providing unidirectional FRP wrapping that can provide the same design shear strength as that of the additionally required ties. The site location makes it difficult to shift lot of steel and concrete (Fig. 13), but FRP based solution requires significantly less qty of material shifting, and additionally, FRP is a very light material in itself. The window of opportunity to access the pier fully is only 2 months/year, as on the other months, the piers would be submerged, as seen in Fig. 3 earlier. A study of rainfall pattern, daylight hours, ambient temperature and river water-levels was required to arrive at that data. There were 182 piers that requires this retrofitting, so methods that would take a long period for execution, such as RCC jacketing, were impractical. FRP application is clean and fast. FRP wrapping has the added advantage of confining the pier uniformly as a continuous layer, as opposed to conventional lateral ties that are discrete steel rings. Taking into account all these factors, the client decided to go with the FRP based solution (Fig. 14).

By using the design guidelines of ACI 440-2R-17, it was found that the piers need 4 layers of 600 GSM unidirectional carbon wraps to generate the equivalent shear strength. However, there was another challenge. The river carries large floating bodies like trees, which may impact the retrofitted piers and thereby damaging the FRP. To prevent this, a thin layer of nominal RCC jacketing could be done on the fully submerged piers and cement plastering can be done in the other piers. The jacketing could be of thickness as small as 40 mm and need to have only nominal reinforcement, hence much less material requirement as compared to traditional jacketing.

Fig. 13 Photograph of the Assam bridge taken before retrofitting

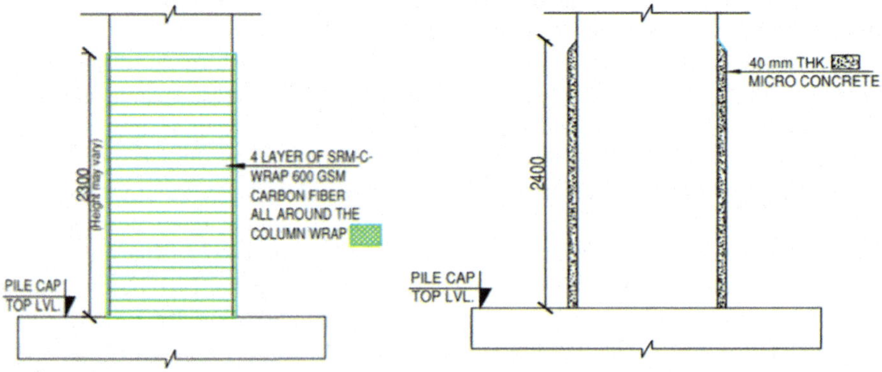

Fig. 14 Schematic sketches of retrofitting scheme to be applied on the piers

The piers were 2300 × 1900 mm in cross section. The FRP contribution to increase in pure axial compressive strength was less, which is typical of rectangular cross sections, especially when dimensions are large. ACI-440-2R-17 recommends the use of FRP for improving compressive strength by FRP wrapping, only to columns of cross-sectional dimensions less than 900 mm, but it can be used for larger columns if the effectiveness is verified experimentally. In this application, the aim is not to improve the compressive strength, but to compensate the shear strength, for which there are no such dimensional restraints in the code.

For most piers, 2.3 m height of FRP wrapping (from top level of pile cap) was required. Since concrete quality was found good, there was no need of structural repair activities like injection grouting and anti-corrosion treatment. The works started with surface grinding and finishing of the piers such that the corners are rounded to at-least 25 mm edge radius. The rounding off is required to prevent stress concentration at the edges which may cause FRP to fail prematurely. The loose dust adhering to the surface after grinding was removed by pressurized air blowing. Structural putty was applied only in small patches to ensure perfect level surface. SRM C-Wrap UD600 is the wrap that was to be installed. It is a 600 GSM unidirectional carbon fiber wrap. It is recommended to use wet lay-up machine for installation of FRP of grade 600 GSM and above. Wet lay-up machine un-wraps the dry roll of FRP, makes the dry fabric pass through a bath of saturant and forces it out through a series of rollers, which ensures complete saturation of the fibers. The saturated fabric forms a wet roll, which is then taken to the site for direct site application (Fig. 15).

On the previously prepared surface of the pier, a layer of primer and saturant were applied one after the other. While the saturant layer was still tacky, the saturated FRP was applied gently without air pockets, all around the pier, using roller brush. This process requires skilled and specially trained manpower. The vertical and horizontal

Fig. 15 Wet layup machine saturating a roll of 600 GSM unidirectional CFRP (left) and the saturated roll of CFRP being applied on the piers after application of primer (right)

overlaps are to be provided as per drawing. Usually, vertical overlaps of 10–20 mm is sufficient, just enough to ensure continuity. Horizontal overlaps are more important. In usual practice, for single layer application, after one complete lap, a horizontal overlap of 150 mm is provided. If multiple layers are applier, the overlap length shall be calculated based on ACI-440-2R-17.

Once the structural requirement is compensated using FRP, the next step is to protect the FRP. In this project, as mentioned earlier, since FRP needs to be shielded from physical impact, a nominal RCC jacketing was provided in the piers on the river. In other locations, a layer of cement plaster was provided. While the surface of the FRP is tacky with the semi-cured saturant, a layer of washed silica-sand is pasted on its surface for ensuring good mechanical bonding between the FRP and the cement based protective layer. For RCC jacketing, steel bars were anchored on the concrete using chemical adhesive, up to a pre-designed depth. After formwork was in place, pre-packaged M50 grade micro-concrete was poured from the carriageway through pipe and fed to the inside of the formwork through gravity pouring (Fig. 16). The cementitious protective layer by itself shields the FRP against its deterioration from UV exposure. In case such cementitious coating is not given, it is necessary to apply a UV protection coat, usually available as a sprayable PU compound. In this project alone, 26,000 m^2 of FRP was consumed, for a total of 182 piers. All the works were completed as planned on time, because of the intensive planning that went into it before execution stage.

Fig. 16 Protection of the FRP wraps on the pier using micro-concrete jacketing

5.2 *Case study 2: Fire damaged hydrocarbon storage tanks in Dhamra*

In the field of structural retrofitting, each project must be treated as unique, since the quality of workmanship in original construction and the distress induced onto the concrete due to environmental factors over the years, are highly variable from project to project. One such project was in Dhamra, where a huge hydrocarbon storage tank that was supported by more than 200 circular columns, caught fire. Once the fire was put out, it was observed that most of the columns were visibly affected by the fire. Spalling was observed in varying intensities, but since the concrete cover was high, no steel was exposed. Non-destructive tests including core-compression tests were carried out on the columns. The consultant reached a conclusion that the cylinder strength of the concrete has declined from 40 to 29 MPa. This was a major cause of concern, since unlike in flexure and shear, reduction in concrete strength directly affects the axial load carrying capacity of the RCC column. Hence, as a first aid, the asset owner performed the first intervention in the form of removal of spalled concrete, application of injection grouting and performing shotcreting.

Based on NDTs and visual observations, the column-wise damage report was documented. For each column, the location and dimensions of spalled portions were identified and tabulated. The retrofitting consultants were then contacted to perform their evaluations and remedial measures. It was found that the injection grouting and shotcreting was done well, thus there was no need to perform structural repair activities and one can concentrate on bringing back the columns to their original design compressive strength of M50 from the current M35 (both values indicate f_{ck}). If a structurally sound surface is obtained, then it is possible to perform this increase using FRP wrapping. Detailed calculations were performed based on ACI-440-2R-17 and found that the columns require four layers of 900 GSM unidirectional carbon fiber for this increment. The structural consultant had indicated that for each column, FRP wrapping shall be performed not along the entire column height, but only 150 mm above and below the distressed portion and directly over the distressed portion itself. This significantly reduced the cost of retrofitting.

Upon approval of the retrofitting scheme, the team set out for execution (Fig. 17). The first target was to obtain a structurally sound and smooth concrete surface for application of FRP. The area to be wrapped was first grinded to level using hand-held electric grinding machine. After sufficiently smooth surface profile was obtained, air blowing was performed for removing the loose dust adhering to the concrete after the grinding activity. This was followed by application of structural putty. These are epoxy-based and are applied in very thin layers, just so the small pitting, minor cracks and undulations are evened out. The putty itself must be of higher compressive strength than the existing concrete strength. After putty application, primer coating is required to ensure that excessive absorption of saturant does not occur. Without primer application, the saturated FRP becomes partially dry after application, leading to ineffective bonding with the substrate.

Fig. 17 Stages of surface preparation before strengthening: **a** Surface grinding, **b** Pressurized air blowing, **c** Structural putty application for smoothening, **d** Primer application

The primed surface is coated with a layer of saturant. Since 900 GSM CFRP is applied, wet layup method was used. The saturated FRP is applied in layers using roller, as soon as the saturant applied on the substrate becomes tacky. The lengthwise (horizontal) overlap required as per calculation was 295 mm. An additional layer of saturant is applied after fully curing the FRP, so that washed silica sand could be pasted on the surface to ensure good mechanical bonding with the subsequent cement-based plastering of 20 mm thickness that was applied for protection. This was followed by curing, using wet gunny bags (Fig. 18). As it was important for the client that such fire incidents do not affect the structure again, the entire area containing the exposed columns were submerged in huge quantities of sand. Even as curing of plastering takes around 15 days, a typical column can be retrofitted from start to finish, from grinding to FRP installation and plastering, within about 3 weeks.

Fig. 18 **a** FRP Wrapping for confinement, **b** Sand pasting for adhesion, **c** Cementitious protective plastering **d** Curing using wet gunny bags for protective plastering

The advantage of using FRP for such applications is that, strengthening of each column can be done in 3–4 days using FRP and then the time-consuming activities such as plastering, can be done by general civil contractors. This project was critical and time sensitive. Just as in all other projects, detailed planning performed and the project was completed on time.

5.3 *Case study 3: Ambet river bridge*

The Ambet river bridge is a 7-span bridge constructed across Savitri River, 200 kms away from Mumbai, connecting Ambet and Mhapral. Each span is 7.5 m wide and supported on hollow concrete structures called hammerheads that are monolithically cast on the top of piers [7]. It was noticed that chunks of concrete were spalling and falling from the hammerheads (Fig. 19) on to the river below, which lead to investigations. The quality of concrete was so low, that the coarse aggregates were getting removed by hand itself (Fig. 20). Delamination survey detected several hollow sounding locations in addition to visible spalling sites with exposed steel, indicating extensive corrosion. Shear cracks were also seen. The precast girders were found to be in sound condition. Other issues such as defunct bearings, jammed expansion joints, honey combing, seepage and instability of slopes of approach roads were also observed.

The original construction was using M20 concrete and Fe 250 steel, which when modelled using FEM, was ok for the current design load requirement. NDT data suggests 30% loss of steel area and in compressive strength of concrete, for which the structure failed for design load. The joint inspection committee comprising of PWD engineers and Sanrachana, concluded that structural distresses such as shear cracks were formed due to loss of steel area and poor concrete quality. The same was getting aggravated due to ongoing corrosion, defunct bearings and jammed expansion joint, as the impact of the vehicular loads were not getting damped. In such cases, the solution must first address the root cause and only after eliminating the same, shall appropriate strengthening measures be incorporated. If FRP based measures are incorporated without addressing corrosion, the structure may continue to deteriorate without showing external warnings and proceed to fail abruptly following the sudden rupture of FRP due to expansive pressure generated through corrosion.

The following were the key targets of this project:

- Compensate for the loss of cross-section of the large diagonal bars which were majorly contributing in shear

Fig. 19 Hammerhead portion, showing corrosion damages at the end articulations

Fig. 20 Loose aggregate and exposed corroded steel

- Compensate for the rebar cross-section loss of vertical walls and bottom inclined bars, which were contributing to shear and flexural strength
- Perform detailed repair of concrete, removing all loose concrete, perform pressure grouting to improve the microstructure of concrete
- Replace the defunct bearings, make the expansion joints functional again
- Repair the pedestals and platform where bearings are installed.

For FRP based activities to be done, the first step was to prepare the concrete. Loose concrete was removed as it was not contributing to strength or stability of the structure. On some areas of the external faces, previously performed shotcrete repairs were also removed, since the steel-mesh used as its reinforcement was rusting. Cracks were opened up into V shaped grooves of 12 mm width and 25 mm depth, filled with non-expansive thixotropic putty. Drill holes of 12 mm dia were made every 250 mm running length of this groove. On the general surface of concrete also, such drills were made at about 250–300 mm c/c spacing.

The honey-combed areas were packed with the same thixotropic putty before holes were drilled. These holes were fitted with non-return valves through which M35 grade low viscosity grout was pressure injected (Fig. 21). Once the grout has cured, exposed steel was treated with alkaline rust converting primer and epoxy based anti-corrosion coating (Fig. 22). Migratory corrosion inhibitor was applied on all accessible concrete surfaces, so as to act as an additional defence mechanism against corrosion.

Fig. 21 Pressure grouting on vertical walls from inside the hammerhead

Fig. 22 Prestressed (top) and non-prestressed (side) carbon laminates, inside hammerhead

Inside the hammerhead, M50 micro-concrete with added 10 mm coarse aggregates and polymer modified mortar were used to extensively repair the concrete. On the external faces, the previously spalled or shotcreted areas were made ready for micro-concrete jacketing as mentioned earlier. For jacketing, at-least 4 shear keys were anchored into parent concrete on every square meter surface area of concrete, which were also corrosion treated. New steel reinforcements were provided as per design and shotcreting was conducted with the free-flow self-compacting micro-concrete. Although 28 days is the full-strength gain period of concrete, it was found to develop over 35 MPa strength in one week itself, hence FRP based activities could be initiated within 10 days of concreting itself. Further, surface preparation was done by application of structural putty and concrete grinding, followed by pressurized air blowing.

On the soffit of the deck portion of the interior of the hammerhead, prestressed carbon laminates of 100 mm width and 1.4 mm thickness were installed, with a nominal 50 kN prestress. In the interior vertical walls of the hammerhead, non-prestressed laminates were installed horizontally along the direction of traffic, after completion of the prestressing activity on the top. All non-prestressed members shall be installed after the installation of prestressed members, otherwise the non-prestressed members will oppose the resultant strains generated in the structure due to prestressing.

On the exterior faces, 900 GSM unidirectional CFRP wraps were provided on the cantilevering portions of the hammerhead (Fig. 23), so as to handle the shear forces and to improve durability by greatly reducing the surface cracking of concrete. The bonding behaviour of the carbon wraps were improved by inserting CFRP anchors at regular intervals. With similar effect, the ends of the carbon laminates are anchored by bolting, using machined aluminium end-plates of H30 grade in case of prestressed laminates and with H9 grade for non-prestressed laminates if required. Carbon laminates were also provided in an inclined manner on the external vertical faces of the hammerhead to mitigate diagonal tension developed due to the loads transferred through the bearings.

The defunct bearings themselves were rested on dilapidated pedestals, which after lifting of the spans, were subject to repair. The pedestals were partly demolished and recast with new reinforcement and micro-concrete, which were then laterally confined with the 900 GSM carbon fibre wraps (Fig. 24), followed by bearing replacement. All exposed FRP were given specialized UV coating after installation.

The strengthening measures taken for the hammerhead where so apt for its condition, that the structure could be transformed from one where from coarse aggregates were getting detached by hand and second layer of rebars were visible at places, to a structure that not-only transfers the load of the 40 m long spans, but also the weight of the huge steel truss, through which dynamic loads of span lifting were transmitted through a different load path! This retrofitting project has been featured in CE&CR journal and it also won their national level award for best retrofitting project in 2022 (Fig. 25).

5.4 *Case study 4: Rashtrapati Bhavan*

This is a case study on where not to use FRP based solutions.

Fig. 23 Anti-corrosion treatment completed on the existing exposed steel and newly added rebars

Fig. 24 CFRP being applied on the cantilevering portions of the hammerhead, **Inset:** Carbon anchors, installed for improving the bonding between concrete and CFRP

Fig. 25 Pedestal after micro-concreting, undergoing CFRP wrapping for additional confinement before installation of bearings

The Rashtrapati Bhavan of India was built in 1929 as the residence of the Viceroy of India. This mansion covers 5 acres on a 330-acre estate with 340 rooms and 2.5 kms of corridor (Fig. 26). Today as the residence and office of the first citizen of India, it is emblematic of the secular and inclusive tradition of the biggest democracy in the world. However, as the effects of ageing were catching up, in the form of severe corrosion of reinforcement bars in the sunshades, also known as chajjas (Fig. 27).

Fig. 26 The front portion of Rashtrapati bhavan photographed while retrofitting was underway

Fig. 27 The lime concrete chajja of Rashtrapati bhavan showing distresses due to corrosion

Made using lime mortar and steel, the chajjas were nearly 2 m wide and inclined downwards from the walls at 4 storey height, following the step-ups and step downs, curves and straight lines along the building perimeter. Upon rusting, corrosion products of significantly larger volume were formed, thereby creating expansive pressure leading to spalling of concrete (Fig. 28).

In any habitable RCC structure, spalling and subsequent falling of chunks of concrete cannot be permitted. Especially when it could be falling from 4 storey height and when it is a building of national importance, occupied by officers of top stature. Using advanced corrosion study instruments such as galvapulse, the exact electrochemical potential generated by corrosion and the concrete resistivity were mapped. Using this data, it is possible to predict the extent of corrosion even

Fig. 28 Dilapidated chajjas showing spalled areas. Temporarily, external supports were given

without making any direct connection with the embedded steel bar. It was found that corrosion was severe throughout the chajja. Also, NDT revealed that the strength of century-old lime concrete was too low for it to be strengthened using FRP as per the pre-requisites laid out by ACI 440-2R-17. In fact, the deterioration was so severe, that the only option left was to demolish and recast the chajja, but this option came with its own challenge.

The underside of the chajja was sculpted with iconic lotus petal shaped hand-made embellishments, each petal slightly different from another. Before demolition, each petal shape must be captured to scale and then recreated with accuracy. For this, 3D lidar scanning of each petal was performed. Corresponding CAD model was generated. Using this, precise 2D drawings were made. FRP based light-weight formwork was manufactured using the drawings and 3D model, with the aid of CNC machine. The CNC machine cut the exact petal shape on plywood. This created a mould for preparing FRP formwork. Using GFRP fabric, saturant and steel bars for added support, the petal shaped FRP formworks were fabricated and sent to site after curing (Fig. 29).

Since the cantilevering length of chajja was 2 m, it was necessary to perform weight reduction. Hence structural light weight concrete was used. However, the underlying issue of corrosion must be addressed first, or otherwise the same issue can recur. After chemical anchorage of sufficient new rebars and anti-corrosion treatment (without epoxy-based compounds) of old and new rebars, hybrid anodes containing in-built power source was installed as per pre-defined circuitry, to provide active cathodic protection. The circuit was such that the potential developed by the anodes can be checked anytime in the monitoring box of each circuit, with the help of the lead wire coming out of an embedded reference electrode. After setting up of the FRP formwork (Figs. 30 and 31), the high early-strength, structural light-weight concrete was cast, followed by water curing, de-shuttering, groove cutting, surface finishing by grinding and finally inorganic painting to match the original colours of the petals (Fig. 32).

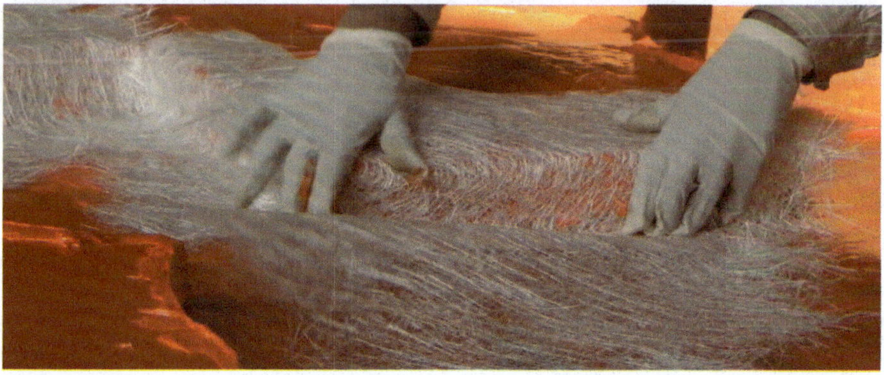

Fig. 29 GFRP fibres used to produce custom light weight formwork based on CNC cuts

Fig. 30 FRP formwork installed for light-weight concreting, after demolition of existing chajja

Fig. 31 FRP formwork installed and additional reinforcements anchored

FRP based solution may not be practical for all projects. In this project, because of the poor quality of concrete, the calculations showed that existing structure did not meet the pre-requisites, but FRP in itself is such a versatile material that in one way or the other, it could be made use of, for executing custom-built solutions such as the uniquely patterned light weight formwork.

Fig. 32 Photos taken before (inset) and after retrofitting the chajjas of Rashtrapati bhavan

6 Key Take-Aways

- FRP based strengthening has the advantage of increasing the load carrying capacity and ductility of a column, without increasing its stiffness, thereby preventing attracting additional loads onto it and behaving desirably under seismic loads.
- Strengthening measures shall not be performed before addressing the underlying structural deficiency or other root cause. Corrosion is an expansive process that may generate excessive internal pressure when not addressed before FRP wrapping, which may cause sudden failure.
- FRP application should be done only after satisfying the structural prerequisites laid down by ACI 440-2R-17
- Surface preparation and quality workmanship are of utmost importance in practical application of FRP. Manufacturer's datasheet must be followed verbatim since FRP systems constantly undergo upgradation through R&D
- FRP cannot be an answer to every strengthening requirement. At times, FRP may be an unsuitable choice. High strength alloy plates, micro-concrete jacketing, addition of structural steel members, span shortening, etc. may be relied upon as and when needed.

References

1. FIB 14 (2001) Externally bonded FRP reinforcement for RC structures. Task Group 9.3 FRP (Fibre Reinforced Polymer) reinforcement for concrete structures
2. Raina VK (2022) Bridges: Past....and....Future
3. ACI 440.2R-17 (2017) Guide for the design and construction of externally bonded FRP systems for strengthening concrete structures
4. IRICEN (2009) Inspection, assessment, repairs and retrofitting of masonry arch bridges. Indian Railways Institute of Civil Engineering, Pune, India.
5. Vivek SV (2019) Consequences of retrofitting on the response of a structure
6. Mirmiran A, Shahawy M (1997) Dilation characteristics of confined concrete. Mechanics of cohesive-frictional. Mater Int J Exp Model Comput Mater Struct 2(3):237–249
7. Ingole P, Gosavi RM, Joshi M (2022) Structural repair and strengthening of road bridge across Savitri River at Ambet, Maharashtra. Indian Highways 38–48

A Study on Ductility and Energy Absorption Capacity of Reinforced Concrete Filled UPVC Columns

P. K. Gupta and P. K. Gupta

Abstract In the present study, Unplastisized Poly Vinyl Chloride (UPVC) tubes filled with concrete are axially loaded to investigate their load carrying capacity and associated ductility. Total twelve specimens of UPVC tubes of diameters 140 and 200 mm with effective length of 900 mm were cast. M20 grade of concrete was filled inside the tubes for the casting of UPVC concrete filled tube (CFUT) column specimens. All these specimens were prepared to investigate the effect of outer diameter to thickness (D/t) ratio on the strength, confinement and ductility of columns specimens. The column specimens were tested for axial loading in the INSTRON machine of capacity 250 tonnes. Their load-deformation curves and modes of deformation were recorded. The test results indicate that as D/t ratio decreases, the confinement and ductility increases. The magnitude of the displacement corresponding to peak load increases as diameter of specimen increases. It can also be concluded from the experiments that concrete filled UPVC tubes can be used as columns.

Keywords Unplastisized Poly Vinyl Chloride (PVC) tubes · Ductility · Confinement · Strength · Buckling

1 Introduction

Concrete Filled Tube (CFT) is an example of blending two materials generally steel and concrete together to avail the compressive strength and stiffness of concrete and ductility of steel in one of the best possible ways. The CFT is prepared by filling the concrete in a steel tube. The CFT structural members such as columns and beams have a number of distinct advantages over an equivalent steel, reinforced concrete, or steel-reinforced concrete members. The orientation of the steel and concrete in the cross section optimizes the strength and stiffness of the section. The steel lies at the

P. K. Gupta (✉) · P. K. Gupta
Structural Engineering Group, Department of Civil Engineering, Indian Institute of Technology Roorkee, Roorkee 247667, Uttarakhand, India
e-mail: spramod_3@yahoo.com

S. B. Singh and C. V. R. Murty (eds.), *RC Structures Strengthened with FRP for Earthquake Resistance*, Composites Science and Technology,
https://doi.org/10.1007/978-981-97-0102-5_10

outer perimeter where it performs most effectively in compression, tension and in resisting bending moment. Also, the stiffness of the CFT is greatly enhanced because the steel, which has a much greater modulus of elasticity than the concrete, is situated farthest from the centroid, where it makes the greatest contribution to the moment of inertia. Additionally, it has been shown that the steel tube confines the concrete core, which increases the compressive strength of concrete core and consequently the axial load carrying capacity and the ductility of CFT structural members. Therefore, it is advantageous to use CFTs as columns [1] in high rise buildings where land space is one of the major concerns due to high price of land. There has been significant research on CFTs over past two decades [2–7]. As a result leading international codes have devised the design procedures and guidelines for CFTs as columns and beams [6]. In the severe environmental conditions the outer steel tube may have weathering effects and as a results land to the lower load carrying capacity of members. UPVC tube can be used as a barrier between the severe environment and the material of the structural element. Therefore, in the present paper an experimental study on Concrete Filled Unplastisized Poly Vinyl Chloride Tubes (CFUTs) as columns is presented. The Unplastisized Poly Vinyl Chloride (UPVC) tubes were used in place of conventional steel tubes. UPVC has certain distinct advantages as a material. Firstly, it is cheaper than the steel. It is a low maintenance material. It does not support combustion and is self extinguishing. It has a high electrical resistivity and has good insulation properties. It is unaffected by most concentrations of acids, alkalis and organic chemicals. It can also withstand huge deformations. Other advantages include its availability in a range of colors and finishes. Due to all these properties it qualifies as a suitable material in building construction.

2 Experimental Setup

Commercially available UPVC tubes generally used for water pipes were chosen for this study. Two diameters of tubes were chosen; they were 140 and 200 mm respectively. All the tubes had a length of 900 mm. Tubes were filled with one grade of concretes, i.e. M20 grade to make 'Concrete Filled UPVC Tubes' (CFUTs). Concrete was prepared as per the guidelines given in IS-10262: 1982, reaffirmed 2004 [8]. Locally available material and OPC 43 grade was used for making the concrete. All the test specimens of similar configuration were casted with one design mix in one go so that their physical properties had a minimum variation. Concrete cube strength of 7 and 28 days were recorded. In total 12 test specimens were prepared for the present study. Six specimens were prepared with UPVC tube having diameter of 140 mm and another six specimens with UPVC tube having diameter 200 mm. The geometrical parameters of UPVC tubes used to prepare three types of specimens are presented in Table 1. The length of all specimens was taken equal to 900 mm to keep the specimen as short column. Figure 1a shows the typical pipes used for preparation of specimens. All these test specimens had their length/diameter (L/D) ratio between 4.5 and 6.4 so that effects of slenderness were not prominent and therefore the effects of confinement and resultant increase in strength could be studied. Their diameter/wall thickness (D/t) ratios were between 35 and 42. All the twelve specimens were prepared with concrete infill of M20 grade and had six TMT bars of 12 mm diameter as longitudinal reinforcement in addition to these two hoop one at top and other at bottom were also welded in the reinforcement (see Fig. 1b). Details of the test specimens are shown in Fig. 1a–f. Figure 1d shows the specimens sealed with polyethylene sheets to provide curing. In order to transfer the axial load without any eccentricity both top and bottom surfaces were prepared using epoxy (Fig. 1e). Figure 1f shows the cross-section of the prepared specimens ready for testing. All the six reinforcement bars are clearly visible in the cross-section. Testing of the specimens was carried out in two ways: first by transferring the load through both concrete and UPVC pipe i.e. both loaded, and second only through concrete core i.e. concrete loaded. For second type of loading i.e. concrete loaded, special steel plates having diameters equal to the inner diameter of UPVC pipes were used. Figure 2a shows these plates. In each diameter specimens two specimens were cured in artificial sea water to study the effect of sea water on their mechanical behavior and tested as concrete loaded. The salient properties of these test specimens are given in Table 1.

The specimens were named on the basis of their diameters, compressive strength of concrete, way of testing and type of curing. In specimen T140M20B-1, 1 signifies the first specimen, T140 signifies external tube diameter of 140 mm, M20 signifies filled with M20 grade of concrete cured with polyethylene sheet and B signifies tested with transfer of load through both tube and concrete.

In specimen T140M20C-2, 2 signifies the second specimen, T140 signifies external tube diameter of 140 mm, M20 signifies filled with M20 grade of concrete cured with polyethylene sheet and B signifies tested with transfer of load through only concrete using the steel plates shown in Fig. 2a.

In specimen D200M20CC-2, 2 signifies the second specimen, D200 signifies external tube diameter of 200 mm, M20 signifies filled with M20 grade of concrete cured with putting in sea water and CC signifies tested with transfer of load through only concrete using the steel plates shown in Fig. 2a.

These specimens were tested on 250 ton capacity UTM available at the Structural Engineering Laboratory of Civil Engineering Department of IIT Roorkee. Load-compression curves were recorded with automatic recorder of the machine. The testing was carried out at 1 mm/m speed of cross head. Figure 2b shows the machine and specimen during testing. In general the specimens failed due development of one

Table 1 Details of UPVC tubes

S. no.	Specimen	Nominal external Dia. (mm)	Mean thickness (t) (mm)	Inner Dia. (mm)	Outer Dia. (D) (mm)	Cross-sectional area (mm^2)	D/t ratio	L/D ratio	Length (L) (mm)
1	T140	140	3.86	134.0	140.39	1377.0	36.3	6.4	900
2	T200	200	5.94	188.0	200.17	3710.2	33.7	4.5	900

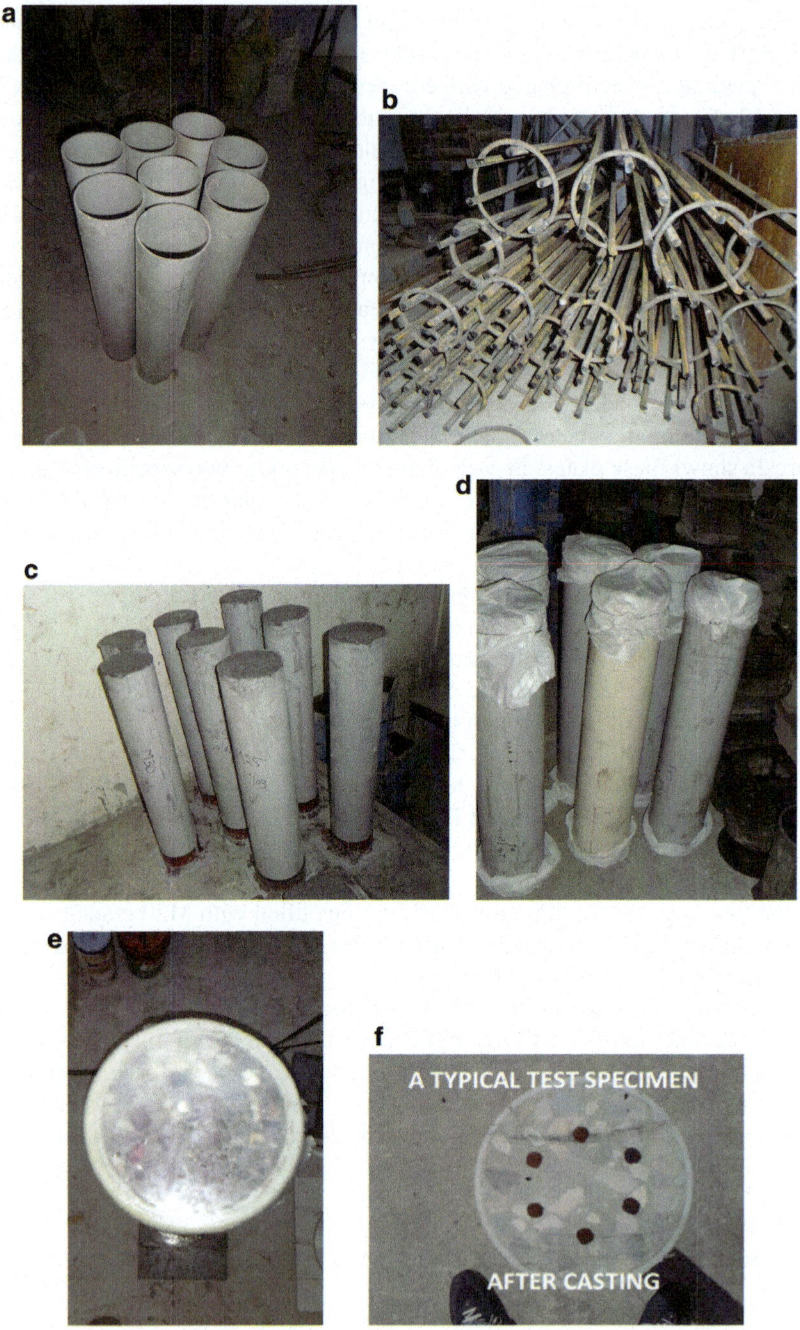

Fig. 1 **a** Unfilled UPVC pipes. **b** Reinforcement for UPVC columns. **c** UPVC pipes filled with concrete. **d** UPVC pipe columns sealed against water loss. **e** Epoxy capping on the test. **f** Cross-section of prepared specimen

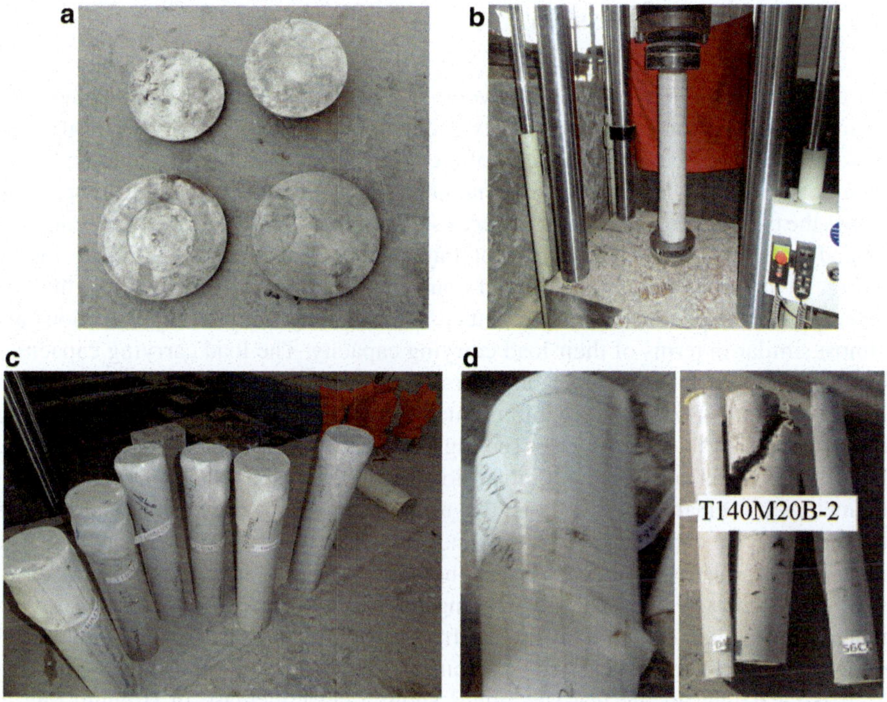

Fig. 2 **a** Steel plates for loading only the concrete core in a CFUT. **b** UPVC pipe columns during testing (only concrete loaded). **c** Typical mode of collapse UPVC pipe columns. **d** Visible shear crack in mode of deformation

shear crack (see Fig. 2c). Figure 2d shows the picture of a typical specimen failed due to development of shear crack.

Two different methods were employed to test the specimens. Four specimens were tested by loading both concrete and UPVC tube and remaining eight were tested by transfer of load through only concrete using the steel plates shown in Fig. 2a.

Eight specimens were tested by loading only concrete core in a CFUT. Remaining four specimens were tested as pure concrete column so that the comparative data is available to judge the increase in strength due to confinement. For loading both concrete and UPVC tubes simultaneously, capping was done on the test specimens using epoxy resin (see Fig. 1e). To load only the concrete steel plates of inner tube diameters were prepared. These details are shown in Fig. 2a.

3 Results and Discussion

Two types of specimens and in total twelve specimens were tested with two methods to investigate the load carrying capacity and their behaviour under axial loading. The Results of the load carrying capacity of the different specimens were obtained from their corresponding load–displacement curves obtained during testing. Figure 3a–b shows the typical variation of load-compression curves for all specimens i.e. concrete loaded and both loaded. The corresponding energy-displacement curves are shown in Fig. 4a–b. After comparison of load-compression curves in Fig. 3 it clear that the behavior of the specimens; in different cases of loadings and curing conditions are almost similar in terms of their load carrying capacity. The load carrying capacities of all specimens having same diameters are almost comparable. It is clear from Fig. 3a–b that UPVC CFTs exhibits ductile behaviour during axial collapse and absorbs considerable energy also (see Fig. 4a–b). Figure 5 shows the comparison of load-compression and energy-compression variations of specimens having different diameters. It is clear from this figure that as the diameter increases and D/t decreases the load carrying capacity and ductility increases due to increase in confinement. The magnitude of the displacement corresponding to peak load increases as diameter of specimen increases. The energy absorbing capacity of specimen also increases with increase in its diameter. It is also clear from Figs. 3 and 4 that curing in sea water does not hamper the load carrying capacity of CFUTs.

It is clear that all the test specimens show a clear increase in strength due to confinement resulting from the use of UPVC tubes. It was seen that the mode of failure was diagonal shear of the concrete core. This was ascertained by cutting the test specimen's outer UPVC cover after the failure of test specimen (see Fig. 2d).

4 Comparison of Load Carrying Capacity of the Test Specimens with Standard Codes of Practice

Results for the load carrying capacity of the CFUT specimens as their ultimate strength (P_u) were obtained from the corresponding load–compression curves in experiments. These values are tabulated in Tables 2. The theoretical values (P_{the}) of load carrying capacity of different concrete filled tubes were predicted using six Internationals codes as per their recommendations and listed in Table 2. To validate the results obtained by the experimental program, they were compared with six standard codes of practice, whereby the predicted strength by these codes was compared with the actual strength of test specimens obtained by the experiments conducted. The comparison results are given in Table 2. Following codes were used for comparison:-

(a) EN 1994-1-1 (Eurocode 4) [9].
(b) ANSI/AISC 360-05 [10].
(c) ACI-318.

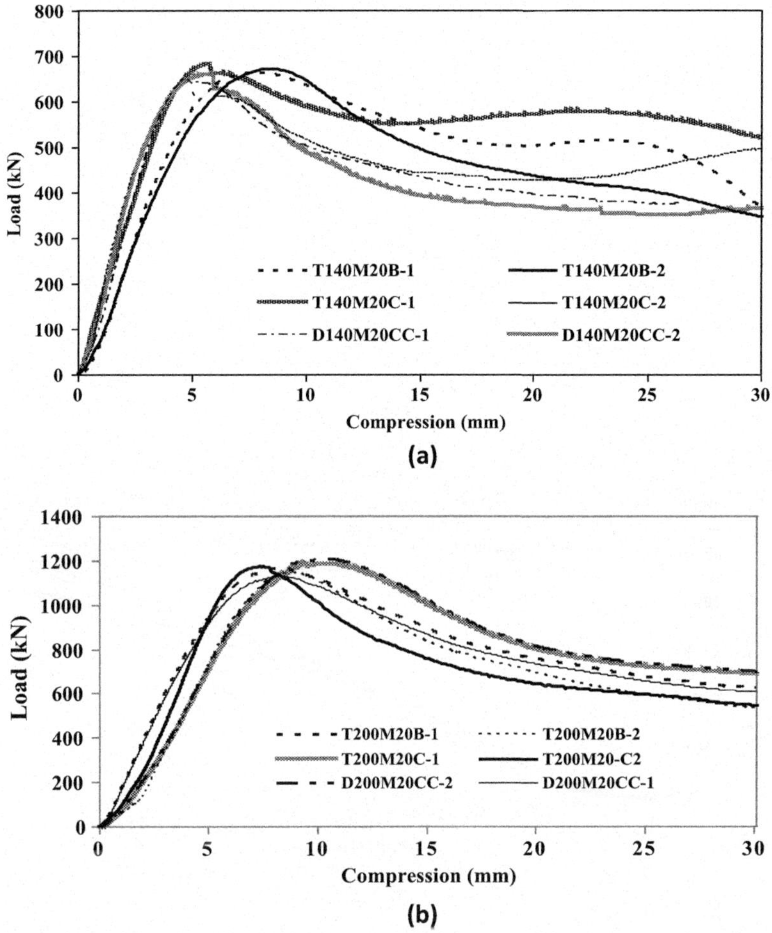

Fig. 3 a–b Comparison of load-compression curves of different specimens tested with different methods **a** diameter = 140 mm, **b** diameter = 200 mm

(d) NBR-8800.
(e) CAN/CSA-S16.1 M 94 [11].
(f) Egyptian Code of Practice.

The data for mechanical properties was as obtained by testing of the tube sample. The other data was taken from the standard UPVC tube manufacturers' websites. Therefore the data used for calculations was as under.

(a) Ultimate tensile strength of UPVC tube = 42.22 MPa
(b) Young's Modulus = 2750.
(c) Compressive strength = 20 MPa.
(d) Shear strength = 39 MPa.

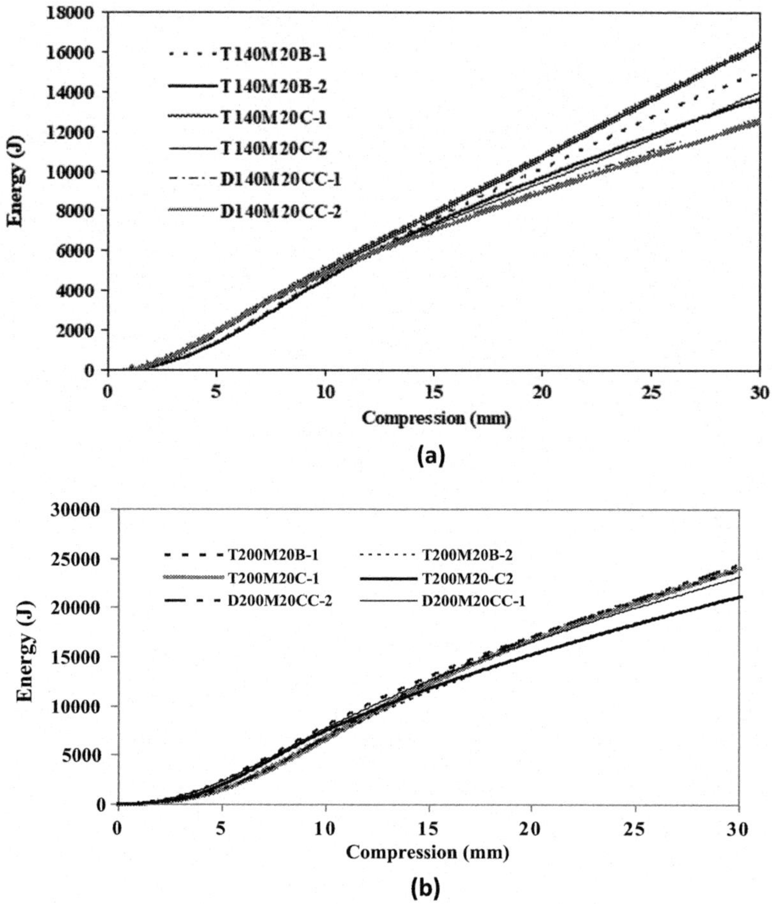

Fig. 4 **a-b** Comparison of energy-compression curves of different specimens tested with different methods **a** diameter = 140 mm, **b** diameter = 200 mm

(e) Poisson's ratio = 0.35–0.38 (Higher value was taken for calculations, where applicable).
(f) Concrete compressive strength (f_{ck}) was taken as 70% of the 28 day cube strength. This was based on the data obtained from testing the pure concrete cores.

To have a balanced comparison only pure CFUTs or the CFUTs having only concrete infill were compared so that the effect of only UPVC tube in generating the confinement could be compared with standard international codes of practice.

It is evident from this table that EC4 and AISC-LRFD give the best estimation of strength for the CFUTs followed by NBR 8800 and ACI. Hence, the use of these standard codes can also be extended to design and strength predictions for CFUTs. However, further studies are recommended on the subject to come out with suitable

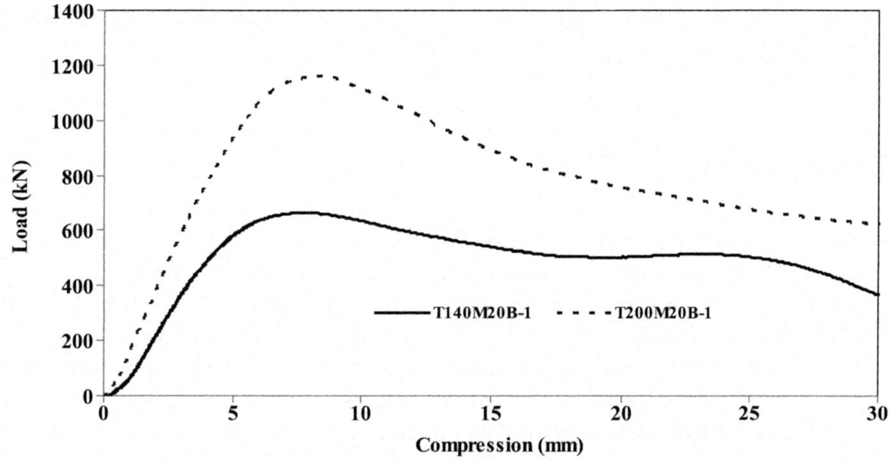

(a)

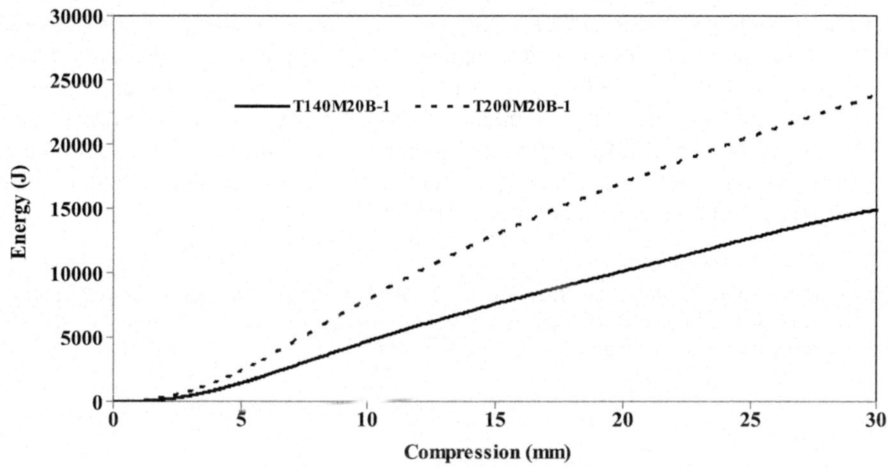

(b)

Fig. 5 a–b Comparison of **a** load-compression and **b** energy-compression curves of specimens having different diameters filled with same grade of concrete tested with both loaded methods

Table 2 Comparison of load carrying capacity (kN) of different specimens

Sp. no.	Type of strength	Exp	EC4	AISC (LRFD)	ACI 318	NBR 8800	Egyptian code	CAN/ CSA-S16.1 M94
T200M20	Ultimate	1150–1180	1058.93	1001.96	933.53	1010.08	238.97	687.10
	Design		785.02	751.47	466.77	795.66	137.62	445.13
T140M20	Ultimate	627–672	669.37	644.35	545.91	657.71	347.91	312.48
	Design		501.95	483.26	272.96	518.04	198.69	201.82

coefficients in these codes' equations for even better strength prediction for CFTs using UPVC tubes in place of conventional steel tubes.

5 Conclusion

- Present study focused on use of an alternate material for steel in conventional CFTs. The purpose of study was to use UPVC due to its certain inherent advantages, i.e. good weather resistance, good chemical resistance and high die-electric strength to provide high electrical resistivity. It also has good insulation properties. It is unaffected by most concentrations of acids, alkalis and organic chemicals. It can also withstand huge deformations. Purpose of the present study was to see whether a UPVC tube can give the comparable increase in strength of concrete core due to confinement vis a vis the steel tube. It is evident from the results that the UPVC tube qualifies as a replacement material for steel tube in CFTs with an addition of other advantages offered by UPVC as listed above.
- The strength prediction in CFUT can be done by standard codes of practice used for the design of CFTs by various nations as is brought out by this study which makes the studies on CFUTs even simpler. With present studies it emerges that CFUTs can be used as column members. It is also suggested that further studies be carried out on CFUTs as structural members. The other variants of PVC such as MPVC and OPVC can also be tried for their suitability in place of UPVC since they have certain other advantages over UPVC.

Acknowledgements Authors are thankful to Building Materials and Technology Promotion Council (BMTPC), Ministry of Housing and Urban Poverty Alleviation, Government of India for providing the financial support for this research work.

References

1. Oehlers DJ, Bradford MA (2004) Composite steel and concrete structural members. Elsevier J. Oxford, Langfordlane, Kindlington, UK
2. Nardin SD (2007) Debs ALHCE. Axial load behaviour of concrete filled steel tubular columns. In: Proceedings of the institution of civil engineers, structures and buildings. SBI160:13–22
3. Giakoumelis G, Lam D (2004) Axial capacity of circular concrete filled tube columns. J Constr Steel Res 60:1049–1068
4. Gupta PK, Sarda SM, Kumar MS (2007) Experimental and computational study of concrete filled steel tubular columns under axial loads. J Constr Steel Res 63:182–193
5. Singh YC (2008) Studies in steel encased RC columns. M. Tech. Dissertation. Department of Civil Engineering, Indian Institute of Technology, Roorkee
6. Zhang J (2004) Design of composite columns. Seminar on steel structures. Department of Civil and Environmental Engineering, Helsinki University of Technology, Finland
7. Raghuvanshi H, Studies on concrete filled circular PVC tubes as columns. M. Tech. Dissertation

8. Department of Civil Engineering, Indian Institute of Technology, Roorkee (2010)
9. IS:10262–1982 (reaffirmed 2004) (2004) Recommended Guidelines for concrete mix design; BIS, New Delhi
10. Eurocode 4: Design of steel and concrete structures (1996) Part 1.1: general rules and rules for buildings. Brussels (Belgium): European Committee for Standardization
11. ANSI/AISC 360–05 (2007) Specification for structural steel buildings. American Institute of Steel Construction Inc., third print
12. CAN/CSA-S16.1 M 94 (2002) Canadian code of practice for composite design

Seismic and Fire Behaviour of FRP Strengthened Reinforced High Strength Concrete Structures—An Overview

Sanket Rawat, Rahul Narula, Prachuryya Kaushik, Divya Prakash Jain, Nitant Upasani, Ashirbad Satapathy, Mansi Bansal, Harish Mulchandani, Shreyas Pranav, and G. Muthukumar

Abstract The development of Fibre Reinforced Polymer (FRP) can be traced to the expanded use of composites after the World War II in the early 1940s, though the use of FRP was considered seriously for use as reinforced concrete until 1960s. Fibre Reinforced Polymers are well recognized as an effective seismic retrofit/fire resistant material for existing concrete buildings. This strengthening domain in civil engineering, a critical part of overall lifecycle aspect of any infrastructure, is more than two decades old and several successful projects have been installed using FRP as reported in several literatures. Many of these retrofitted buildings have experienced significant earthquakes and performed as designed, validating the effectiveness of the FRP and technology. Extensive laboratory testing and actual earthquakes have led to the growth of dependable design methodologies and guidelines for FRP to be used by the engineering fraternity. FRP materials have a high strength-to-weight ratio, which make them a perfect material for seismic retrofit. Although they do not add significant mass to a structure, they certainly enhance the capacity of various structural components. This also avoids the mandate of performing the analysis again without appreciable weight change and further consequential effect on foundation after due strengthening. FRP possesses innate characteristics to deal with fire and heated environment. This article is an attempt to highlight some of the features of FRP Reinforced High-Strength Concrete Structures from both Seismic and Fire viewpoints.

Keywords Concrete · Ductility · Design · Materials · Seismic performance · Fire · FRP

S. Rawat · R. Narula · P. Kaushik · D. P. Jain · N. Upasani · A. Satapathy · M. Bansal · H. Mulchandani · S. Pranav · G. Muthukumar (✉)
Department of Civil Engineering, BITS Pilani, Pilani Campus, Pilani 333 031, Rajasthan, India
e-mail: muthug@pilani.bits-pilani.ac.in

S. B. Singh and C. V. R. Murty (eds.), *RC Structures Strengthened with FRP for Earthquake Resistance*, Composites Science and Technology,
https://doi.org/10.1007/978-981-97-0102-5_11

1 Background and Introduction

The need for new materials and new versions of existing materials has been growing at an accelerated rate. Out of all characteristics, the most sought after and indispensable material characteristics in the construction industry are high workability and high strength, as far as concrete is concerned. Also, the growing need of tall buildings requires that the structural designs are optimized to perfection (no more and no less), thereby instilling the need for research in search of high-performance concrete. Consequently, new versions of concrete such as High Strength Concrete, High-Performance Concrete, and Ultra-High-Performance concrete have emerged. The primary reason behind the application of such performance concrete is to build structures with minimum material use and maximum performance level. Under severe earthquake loading or fire, ductility is highly desirable and is focussed on profoundly during the design stage itself.

Failures due to severe heating environment: It is reported in literature that about 25,000 persons die due to fires and related causes in India. According to literature, about 45% of the claims are due to fire losses and about Rs. 1000 crores are lost every year due to fire. Electric defaults are the one of the primary reasons in initiating fire. Many a times, beams with openings meant to carry electric lines and AC ducts are susceptible to such failure. In terms of the potential risk, 'Fire' has been ranked fifth and according to National Crime Record Bureau (NCRB), a total of 18,450 cases of fire accidents were reported in India in 2015, with 1,193 persons injured and 17,700 killed during the disaster. After the 9–11 attack on the World Trade Centre, interest in the design of structures for fire has greatly increased.

However, in May 2018, a 26-storey high-rise building has collapsed in Brazil in 90 min as building was unable to resist the severe heating environment. In the year 2017, a two-storeyed building in West Delhi collapsed after fire supposedly due to faulty wiring in the air-conditioning system. In June 2018, major fire broke in G + 5 building and subsequently building collapses. Incidentally, this building was already considered dangerous and such un-demolished building poses severe danger for the pedestrians who are passing nearby the dangerous buildings. The schools where mid-day meals are cooked are potential locations of fire hazards. In commercial establishments, major fires start in storage area and warehouses than production areas. The threat posed by fire following an earthquake has been highlighted by the occurrence of a number of past earthquakes [27], notably those witnessed in San Francisco (1906), Northridge (1994), Los Angeles (1994), and Kobe, Japan (1995).

Incidence of a fire may be pointedly riskier under post-earthquake circumstances rather than under usual conditions and thereby creating a greater need for Ductile design. Ductile detailing requires good amount of reinforcing bar to create good confining effect but has the potential to naturally creating congestion in the structural member for concreting to take place. Under such circumstances, the need for high workability and flowability of concrete using high-range water reducers is mandated to meet the construction requirement.

Concrete is widely appreciated in the construction industry. Indeed, it is the most consumed material in the world after water but its major disadvantage is its brittle nature, which is attributed to its poor resistance to crack formation, low tensile strength, and strain capacities. In the last two decades, the construction of high-rise buildings using high-strength concrete has grown significantly in many residential and commercial sectors due to paucity of land. Though HSC is preferred from strength and durability point of view in ambient temperature conditions, normal strength concrete exhibits good performance under fire situations. However, the use of high-strength concrete in RC members cannot be avoided as it keeps the structural size to the optimum level. On the other hand, with the introduction of high strength concrete, structural members have become very thin and hence less fire resistant. The concrete's physical, chemical and mechanical properties do undergo extraordinary modifications when subjected to elevated temperatures and a considerable loss in strength occurs when concrete is heated above 300 °C [16]. Instead of conventional steel reinforcement, this article attempts to highlight the features of FRP reinforced high-strength concrete structures.

2 Firm Choice of Comfort in FRP

The retrofitting of existing concrete structures to counteract high design loads, correct strength loss due to deterioration, correct design or construction deficiencies, or increased ductility has traditionally been accomplished by the use of established construction materials and techniques. Steel plates with external bonding, use of steel/concrete jackets, and external post-tensioning are a few such techniques currently being deployed. Orthodox methods of seismic assessment and design of buildings with RC walls were challenged after the peculiar failure modes observed in RC structural walls in the 2010 Chile and February 2011 New Zealand (NZ) earthquakes. While some of the existing methods are also good at improving the specific building performance, these methods lack overall resilience in terms of completing the task with minimum disturbance in day-to-day operations. Composites made of fibres in a polymeric resin, also known as fibre-reinforced polymers (FRPs), have been developed as an alternative to conventional materials for repair and rehabilitation (ACI 440–2R3). In the last decade, the acceptance of the usage of FRP as a structural repair is beyond doubt.

3 Few Inherent Capabilities of FRP (As Per ACI–440)

- The greater interest in strengthening of RC columns by external wrapping since 1980s have led to the investigation in the direction towards the confinement of concrete with FRP laminates.

- Ductility has also been a concern where FRP is primarily used. It was highlighted that the strength gain is often complimented by a reduced deflection at failure (ductility). Nevertheless, the proper design is extremely essential to ensure that the steel reinforcement in the beam yield before failure, giving advanced cautioning prior to failure of the beam.
- The fact that shear strength in concrete is resisted by diagonal tension means that the FRP can be used as shear reinforcement. The design guidelines with respect to limiting low value of the allowable strain in the FRP for shear ensures that shear crack does not become too wide and thereby ensuring the keeping intact of aggregate interlock.
- Complete wrapping of the FRP laminate around the section is the most efficient wrapping technique and is most frequently used in column applications where all four sides of the column can generally be accessed. It is impractical to completely wrap beam members due to the presence of an integral slab; therefore, the shear strength can be enhanced by wrapping the FRP laminate on three sides of the member (U-wrap) or pasting on two opposite sides of the member.
- The existing method of formulating FRP RC design guidelines by adjusting conventional RC guidelines may seem reasonable, but it is not fully appropriate. The basis of this statement is that conventional RC design guidelines assume that the principal mode of failure is always ductile, due to flexural reinforcement yielding. However, the same cannot be assumed for FRP RC; FRP RC guidelines assume that brittle failure would be sustained because of either concrete crushing or FRP reinforcement rupture.
- FRP bars are anisotropic in nature and can be manufactured using a variety of techniques such as pultrusion, braiding, and weaving. The characteristics of FRP bar are largely influenced by fiber volume, fiber type, resin type, fiber orientation, dimensional effects, and quality control during manufacturing. Hence, it is suggested that manufacturer's material data should be consulted for specific product properties.

4 Literature Review—Fire Resistance

An analytical model, developed by [10], based on extensive experimental investigation of the structural response of beams subjected to the elevated temperature simulating the fire loading indicated that the significant shear cracks were developed much early near the continuous support, thus highlighting the importance of considering fire in the design of concrete structures. However, the concrete reported in the study was of normal-weight concrete.

Zhang et al. [35] investigated the performance of retrofitted RC deep beams using Carbon Fibre Reinforced Polymer (CFRP) laminates and observed the significant improvement in ductility and shear strength.

Evaluation of the fire endurance of FRP reinforced concrete slabs by [17] highlighted the factors to be considered for the fire resistant design. The type of reinforcement, concrete cover and type of aggregate were the important factors reported in the study. The differences in the critical temperature of the steel reinforcement and FRP reinforcement were also reported in this paper. It was also highlighted that the aggregate type has moderate influence on the fire resistance of FRP reinforced concrete slabs. It was found that concrete made of carbonate aggregate has 10% greater fire resistance as compared to siliceous aggregates; this was seen in terms of the degree of temperature rise in specimen concrete slabs. Moreover, with higher concrete cover, the higher fire resistance can be obtained as advocated in the study and hence spalling characteristics are considered to be important.

In the past, engineers have endorsed the use of advanced analytical models to determine fire growth within a compartment using the finite element models of structural components to evaluate the temperatures of many structural elements within a component by heat transfer analysis [6]. The literature advocates that the input heat, thermal expansion, degradation in strength and stiffness of materials at elevated temperatures are required to be factored into the design of building components [9].

Another work highlighting the influence of reinforcement type on fire resistance was reported in [20] and it was widely considered that FRP reinforced concrete elements have lower fire resistances than conventional steel reinforcement due to the sensitivity of FRP to the change in tensile and bond strength under severe elevated temperature conditions. Kabay (2014) reported that the addition of basalt fibres in normal as well as in high strength concrete can lead to massive improvement in flexural strength (9–13%), fracture energy (126–140%) and reduction in abrasion wear (2–18%). However, it may also be noted that the compressive strength of concrete generally decreases on addition of basalt fibres.

It was also reported in [24] that at elevated temperatures, the moisture evaporation takes place and moisture transportation in concrete has been reported to be a complex phenomenon, not essentially controlled by temperature, pore pressure and vapor content

Zaidi et al. [33] highlighted the influence of FRP properties on the numerical deformations in FRP bars-RC elements in heated temperature and observed that circumferential thermal deformations increase, profoundly with the increase of elasticity modulus in the transverse direction of concrete especially in high temperature zone. It was also highlighted that the concrete cover thickness has no big effect on the circumferential thermal deformations of FRP bar-reinforced concrete cylinders.

The detailed study was reported by [5] by investigating the influence of cracking on the heat propagation in RC structures (beams, columns and frames) experimentally and analytically and predicted that the cracked regions are susceptible to the increase in the rate of heat propagation. It was concluded that the temperature distribution needs to be different for different nature of sections in order to arrive at the safe design.

The demand for high-strength concrete (HSC) has increased in recent years with the increase of modern construction projects (e.g., long-span bridges, super high-rise buildings, offshore platforms and underground structures). It is well understood that improving the performance of concrete can enhance the lifespan of concrete structures (Wang et al. 2017).

The growing global population and the enormous economic development in areas prone to calamities have amplified the chance of several catastrophic incidents, which lead to disruption of buildings and infrastructures. Marasco et al. [21] discussed the importance of hazard analysis of hospital building in an effort to improve structural safety and resiliency and also to reduce the building life-cycle costs.

In the recent past, it was reported that development of Steel-Carbon FRP Composite Bars (SCFCB) actually makes up for the lack of ductility of FRP rebars to certain extent and thereby reducing engineering costs [30]. The results also indicated that under the same conditions of reinforcement ratios, the SCFCB-reinforced beams exhibit better performance than CFRP-reinforced beams, and stiffness is slightly lower than that of steel-reinforced beams.

Tariq and Bhargava [28] highlighted the importance of dealing with aggressive corrosive environment and subjected to accidental fire. It was highlighted that accidental fires can cause irreparable damage to the construction. The load–deflection characteristics of structural elements were studied under the influence of corrosion-temperature interaction.

Gedam [13] proposed the performance-based fire resistance design method for evaluating the flexural carrying capacity of reinforced concrete beams. It was observed that the developed method is capable of predicting fire-resistance rating of RC beams. The type of aggregate plays a crucial role in the flexural carrying capacity at fire lad.

Cao and Nguyen [7] highlighted the importance of flexural performance of post-fire Reinforced Concrete Beams as a part of retrofitting strategy. It was observed that the ductility of post-fire specimens was decreased by up to 61.1%. Apart from ductility, yield strength, yield stiffness and yield deflection were also affected.

5 Literature Review—Earthquake Resistance

Erkmen and Saatcioglu [11] studied the seismic performance of carbon fiber reinforced polymer concrete frame buildings. It was found that upon applying cyclic dynamic loading, seismic force demand can be lesser than that computed through equivalent static load analysis using experimentally determined period values. This highlights the significance of FRP as a material as a choice for earthquake resistance as well.

Eslami and Ronagh [12] performed a numerical examination of the enhancement in seismic performance of RC buildings having glass and carbon FRP wrapping: both code-complying and poorly-confined reinforced concrete buildings were considered. Non-linear analysis showed that the GFRP and CFRP wrappings are incapable of enhancing the ductility of code-complying structures, but they significantly enhanced the ground motion resistance of poorly-confined buildings.

A study reported by [27] highlighted the influence of ductile detailing on the performance of RC building frame subjected to earthquake and fire and concluded that the ductile detailing was found essential in arresting the fire damage and the extensive spalling. A full-scale fire test was conducted on an already damaged RC frame. A non-ductile detailing resulted in the higher temperatures and thermal damages in the RC frame and its constituent elements. It was also highlighted that the threat posed by post-earthquake fire scenario such as San Francisco (1906), Northridge, Los Angeles (1994), and Kobe (1995) resulted in massive failure of buildings and loss of human lives. Until the mid-1970s, the engineering community knew little about how to design structures to resist earthquake loads safely. As a result, the entire global inventory of buildings and bridges that were constructed prior to those dates are unsafe for resisting earthquake loads. This is evident by the large number of older buildings that have collapsed or were severely damaged in recent earthquakes.

Salem and Issa [26] performed non-linear finite element analysis of high and ultra-high strength concrete beams reinforced with FRP bars and observed that concrete strength has a small effect on the ultimate capacity. However, the deflection is higher for beams with lesser strength. It was also observed that CFRP reinforced beams showed higher capacity and lesser ductility, while GFRP reinforced beams showed lesser capacity and higher ductility.

Del Zoppo et al. [8] observed from his experience on earthquakes and also through experimental analysis that use of external reinforcement systems, such as CFRP, is herein experimentally investigated as a reliable method for enhancing the capacity of short RC members and also helps in preventing the undesirable brittle shear failure. It was also highlighted that L'Aquila earthquake (2009) confirmed the high incidence of columns brittle failure due to shear, especially in case of reduced shear length (i.e. band-type windows, semi-basement, etc.). 56 out of 284 heavily damaged RC buildings were susceptible to shear failure of columns. Also, shear failures were also observed in slender columns for those structures designed using "first-generation" seismic codes, non-conforming to present day requirements.

Lee et al. [18] proposed a novel sprayed FRP strengthening technique for RC columns. Different permutations and combinations of chopped glass and carbon fibers with epoxy/vinyl ester resin were gushed onto the RC columns and dynamic loading was applied. The shear strength and deflection capacity of the column increased pointedly upon the usage of FRP; the most optimum mix was found to be the one having glass and carbon fibers mixed with resin in a 1:2 proportion. The shear strength of the FRP-strengthened specimens, on average, was found to be 31% greater than the control columns.

Zeng et al. [34] studied the seismic behavior of basalt **F**ibre reinforced Polymer-**R**ecycled Concrete Aggregate-**S**teel **C**olumns (FRSCs) having shear connectors. FRSCs are capable of withstanding significant lateral loads, have very high ductility. The presence of shear connectors on the inner steel tube was observed to be instrumental in opposing the buckling of steel, and enhancing the lateral load capacity.

Mincigrucci et al. [23] conducted a comparative study between GFRP, RC and steel frames subjected to seismic loading. Using GFRP decreased the base shear by approximately 40% in comparison to steel and by 88.5% in comparison to reinforced concrete. The Von Mises equivalent stress time histories reveal that the FRP frames exhibit more regularity in behaviour than steel and RC. However, the FRP frames show a less uniform stress distribution than steel and RC.

6 Literature Review—High Strength Concrete

ASCE [3] defines sustainability as a combination of economic, environmental, and social circumstances in the society has the capability and possibility to maintain and enhance its quality of life for the upcoming generations without deteriorating the quantity, quality, or accessibility of resources. Even though the embodied energy of concrete is among the lowest compared to other engineering materials [4], the concrete industry is still one of the significant industrial pollutants. The cement industry alone is accountable for approximately 5–8% of the world's CO_2 emissions [22]. It is also to be noted that most materials used in construction are still natural and non-renewable.

Sustainable Ultra High-Performance Concrete (UHPC) is defined as that UHPC whose fresh-state and hardened-state properties can be tailored to meet design specifications including sustainability [31]. The partial utilization of agricultural and industrial waste as alternatives for cement, aggregate, and rebar materials in the manufacture of sustainable UHPC can be the solution. ACI 239R (ACI 2018) defines UHPC as a fiber-reinforced concrete that has a compressive strength of at least 150 MPa with specific durability, ductility, and toughness requirements. Also, according to ACI 363R (ACI 2005), high-strength concrete is concrete having a compressive strength of 55 MPa or greater.

Though UHPC is having many benefits in terms of certain characteristics such as impermeable, ductility, energy absorption, it may also potentially cause shrinkage problems and can be very fragile without any steel fibre [29]. There is a great need to reduce the cost of such HPC by looking for materials that not only satisfy technical needs, but also satisfy the budgetary constraints. UHPC certainly has teething issues and needs further refinement before being put into practice in masses.

7 Literature Review—Basalt Fibre—Sustainable Construction Material of the Future

Basalt fibre was originally developed in the Soviet Union during the 1960–1980s. Basalt fiber is a material made from extremely fine fibers of basalt, which is composed of the minerals plagioclase, pyroxene, and olivine. Basalt is usually black or dark gray

Fig. 1 Chopped basalt fibers

and relatively featureless. It is composed of mineral grains which are mostly indistinguishable to the naked eye. Basalt may also contain volcanic glass (Fig. 1) (based on Zhang et al. [36]).

Basalt fiber is an inorganic material produced from volcanic rock called basalt. The production of basalt fibers does not create any environmental waste and it is nontoxic and biodegradable. Basalt fiber has been considered to be environmentally safe and noncorrosive with good insulating characteristics and thermal endurance. It is stable in all aggressive environments, does not conduct electricity, and possesses a high tensile strength. It is believed that basalt fibre-reinforced concrete (BFRC) takes the pole position in the construction industry because it is cheaper, greener, lighter, and eliminates the problem of corrosion of reinforcement bars and corrosion-led damages in concrete structures. Currently, basalt chopped fibres are available in various lengths ranging from 12 to 100 mm and various diameters ranging from 10 to 20 μm.

The use of high-performing materials (HPM) such as high strength concrete (HSC) and fibre-reinforced polymers (FRP) have sneaked their way into the construction world in the recent years. The reason for the emergence of FRP is primarily from the durability (non-corrosive) point of view. Carbon and Glass fibres are two of such materials which have gained significant attention due to their exceptional mechanical performance and durability. However, their use is still a concern from environment perspective as incineration of discarded fibres generates plenty of smoke and unhealthy odours. To overcome such environmentally hazardous concerns, basalt has emerged as a suitable type of fibre, which has found multiple applications in many fields. In 1990s, Basalt found its way in multiple applications of Civil Engineering field and is now recognized as 'The twenty-first century non-polluting green material'. The use of continuous basalt fibre, considered to be the potential fire resistant material, has already been explored as an alternative to the steel reinforcement in achieving the desired strength and ductility characteristics under very severe temperature conditions.

Table 1 Chemical composition of basalt fibre

Name of the chemical compound	Percentage
SiO_2	49.58
TiO_2	2.08
Al_2O_3	14.48
Fe_2O_3	4.42
FeO	9.43
MnO	0.17
MgO	5.10
CaO	8.50
Na_2O	1.89
K_2O	1.12
P_2O_5	0.35

Basalt is an igneous rock that contains more than 45 and less than 52% of SiO_2 and less than five percent of total Alkalies ($K_2O + Na_2O$) (Table 1) (based on Yuvaraj et al. [32]).

Yoder and Tilley (1962) developed Basic Tetrahedron model (Dhand et al. 2015) for the classification of minerals that constitute basalt rocks. The four corners of the tetrahedron represented the four major constitutive minerals i.e. Forsterite (Olivine), Diopside (Pyroxene), Nepheline (Feldspathoid) and Quartz. Chemically, basalt rocks cover almost 70% of the earth's surface and mainly contain Silica (SiO_2), Alumina (Al_2O_3), Ferrous Oxide (FeO), Calcium Oxide (CaO), Magnesium Oxide (MgO) etc. These rocks are classified as alkaline, mild-acidic and acidic basalts according to the variation in the SiO_2 content. Generally, only acidic basalt (SiO_2 content > 46%) qualifies as suitable for the preparation of basalt fibre, as high silica content is required to develop glass network. Moreover, SiO_2 is also responsible for strength, hardness and thermal characteristics of basalt. The brittleness of basalt is due to the crystalline nature of SiO_2. Al_2O_3 and FeO, on the other hand, are responsible for imparting characteristics like wear resistance, alkali and acid resistance, thermal stability, stiffness etc. [25]. The brown colour of basalt fiber is due to the presence of FeO [19]. CaO provides strength and bonding nature to basalt. MgO is responsible for Its chemical stability at elevated temperatures and moisture resistance.

8 Manufacturing of Basalt Fibres

Basalt rock is principally composed of silica, alumina, with lime, magnesium oxide and ferric oxide found in lesser percentages. For fabrication of continuous basalt fibre (CBF), the quantity of each material needs to be controlled. Mineralogically speaking, basalt is primarily constituted of plagioclase, pyroxene and olivine. To create basalt fibre, the basalt rock is mined and crushed into basalt fractures. Batches of basalt fractures are sorted and mixed in order to achieve the desired composition. These blended fractures are then melted in a furnace. Once the basalt fractures are heated to an optimal temperature of between 1400 and 1600 °C, the molten basalt is extruded into continuous filaments with a diameter of 12–18 μm. CBF may be formed into chopped basalt fibre strands, basalt fabrics, basalt wires or meshes, which can then be used in a wide range of application areas. It is similar to fiberglass, having better physico-mechanical properties than fiberglass, but being significantly cheaper than carbon fibre.

9 Summary and Conclusions

This paper presents overview of the importance of considering various aspects of Fire and Seismic behavior and also the usage of FRP, especially the Basalt Fibre. The high-strength and high-performance concrete are going to be unavoidable in future because of multiple constraints and expectations in the construction industry.

FRP as a material has a potential to excel in many areas including durability, corrosiveness, fire resistance etc. Also, basalt, which is considered as a green and sustainable material has excellent capability in dealing with fire. It also possesses better physico-mechanical properties and hence suitability is non-negotiable.

Strength, durability, stiffness, and ductility are few of the capabilities that are sought in structural members. The design of FRP can be done in such a way that it satisfies good number of characteristics and thus making it a viable material to be used in the construction of buildings. A high strength to weight ratio is very promising, and hence the issue of weight increase due to FRP does not emerge as a discussion point.

References

1. ACI Committee 363 (2005) ACI 363R: high-strength concrete. 228:79–80. https://doi.org/10.14359/14461
2. ACI Committee 440 (2008) ACI 440.2R-08: guide for the design and construction of externally bonded FRP systems for strengthening concrete structures. https://edisciplinas.usp.br/pluginfile.php/3435659/mod_resource/content/1/4402r_08.pdf
3. ASCE. 2017. Minimum design loads and associated criteria for buildings and other structures. ASCE/SEI 7–16. Reston, VA: ASCE.
4. Ashby M, Shercliff H, Cebon D (2007) Materials: engineering, science, processing and design. Elsevier, Oxford
5. Ba G, Miao J, Zhang W, Liu C (2016) Influence of cracking on heat propagation in reinforced concrete structures. ASCE J Struct Eng 142(7):11
6. Bilow DN, Kamara ME (2008) Fire and concrete structures. In: Structures congress. Vancouver, British Columbia, Canada
7. Cao VV, Nguyen VN (2022) Flexural performance of post fire reinforced concrete beams: experiments and theoretical analysis. J Perform Constr Facil 36(3):1–12

8. Del Zoppo M, Di Ludovico M, Balsamo A, Prota A, Manfredi G (2017) FRP for seismic strengthening of shear controlled RC columns: experience from earthquakes and experimental analysis. Compos B Eng 129:47–57. https://doi.org/10.1016/j.compositesb.2017.07.028
9. Ellingwood BR (2005) Load combination requirements for fire-resistant structural design. J Soc Fire Prot Eng 15(1):43–61
10. Ellingwood B, Lin TD (1991) Flexural and shear behavior of concrete beams during fires. ASCE J Struct Eng 117(2):440–458
11. Erkmen C, Saatcioglu M (2008) Seismic bahviour of FRP reinforced concrete frame buildings. In: The 1st world conference on earthquake engineering
12. Eslami A, Ronagh HR (2013) Effect of FRP wrapping in seismic performance of RC buildings with and without special detailing–a case study. Compos B Eng 45(1):1265–1274. https://doi.org/10.1016/j.compositesb.2012.09.031
13. Gedam BA (2021) Fire resistance design method for reinforced concrete beams to evaluate fire-resistance rating. Structures 33:855–877. https://doi.org/10.1016/j.istruc.2021.04.046
14. Heintz JA (2014) Recommendations for seismic design of reinforced concrete wall buildings based on studies of the 2010 Maule, Chile Earthquake. https://tsapps.nist.gov/publication/get_pdf.cfm?pub_id=916206
15. Kam WY, Pampanin S (2011) Revisiting performance-based seismic design in the aftermath of the Christchurch 2010–2011: raising the bar to meet societal expectations
16. Khoury GA (1992) Compressive strength of concrete at high temperatures: a reassessment. Mag Concr Res 44(161):291–309. https://doi.org/10.1680/macr.1992.44.161.291
17. Kodur VKR, Bisby LA (2005) Evaluation of fire endurance of concrete slabs reinforced with fibre-reinforced polymer bars. ASCE J Struct Eng 131(1):34–43
18. Lee KS, Lee BY, Seo SY (2016) A seismic strengthening technique for reinforced concrete columns using sprayed FRP. Polymers 8(4):107. https://doi.org/10.3390/polym8040107
19. Li M, Gong F, Wu Z (2020) Study on mechanical properties of alkali-resistant basalt fiber reinforced concrete. Constr Build Mater 245:118424. https://doi.org/10.1016/j.conbuildmat.2020.118424
20. Maluk C, Bisby L, Terrasi GP, Green M (2011) Bond strength of CRFP and steel bars in concrete at elevated temperature. In: ACI SP 279. American Concrete Institute, Detroit
21. Marasco S, Noori AZ, Cimellaro GP (2017) Cascading hazard analysis of a hospital building. J Struct Eng 143(9):1–15
22. Mehta PK, Monteiro PJM (2006) Concrete: microstructure, properties, and materials, 3rd edn. McGraw-Hill, New York
23. Mincigrucci L, Civera M, Lenticchia E, Ceravolo R, John M, Russo S (2023) Comparative structural analysis of GFRP, reinforced concrete, and steel frames under seismic loads. Materials 16(14):4908. https://doi.org/10.3390/ma16144908
24. Peter A, Murugesan K, Sharma U, Arora P (2014) Numerical study of heat and moisture transport through concrete at elevated temperatures. J Mech Sci Technol 28(5):1967–1977
25. Raj S, Kumar V, Kumar B, Iyer NR (2016) Basalt: structural insight as a construction material. Sādhanā 42(1):75–84. https://doi.org/10.1007/s12046-016-0573-9
26. Salem M, Issa MS (2023) Nonlinear finite element analysis of high and ultra-high strength concrete beams reinforced with FRP bars. HBRC J 19(1):15–31. https://doi.org/10.1080/16874048.2023.2170765
27. Shah AH, Sharma UK, Kamnath P, Bhargava P, Reddy GR, Singh T (2016) Effect of ductile detailing on the performance of reinforced concrete building frame subjected to earthquake and fire. ASCE J Perform Constr Facil 30(5):17
28. Tariq F, Bhargava P (2021) Flexural behavior of corroded RC beams exposed to Fire. Elsevier Struct 33:1366–1375
29. Wang D, Shi C, Wu Z, Xiao J, Huang Z, Fang Z (2015) A review on ultra-high performance concrete: Part II. Hydration, microstructure and properties. Constr Build Mater 96:368–377
30. Wang L, Zhang J, Huang C, Fu F (2020) Comparative study of steel-FRP, FRP and steel-reinforced coral concrete beams in their flexural performance. Materials (Basel) 13(9):2097

31. Wang X, Wu D, Zhang J, Yu R, Hou D, Shui Z (2021) Design for sustainable ultra-high performance concrete: a review. Constr Build Mater 307:1–24
32. Yuvaraj M, Rajmohan M, Naveen G, Mohanraj S (2014) Mechanical characterisation of basalt based composite materials. In: International conference on recent trends in engineering and management. https://doi.org/10.13140/2.1.2554.3369
33. Zaidi A, Brahim MM, Mouattah K, Masmoudi R (2016) FRP properties effect on numerical deformations in FRP bars-reinforced concrete elements in hot zone. In: International conference on materials and energy
34. Zeng L, Li L, Yang X, Liu F (2019) Seismic behavior of large-scale FRP- recycled aggregate concrete-steel columns with shear connectors. Earthq Eng Eng Vib 18(4):823–844. https://doi.org/10.1007/s11803-019-0538-1
35. Zhang Z, Hsu CTT, Moren J (2004) Shear strengthening of reinforced concrete deep beams using carbon fiber reinforced polymer laminates. ASCE J Compos Constr 8(5):403–414
36. Zhang W, Shi D, Shen Z, Zhang J, Zhao S, Gan L, Li Q, Chen Y, Tang P (2023) Influence of chopped basalt fibers on the fracture performance of concrete subjected to calcium leaching. Theor Appl Fract Mech 125:103934–103934. https://doi.org/10.1016/j.tafmec.2023.103934

Retrofitting Methods for Shear Strengthening of Reinforced Concrete Beams Using CFRP

R. Arvindh Raj and R. Senthilkumar

Abstract Due to the ageing of infrastructures, there is a need for repair and rehabilitation of the structures at later stages. The usage of carbon fiber-reinforced polymer for strengthening is one such practice that has become common in the Indian scenario. This paper specifically discusses the comparison of different application methods of carbon fiber-reinforced polymers for enhancing the shear capacity of the reinforced concrete beams. Near surface mounted carbon fiber-reinforced polymer strips, externally bonded carbon fiber-reinforced polymer strips and externally bonded carbon fiber-reinforced polymer strips with U–wrap are the methods adopted for strengthening reinforced concrete beams for shear. Three-point loading flexure tests were conducted on the strengthened beams, and it was found that the near surface mounted method exhibited a 28.5% increase in the shear capacity of the beam. In contrast, the externally bonded method exhibited an 18.5% increase, and externally bonded with U–wrap exhibited a 142% increase in the shear capacity of the beam compared to the control beam.

Keywords Retrofitting · Shear · Near-surface mounted · Externally bonded · Reinforced concrete beams

1 Introduction

During the life of a structure, various situations would arise that may not have been considered during the initial design stage. One such example is changing the structure's functionality leading to heavier loads that may not have been considered in the initial stage. For such conditions, demolishing and constructing a new building will incur more time and cost. Thus, retrofitting and strengthening have become the ideal solution to these situations. Though various methods of strengthening techniques are available, strengthening using Carbon Fibre Reinforced Polymer Composites

R. Arvindh Raj · R. Senthilkumar (✉)
Department of Civil Engineering, NIT, Tiruchirapalli 620015, India
e-mail: senthilr@nitt.edu

S. B. Singh and C. V. R. Murty (eds.), *RC Structures Strengthened with FRP for Earthquake Resistance*, Composites Science and Technology,
https://doi.org/10.1007/978-981-97-0102-5_12

(CFRP) has become a prominent solution because of its strength and application time. Based on the intended applications of the CFRP, they are classified as Bond-critical and Contact-critical applications. The bond-critical application requires an adhesive bond between the FRP and concrete, which can be used for flexural or shear strengthening of beams, columns and slabs. The contact-critical application requires intimate contact between FRP and concrete, such as column confinement.

Here in this study, discussions on the bond critical applications and the methods of strengthening using CFRP laminates which are Near Surface Mounted CFRP (NSM), Externally Bonded CFRP (EB), and Externally bonded CFRP strips with U–Wrap CFRP (EBU), are critically presented. Briefly, the NSM method involves cutting grooves in the concrete for the required depth to install the CFRP laminates with epoxy adhesives. In the EB method, the CFRP laminate is glued with epoxy adhesive on the surface of the Reinforced Concrete (RC) members. Similarly, the U-Wrap involves the installation of CFRP laminates on the entire member for the required length except the top surface using epoxy resins. For the installation of CFRP, ACI 440 2R.17 (2017) suggests various parameters to be adhered to. One such parameter is that the minimum tensile strength and compressive strength of concrete should be 1.4 MPa and 17 MPa, respectively, to carry out the bond critical applications. Whereas the same is not required for contact-critical applications.

Further profiling of the concrete surface for CFRP laminates is stipulated by ACI 546R (2017) and ICRI 03,730 (2008). They need to be compatible with adhesion to concrete surfaces as well as FRPs and should be able to resist environmental impact and be highly workable. Various studies have been done in the field of retrofitting using CFRP. Barros (2006) concluded that the NSM method is an effective method of shear strengthening of beams in comparison with the results of three variations in the specimens—without shear reinforcement, with steel shear reinforcement and with NSM CFRP laminate shear reinforcement. As a result, it was found that steel shear reinforcement exhibited an 85% increase in the shear capacity, whereas NSM CFRP exhibited a 102% increase in the shear capacity than the control specimen. Further, Zhang (2017) found that the bond between the epoxy and CFRP laminate was better in the NSM method than in the EB method. This was because the area of contact is more in the NSM method, and hence the separation failure of the concrete cover was prominent in the NSM method. In a recent study related to the groove spacing of CFRP laminates, Zhang (2017) observed that the bond strength of the CFRP laminate increases with an increase in the groove spacing. Comparing the different factors affecting the bond strength, Sharaky (2013) suggested that in the NSM method of CFRP, the failure load can be increased by 17% with an increase in 25% bond length. From the above discussions, the NSM method is observed to be a viable method for the shear strengthening of RC beams. Various past studies have established the enhancement in the flexural strength of the strengthened RC beam with different methods of CFRP, but further studies are required to interpret the performance of shear strengthening of RC beams using different techniques. Therefore, this study compares the shear strength enhancement of RC beams strengthened using three existing approaches such as NSM, EB and EBU, which can be effectively practiced at the site.

2 Experimental Program

2.1 Design and Casting of Beam Specimens

A total of 8 RC beams (4 pairs) were cast, along with 3 cubes and 3 cylinders. The beams cast were control beam specimens, beam specimens strengthened by the NSM method, beam specimens strengthened by the EB method and beam specimens strengthened by the EBU method. Cubes and cylinders were cast to ascertain the mechanical properties of concrete. All beams were of length 1m, width 0.1 m and depth 0.2 m. The grade of concrete adopted was M40. The reinforcement details for all the beams adopted were Fe 415 grade steel with 2 numbers of 12 mm diameter bars as the main reinforcement with 2 numbers of 12 mm diameter bars as hanger bars and stirrups of 8 mm diameter bars at 125 mm c/c. Figures 1a, b and 2 show the cross-section details, reinforcement arrangement and casting of RC beams.

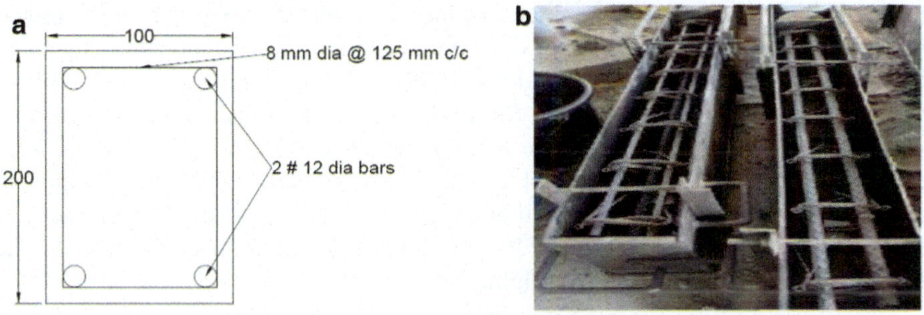

Fig. 1 **a** Cross-section details. **b** Reinforcement details

Fig. 2 Cast RC beams

2.2 Preparation of RC Beam Specimens

The preparation procedure for strengthening the beam specimens using the three methods is explained below. The properties of the CFRP laminates and ply are stipulated in Table 1. A thixotropic epoxy adhesive resin named Sikadur-30 LP which is tested according to EN 1504-4 was utilised for the study.

2.2.1 Near Surface Mounting Method (NSM)

Near Surface Mounting method involves cutting a groove through the concrete (Fig. 3) and placing CFRP laminate inside the groove using an epoxy adhesive (Fig. 4). On a trial basis, two grooves were cut for the CFRP installation. Care was taken while cutting to ensure that no damage was caused to the steel reinforcement. As per ACI 440 2R.17 (2017), the groove's size should be at least 3 times the width and 1.5 times the depth of CFRP laminate. Therefore, in order to place a 1.4 mm thick and 25 mm wide CFRP laminate, a 4 mm wide and 35 mm deep groove cut was made. This was due to the fact that, since it was a bond-critical application, the bond between the FRP and concrete should be 100% and also should ensure perfect load transfer between FRP and concrete. Similarly, the spacing between the two grooves was kept at twice the depth of the groove, which is 80 mm. The grooves were cleaned completely by removing all the dust from the groove. The two-component epoxy adhesive was mixed and filled in the groove completely, and further adhesive was applied on the surface of the CFRP ply and firmly placed inside the groove. The adhesive was allowed to cure overnight.

2.2.2 Externally Bonded Method (EB)

The externally bonded method involves preparing the surface and bonding the laminate. Being a bond-critical application, the bond between the FRP and concrete is very vital. The surface was prepared as stipulated by ACI 546R (2017), Concrete Surface Preparation Level 3 (CSP 3). Grinding of the surface was done extensively by using an electrically operated grinding machine to remove any form of lattice and dust. On a trial basis, it was decided to place CFRP laminates for the entire beam

Table 1 Properties of CFRP ply and laminates

Property	CFRP ply	CFRP laminates
Weight of the fiber	450 GSM	–
Thickness	0.25 mm	1.4 mm
Tensile strength	4900 MPa	2800 MPa
Elongation	1.5%	1.7%
Fiber orientation	Unidirectional	Unidirectional

Fig. 3 Grooves for NSM

Fig. 4 Installation of CFRP

width by stacking together two laminates having a width of 50 mm and a thickness of 1.4 mm. Before attaching the laminate, the undulations on the concrete surface were leveled using thixotropic epoxy putty to ensure perfect bonding of the FRP and concrete. The adhesive was mixed thoroughly until a uniform colour was obtained using a mixer machine. The surplus adhesive was applied on the soffit of the beam, and the laminates were installed effectively using rollers to remove any entrapped air bubbles and to ensure that the contact area of FRP was completely coated with the epoxy adhesive, as shown in Figs. 5 and 6.

Fig. 5 Applying Epoxy adhesive for installation of CFRP

Fig. 6 CFRP laminate installed to the beam using EB Method

2.2.3 Externally Bonded with U—Wrap (EBU)

CFRP laminates were installed as described in EB, and the CFRP plies were applied as U-wraps to compare the performance of the combined action of externally bonded laminates and U-wraps. Additionally, the corners of the beams were rounded to an average radius of 15 mm for reducing the stress concentration in the FRP system as per ACI 44-2R.17 (2017) after the installation of the U-wrap. All the undulations on the concrete surface were leveled using thixotropic epoxy putty to ensure a uniform level surface for the U-wrap installation. Post installation of the CFRP laminate by EB method, epoxy primer was applied uniformly to the surface and allowed to cure for 24 h. A wet layup system of application was adopted for the installation of the CFRP U-wrap plies. 450 GSM CFRP fiber layers were saturated with epoxy matrix and laid over the concrete surface using a hand layup process. Rollers were used to ensure that the saturant passed through the fibers to provide a uniform layer of

Fig. 7 CFRP laminate installed to the beam using the EBU Method

wrapping over the beam. The beam was left undisturbed for a period of 48 h. Figure 7 shows the beam specimen strengthened using the EBU method.

2.3 *Experimental Setup*

The three-point loading test was carried out using a Computerized Universal Testing Machine (UTM) of 100 ton capacity. Tests were carried out on the 28th day of casting and curing for all specimens. The three-point loading has been performed on the beam specimens to ascertain the shear capacity of the control beam and the strengthened beams, as shown in Fig. 8. The displacement was recorded at the mid-span of the beam corresponding to every load increment. The displacement readings were noted based on the UTM values as there were constraints in fitting LVDT or dial gauge. The deflection pertaining to the beam was based on the column movement in the UTM. The working length of the beam was taken as 0.8 m, and the specimens were loaded till the failure at a loading rate of 1.8 kN/min.

3 Test Results and Discussions

The load–deflection plots of the tested RC specimens are shown in Fig. 9. It was observed that the control specimen exhibited diagonal cracks of 45 degrees from a distance of L/3 from the support region, which implies pure shear failure and the

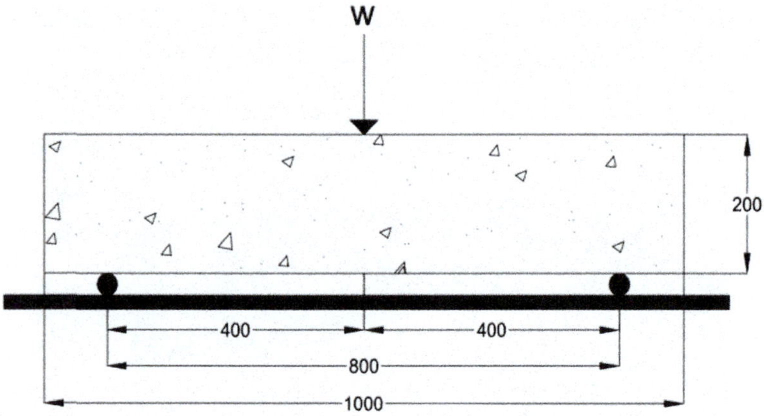

Fig. 8 Schematic diagram of three-point loading setup

crack width increased with an increase in the applied load. The specimen failed in a brittle manner at a load of 70 kN. The NSM specimen exhibited the same cracking and failure pattern with a relative increase of 28.5% in the load-carrying capacity.

The increase in the shear resistance is due to the dowel action of CFRP laminates. Concrete crushing was noticed in both specimen types at the loading point. Figures 10, 11, 12 and 13 showcase the different failure modes observed in the tested specimens. In the case of the EB strengthened beam, the increase in shear capacity was observed as 18.5%, which is marginally more than the control but lesser than NSM strengthened specimens. The EBU specimen showed a very high increase in the shear capacity. During testing, delamination occurred in the CFRP laminates, and cohesive failure was witnessed in the concrete layer that was bonded to CFRP laminates at the failure stage. The very high increase of 142% in shear was due to the

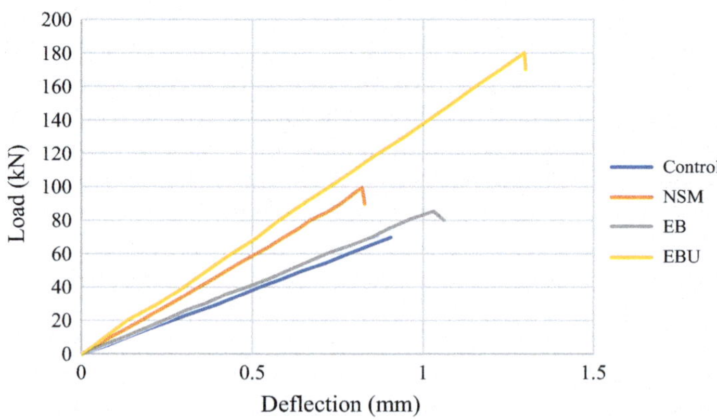

Fig. 9 Load versus Deflection for all specimens

Fig. 10 Concrete Crushing in control beam

Fig. 11 Cracking in NSM specimen

confinement provided by the CFRP U-wraps, which in turn increased the compressive strength of the concrete by accounting for the increase in the shear strength of the specimen and also the presence of the CFRP laminate placed through EB method would account to the increase in the shear capacity of the EBU specimen.

4 Summary and Conclusions

In this study, three different types of retrofitting techniques focused on enhancing the shear capacity of an RC beam are discussed. Based on the test results, the following conclusions are drawn.

Fig. 12 Shear crack in EB specimen

Fig. 13 CFRP de-lamination and concrete crushing in EBU specimen

- The EBU method of retrofitting showed a very prominent increase in the shear capacity of the beam compared to other techniques primarily due to the confinement provided by the CFRP wrap systems.
- The NSM is another efficient method that can be adopted to improve the shear capacity, which enhanced the performance by 25% due to its dowel action and higher stiffness (vertical orientation).
- The EB method did not show much increment in the shear strength compared to other methods because of delamination and lower stiffness (horizontal orientation) of CFRP laminates. Hence, EBU and NSM are better techniques to improve the shear strength of the RC beam.

- Concrete crushing and de-lamination of CFRP is the most prominent failure mode noticed in all cases. Surface preparation of the substrate is critical for the efficient performance of CFRP to prevent bond failure between concrete and CFRP.

References

1. Sabol P (2013) Shear strengthening of concrete members using NSM method. In: Concrete and concrete structure conference
2. Zhang SS, Yu T, Chen GM (2017) Reinforced concrete beams strengthened in flexure with near surface mounted (NSM) CFRP strips: Current status and research needs. Compos Part B 131:30–42
3. Zhang SS, Yu T, Chen GM (2017) Effect of groove spacing on bond strength of near surface mounted (NSM) bonded joints with multiple FRP strips. Constr Build Mater 155:103–113
4. Sharaky IA, Torres L, Baena M (2013) An experimental study on different factors affecting the bond of NSM FRP bars in concrete. Compos Part 99:350–365
5. Seo S-Y, Lee MS (2016) Flexural analysis of RC bean strengthens by partially de-bonded NSM FRP strip. Compos Part B-21-30
6. Yang J, Wang L (2019) Experimental research on flexural behaviors of damaged PRC beams strengthened with NSM CFRP strips. Constr Build Mater 190:265–275
7. Parretti R, Nanni A (2004) Strengthening of RC members using near surface mounted FRP composites: design overview. Adv Struct Eng 7(6):469–483
8. De Lorenzis L, Nanni A (2001) Shear strengthening of reinforced concrete beams with NSM rods. ACI Struct J 98(1):60–68
9. Barros JAO, Dias SJE (2006) Near surface mounted CFRP laminates for shear strengthening of concrete beams. Cement Concr Compos 28(3):276–292
10. Barros JAO, Dias SJE (2005) Shear strengthening of RC beams with NSM CFRP laminates. In: 7th international symposium on Giber Reinforced Polymer (FRP) reinforcement for concrete structures (FRP7RCS). Kansas, USA, DP-230-47, pp 807–824
11. Externally bonded FRP systems for strengthening concrete structures—ACI 440. 2R-17

Use of FRP on Concrete Specimen, RC Elements and Components for Higher Load Resistance

G. R. Reddy, Milan Savji Nakum, F. K. Muhammad Khizar, and Lakhani M. Salman

Abstract Most of the RC structures exposed to corrosion environment are getting distressed and there is a loss of capacity of structural elements and components. In addition, existing RC structures are generally weak at joint levels and causing failures due to external events like earthquakes. Present chapter is made explaining the detailed calculations to improve the concrete properties by FRP confinement. A detailed procedure is provided to evaluate the improved strength of RC elements strengthened with FRP plate. Also procedure explained to evaluate improved strength of RC structural component (joint). All these procedures are validated with published related experimental data.

1 Introduction

Structures are subjected to normal loads such as dead weight, live loads, imposed loads, summer and winter temperatures, normal machine/vehicle vibration loads, etc. It addition it is also subjected to accidental loads induced due to earthquake, extreme wind, blasts, etc. The demand on the structure due to natural hazards such as earthquakes keep changing with time. One example is recent Turkey earthquake, 2023 and past Kobe earthquake [1–3]. The failures are same as shown in Fig. 1 and there is a big gap to implement the corrections needed in the structures.

The earthquake amplitudes such as peak ground acceleration has been exceeded the design basis earthquake levels causing the damages of the structure. Some are partially damaged and some are collapsed. To take care of these aspects, hazard levels have to be improved with new data and structures have to be revisited frequently and assessed for its strength and serviceability status. If the structures are not meeting the new hazard or demand, the structure needs to be rehabilitated to meet the initial design intent and retrofitted if the load demand increases. Conventionally after proper repair, steel jacketing, concrete jacketing, bracing etc. are adopted to rehabilitate

G. R. Reddy (✉) · M. S. Nakum · F. K. Muhammad Khizar · L. M. Salman
Structural Engineering Division, VJTI, Mumbai, India
e-mail: grrddy@yahoo.com

S. B. Singh and C. V. R. Murty (eds.), *RC Structures Strengthened with FRP for Earthquake Resistance*, Composites Science and Technology,
https://doi.org/10.1007/978-981-97-0102-5_13

a. Failure of columns of metrotrack deck columns

b. Column failures in weak axis Direction

Fig. 1 Failure of columns during Kobe (1995) and recent (2023) Turkey Earthquake

and retrofit the structure. However, recently fiber reinforced polymers (FRP) due to its mechanical properties as shown in Table 1 is taking the lead material for rehabilitation and retrofitting of structures especially reinforced concrete structures. Wrapping the concrete structural elements will improve the confinement and hence increase in strength as shown in Fig. 2a and is similar to effect of the confined steel. It is essential to note that the wrapping will be effective only when the concrete has lateral deformation the obvious behaviour of concrete when start loading. When there is a need to increase the flexural capacity, it is achieved by using FRP laminates similar to the reinforcement. It is located generally near the surface of the concrete of structural elements. Both the confinement and flexure effects [4] are clearly shown in Fig. 2b, c. Considering these characteristics of FRP wrapping and strengthening, this chapter is developed explaining the procedure of evaluating the enhanced mechanical properties of concrete due to wrapping, flexure properties of concrete elements such as beam and columns, components (combination of beam an column) and structural system.

Failure Modes of RC Components with FRP Plates (Laminates)

While estimating the capacity of RC structural element, components, it is important to consider the proper failure modes. There are four modes of failures as explained in Fig. 3 [5].

Table 1 Mechanical properties of fibres and steel

Property	E-glass	Carbon-HS	Basalt	Aramid Kevlar-29	Steel
Density (g/cc)	2.55	1.75	2.7	1.44	7.8
Young's modulus (GPa)	69	221	89	83	210
Tensile strength (MPa)	3450	3100	4840	3620	500[a]

[a]Fe500 reinforcement steel

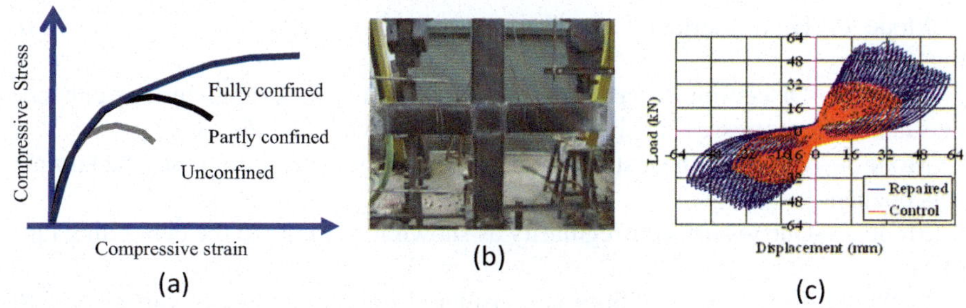

Fig. 2 Effect of FRP rehabilitation on RCC structural component

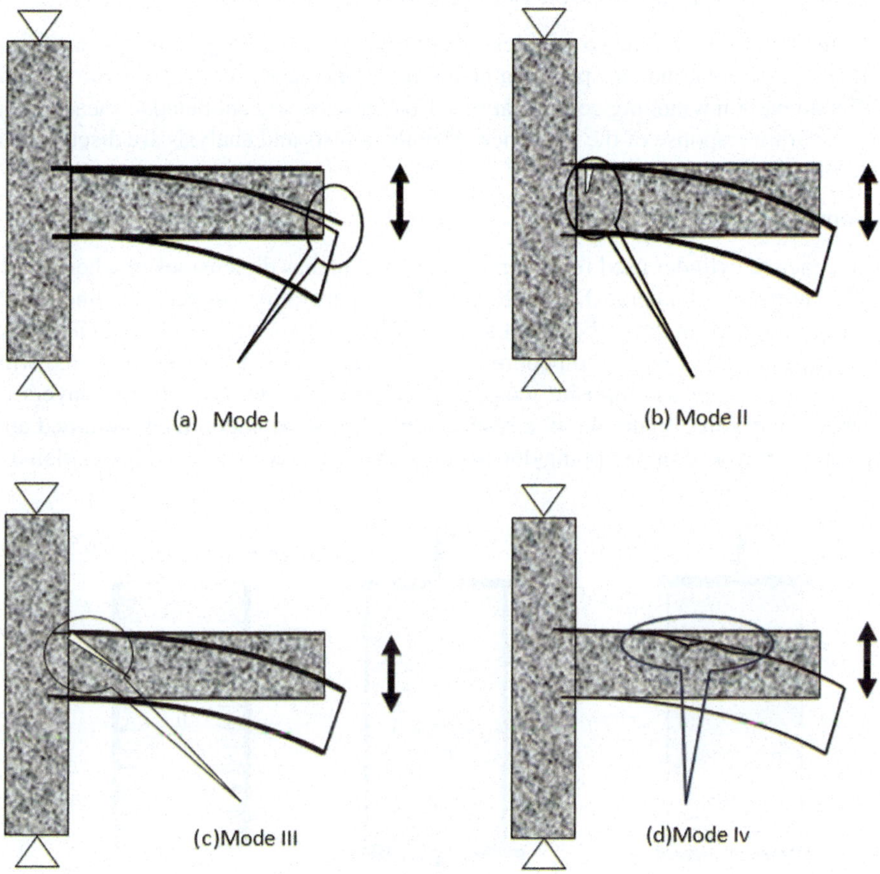

Fig. 3 Failure modes of FRP strengthened RC elements

Mode 1—High interfacial shear stresses causes the debonding as shown in the Fig. 3a.

Mode 2—Discontinuity generated due to flexure cracks that further enhances interfacial shear stresses causing debonding as shown in the Fig. 3b.

Mode 3—When the shear stresses are predominant to flexure stresses the laminate gets opened.

Mode 4—Due to surface irregularity as shown in the Fig. 3d the debonding takes place.

In this chapter mode 2 failure is considered for estimating the capacity of RC elements and components strengthened with FRP.

Effect of FRP Confinement on Concrete Material Properties

Lin and Chen [6] conducted a series of experiments to investigate the stress–strain characteristics of concrete specimen under unconfined and confined conditions with different carbon wrapping configurations. Load tests were conducted to measure the stress–strain response of the specimen. Details of tests and analysis are discussed in the following sections.

Geometric Properties:

The concrete cylinder used for theoretical evaluation has dimensions of a height (H) of 240 mm and a diameter (D) of 120 mm. The aspect ratio of concrete cylinder is 2.

The fibre used in the carbon composite is T300, and the matrix material is epoxy. The thickness (t) of the carbon composite is 0.5 mm. One specimen was tested with one layer of carbon composite wrapping and one more tested with two layers of carbon composite. Figure 4a, b, c respectively shows the unconfined, confined and pressure developed on wrapping due to compressive loads on cylinder are explained.

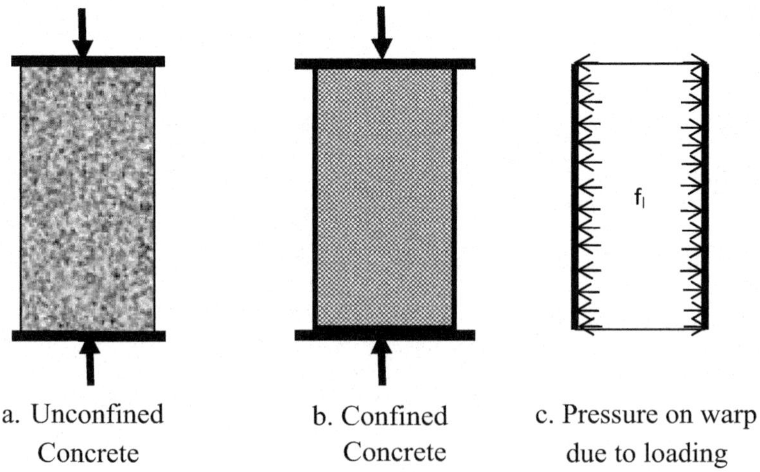

Fig. 4 Concrete cylinder without/with carbon composite sheet wrapping

Material Properties:

Unconfined compressive strength of concrete (f_{cd}) = 32.7 MPa.

Modulus of elasticity of concrete (E_c) = 26.29 GPa.

Compressive strength of carbon composite (f_f) = 770 MPa.

Modulus of elasticity of carbon composite (E_f) = 157.54 GPa.

Evaluating the Stress and Strain Characteristics of Unconfined and Confined Concrete

As per the National Research Council [5] the strength of confined concrete is evaluated using Eq. 1.

$$\frac{f_{ccd}}{f_{cd}} = 1 + 2.6\left(\frac{f_{1,eff}}{f_{cd}}\right)^{2/3} \quad (1)$$

where

f_{cd} is the design strength of unconfined concrete.
$f_{l,\,eff}$ is the effective confinement lateral pressure as explained in Fig. 4 is given as:

$$f_{1,eff} = k_{eff} \cdot f_1 \quad (2)$$

where k_{eff} is a coefficient of efficiency (≤ 1), defined as the ratio between the volume of the effectively confined concrete $V_{c,\,eff,}$ and the volume of the concrete member Vc.

The confinement lateral pressure as shown in Fig. 4c shall be evaluated as follows:

$$f_l = \frac{1}{2}.\rho_f.E_f.\varepsilon_{fd,rid} \quad (3)$$

where ρ_f is the geometric strengthening ratio as a function of section shape (circular or rectangular) and FRP configuration (continuous or discontinuous wrapping) as shown in Fig. 5, E_f is Young modulus of elasticity of the FRP in the direction of fibers, and $\varepsilon_{fd,\,rid}$ (0.004) is a reduced FRP design strain.

The coefficient of efficiency, k_{eff}, shall be expressed as:

$$k_{eff} = k_H.k_v.k_\alpha \quad (4)$$

The coefficient of horizontal efficiency, k_H, depends on the cross-section shape. For cylinder $k_H = 1$ and $k_\alpha = 1$ for fibers normal to the axis.

The coefficient of vertical efficiency, k_V, depends on FRP configurations as shown in Fig. 5.

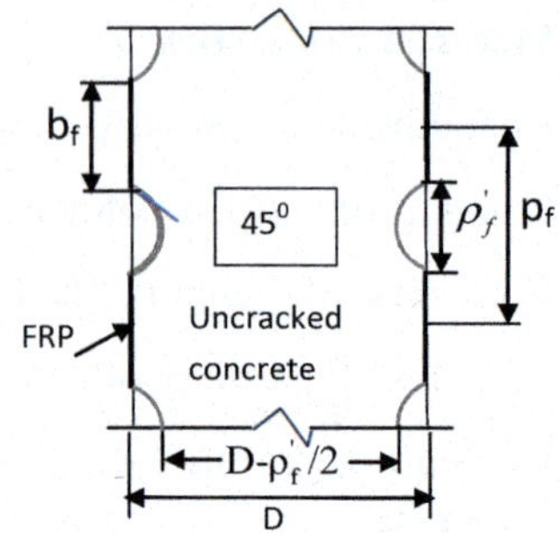

Fig. 5 Schematic of confined concrete with FRP

For RC confined members with continuous FRP wrapping, it is assumed $k_V = 1$.

$$\rho_f = \frac{4.t_f.b_f}{D.p_f} \tag{5}$$

where t_f, b_f and p_f represent FRP thickness, width, and spacing, respectively, and D is the diameter of the circular cross section as shown in Fig. 5.

In the case of continuous wrapping, ρ_f becomes

$$\rho_f = \frac{4 \times t_f}{D} \tag{6}$$

Geometric Strengthening ratios for one layer and two layers wrapping are evaluated using Eq. 6 as follows.

$$\rho_f = \frac{4 \times 0.5 \times 10^{-3}}{120 \times 10^{-3}} = 0.0167 \text{ (for one layer)}$$

$$\rho_f = \frac{4 \times 2 \times 0.5 \times 10^{-3}}{120 \times 10^{-3}} = 0.0334 \text{ (for two layers)}$$

Confinement lateral pressure for one layer and two layers are evaluated using Eq. 3 as follows.

$$f_l = \frac{1}{2} \times 0.0167 \times 157.54 \times 10^9 \times 0.004 = 5.26 \text{ MPa (for one layer)}$$

$$f_l = \frac{1}{2} \times 0.0167 \times 157.54 \times 10^9 \times 0.004 = 10.52 \text{ MPa (for two layers)}$$

Effective confinement lateral pressure is evaluated using Eqs. 2 and 4 as follows.

$f_{l,eff} = k_{eff}.\ f_l = 5.26$ MPa for one layer.

$= 10.52$ MPa for two layers.

Using effective confined lateral pressure and unconfined strength of concrete in Eq. 1, confined strength of concrete is evaluated as follows.

$$f_{ccd} = 32.7 \times 10^6\left(1 + 2.6\left(\frac{5.26 \times 10^6}{32.7 \times 10^6}\right)\right) = 57.85\ \text{MPa}\ \ (\text{withonelayer})$$

$$f_{ccd} = 32.7 \times 10^6\left(1 + 2.6\left(\frac{10.52 \times 10^6}{32.7 \times 10^6}\right)\right) = 72.62\ \text{MPa (with two layers)}$$

Cylindrical strength of the concrete is also estimated using the empirical relation given by Huei-Jeng Lin and Chin-Ting Chen [6, 7] as follows.

$$\mathrm{f_{ccd}} = \mathrm{f_{cd}} + 2 \times 2 \times \frac{\mathrm{f_{cm}}}{\mathrm{D}} \times \mathrm{t} \tag{7}$$

D diameter of concrete cylinder
$\mathrm{f_{cd}}$ strength of unconfined concrete
$\mathrm{f_{ccd}}$ strength of confined concrete
$\mathrm{f_{cm}}$ tensile strength of composite

$$\mathrm{f_{ccd}} = 32.7 \times 10^6 + 2 \times 2 \times \frac{770 \times 10^6}{120 \times 10^{-3}} \times 0.5 \times 10^{-3} = 45.53\,\text{MPa (with one layer)}$$

$$\begin{aligned}\mathrm{f_{ccd}} &= 32.7 \times 10^6 + 2 \times 2 \times 2 \times \frac{770 \times 10^6}{120 \times 10^{-3}} \times 0.5 \times 10^{-3} \\ &= 58.36\,\text{MPa (with two layers)}\end{aligned}$$

and compared with estimates made above using CNR-DT 200 R1/2013 [5] and experiments in Table 2.

Evaluation of Stress–strain Relation of Unconfined and Confined Concrete

The Hognestad stress–strain relation [9] for unconfined concrete is given as follows.

$$f_c = f_{cd}\left[\frac{2\varepsilon_c}{\varepsilon_o} - \left(\frac{\varepsilon_c}{\varepsilon_o}\right)^2\right] \tag{8}$$

where $\varepsilon_0 = \frac{2f_{cd}}{E_c} = \frac{2\times 32.7\times 10^6}{26.29\times 10^6}0.0025.$

Table 2 Comparison of unconfined and confined concrete strengths (MPa)

S. no.	Experiments		CNR-DT200R1/2013 [9]	Lin and Chen [6]
	Unconfined	Confined		
1 (One layer)	32.7	50.3	57.85 (45.13)[a]	45.53
2 (Two layers)	32.7	66.6	72.63 (48.41)[a]	58.36

Note [a]Values in the bracket are at 0.4% strain

Using Eq. 8 for different values of strains, stresses are evaluated and given in Table 3 and also shown in Fig. 7.

Schematic of stress–strain relation of unconfined and confined concrete is shown in Fig. 6.

For OA

$$f_c = f_{cd}\left(a - \overline{\varepsilon} - \overline{\varepsilon}^2\right) \quad 0 \leq \overline{\varepsilon} \leq 1$$
$$\overline{\varepsilon} = \frac{\varepsilon_c}{\varepsilon_{co}} \tag{9}$$

where εcounconfined compressive strain of concrete = 0.0025 (Fig. 7).

For linear branch (AB)

$$f = f_{cd}(1 + b \times \overline{\varepsilon}) \quad 1 \leq \overline{\varepsilon} \leq \frac{\varepsilon_{ccu}}{\varepsilon_{co}}$$
$$\overline{\varepsilon} = \frac{\varepsilon_c}{\varepsilon_{co}} \tag{10}$$

Table 3 Stress–strain data of unconfined concrete

Steps	Unconfined concrete	
	f_c	ε_c
1	0.000	0.00000
2	11.772	0.00050
3	20.928	0.00100
4	27.468	0.00149
5	31.392	0.00199
6	32.700	0.00249
7	31.474	0.00282
8	30.248	0.00314
9	29.021	0.00347
10	27.795	0.00380

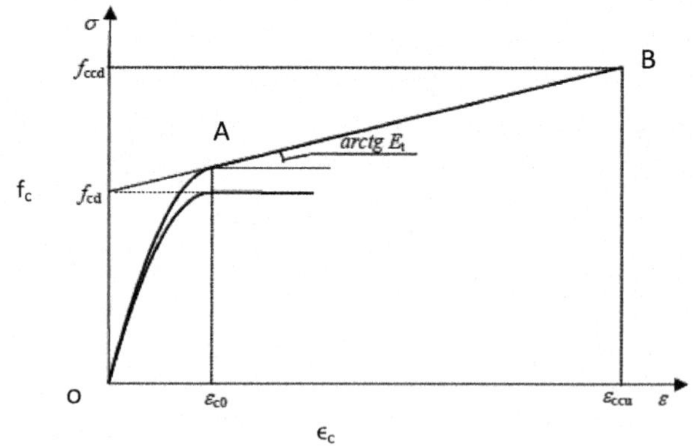

Fig. 6 Schematic of Stress-strains for unconfined and confined concrete

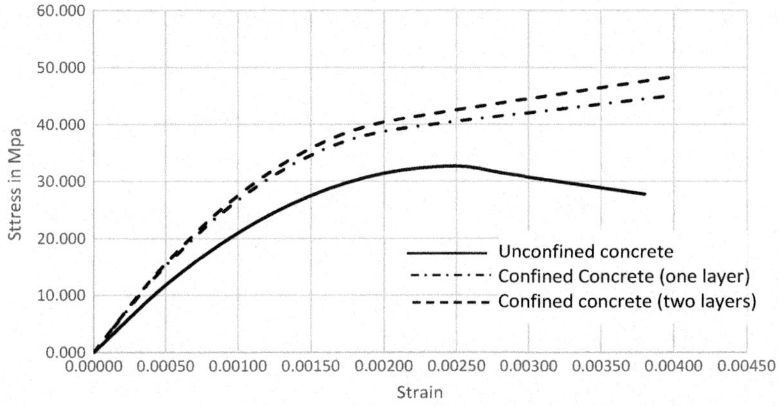

Fig. 7 Comparison of compression stress–strain characteristics of confined and unconfined concrete

where a and b are evaluated using tangent modulus as sown in Fig. 6 and using unconfined strength and strains of concrete as follows.

$$\mathrm{a} = 1 + \gamma; \quad \mathrm{b} = \gamma - 1$$

$$\gamma = \frac{f_{cd} + E_t.\epsilon_{co}}{f_{cd}} \tag{11}$$

$$E_t = \frac{f_{ccd} - f_{cd}}{\epsilon_{ccu}} \tag{12}$$

where

$$\varepsilon_{ccu} = 0.0035 + 0.015\sqrt{\frac{f_{1eff}}{f_{cd}}} \tag{13}$$

Tangent modulus as shown in Fig. 6 for one layer and two layers wrapping concrete using Eq. 12 are evaluated as follows.

$$E_t = \frac{57.85 \times 10^6 - 32.7 \times 10^6}{0.00952} = 2642\,\text{MPa For one layer}$$

$$E_t = \frac{72.63 \times 10^6 - 32.7 \times 10^6}{0.00952} = 3327\,\text{MPa For two layer}$$

Constants 'a' and 'b' of the confined material models are evaluated for single and two layers confinement using Eq. 11 as follows.

$$\gamma = \frac{32.7 \times 10^6 + 2642 \times 10^6 \times 0.0025}{32.7 \times 10^6} = 1.201 \text{ for one layer}$$

$$\gamma = \frac{32.7 \times 10^6 + 3327 \times 10^6 \times 0.0025}{32.7 \times 10^6} = 1.254 \text{ for two layers}$$

a = 2.201 and b = 0.201 for one layer

a = 2.254 and b = 0.254 for two layers.

Using Eq. 13 ultimate confined compression strains are evaluated as follows for one layer and two layers confinement.

$$\varepsilon_{ccu} = 0.0035 + 0.015\sqrt{\frac{5.26 \times 10^6}{32.7 \times 10^6}} \text{ for one layer}$$

$$\varepsilon_{ccu} = 0.0035 + 0.015\sqrt{\frac{10.52 \times 10^6}{32.7 \times 10^6}} \text{ for two layers}$$

Using above data and Eqs. 8–10, stresses are evaluated for different strains in the concrete and tabulated in Table 4. Comparison of stress and strain characteristics of confined and unconfined concrete is shown in Fig. 7 for one layer and two layers confinement.

Table 4 Stress and strains of concrete with one layer and two layers confinement

Steps	One layer		Two layers	
	ε_c	f_c	ε_c	f_c
1	0.00000	0.00	0.00000	0.00
2	0.00042	13.09	0.00042	13.43
3	0.00085	23.55	0.00085	24.25
4	0.00127	31.41	0.00127	32.45
5	0.00169	36.65	0.00169	38.04
6	0.00211	39.27	0.00211	41.00
7	0.00400	45.13	0.00400	48.41
8	0.00706	54.65	0.00706	60.43
9	0.00953	62.32	0.00953	70.13
10			0.01200	79.84

Effect of FRP Laminates on Flexure Properties of RCC Beam Geometric Properties of Beam Tested

A simply supported reinforced cement concrete beam of span 2400 mm has a rectangular cross-section of dimension 150 mm × 250 mm [8] is shown in Fig. 8 was tested by Wenwei and Guo [8]. The main flexural reinforcement consisted of three 14 mm diameter deformed bars with a sectional area of 462 mm^2. Two 8 mm diameter round bars with a sectional area of 100.5 mm^2 were used as compression reinforcement. Shear reinforcement consisted of 6 mm diameter round steel stirrups spaced at 100 mm centre to centre. The overhang on either side of the beam is 150 mm as shown in Fig. 8. Table 5 provides the details of mechanical properties of steel, concrete and CFRP. Four point bending tests were performed [8] on Beams. One beam was used as a control specimen and the other beam was strengthened in flexure with CFRP.

The CFRP material 150 mm wide and 0.111 mm thick was externally bonded to the tension face of the concrete beam. All parameters mentioned above are used for theoretical calculation for estimation of the M–φ, M–θ, and P–Δ curve of beams with and without CFRP strengthening.

For analysis, the beam is modelled as shown in Fig. 9. Elements 1 and 6 are linear beam elements, elements 2 and 5 are rotational spring elements and elements 3 and 4 are shear springs. The left and right portions where concrete is confined with shear reinforcement are modelled considering linear behaviour. Its 2-D stiffness matric is shown in Annexure A1. The portion subjected to constant bending moment causing flexure failure is modelled with nonlinear flexure rotational stiffness which is obtained using the following procedure.

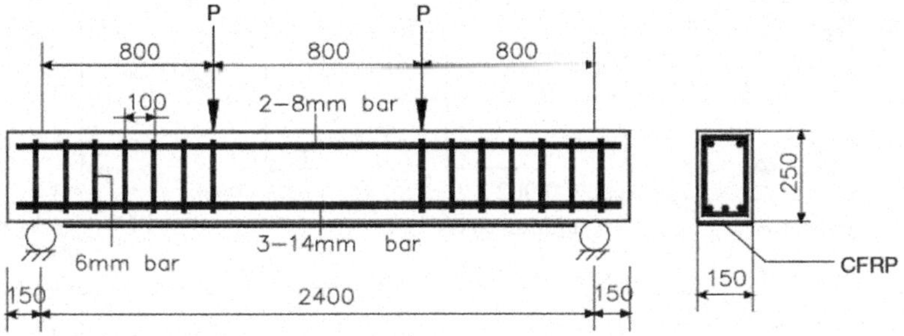

Fig. 8 Geometry details of beams tested

Table 5 Material properties of beams tested [8]

Material		f_y (MPa)	ε_y (%)	f_u (MPa)	ε (%)	E (GPa)
Steel	8 mm bar	352.1	0.168	523.9	–	210
	14 mm bar	364.9	0.183	535.9		200
Concrete	C30	–	–	40.3	1.58	32.7
CFRP				3350		212

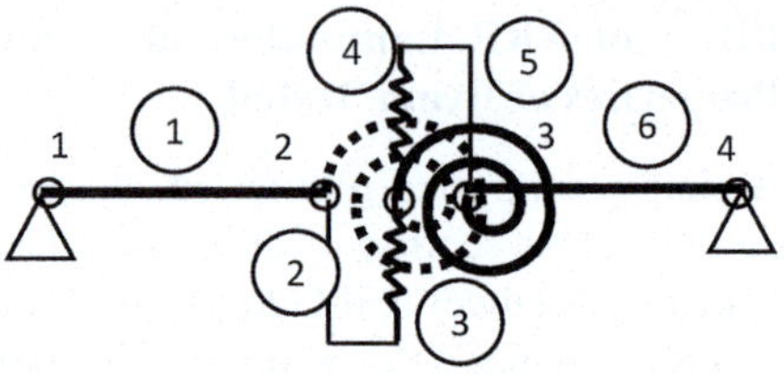

Fig. 9 Mathematical model of the beam

Steps for Estimating M–φ, M–θ, and P–Δ Diagram of the Simply Supported Beam with and Without Carbon Fiber-Reinforced Polymer Strengthening

The following steps [5, 9] shall be carried out to determine moment–curvature, moment-rotation, and load–deflection characteristics for a given section.

i. Assume a value of concrete strain at the extreme compression fiber, ε_{cm}.
ii. Assume a value of neutral axis depth, kd.
iii. Calculate stress block parameters α and γ for the assumed value of ε_{cm} as shown in Fig. 10 for the cases of without and with fibers using an appropriate equation based on the region in which ε_{cm} lies and the assumed stress–strain model as explained in Annexure A2 considering applicable cases of with and without confinement.
iv. Calculate the total compressive force in concrete, C_{con}.

$$C_{con} = \alpha f_c' bkd$$

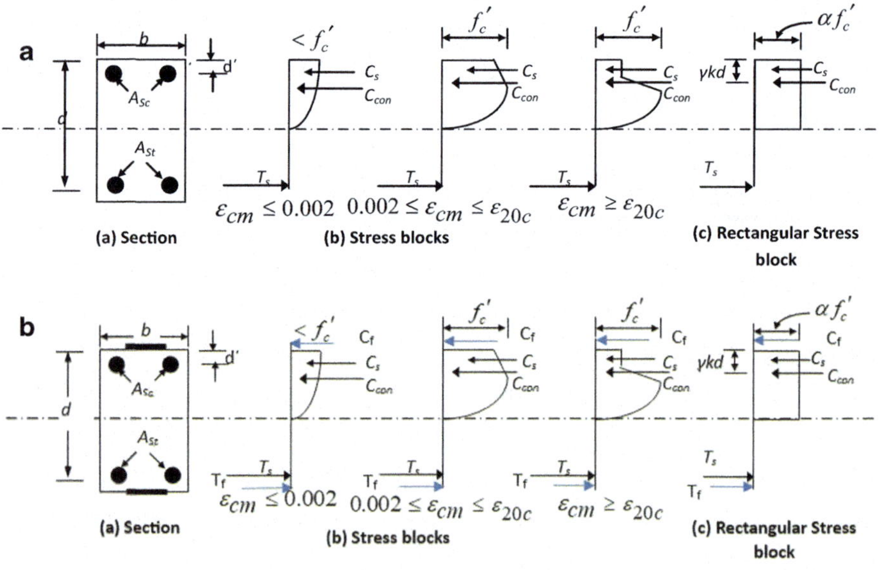

Fig. 10 **a** Stress blocks at different extreme compression fiber strain, **b** Stress blocks at different extreme compression fiber strain with CFRP

v. Calculate strains at different levels of steel and fiber and find the corresponding stresses in reinforcement bars using the stress–strain curve for steel and fiber.

$$\varepsilon_{si} = \varepsilon_{cm} \frac{kd - d_i}{kd}$$

where d_i is depth of location of steel reinforcement or fiber.

vi. Calculate the compressive (Csi) in steel and fiber C_f, tensile (Tsi) forces in reinforcement bars and in fibers T_f.

$$C_{si} \text{ or } T_{si} = f_{si} A_{si} \text{ and } C_f \text{ or } T_f = f_f A_f$$

Kindly note that for beam problem, since there are no fibers no fiber force components.

vii. Check whether the force equilibrium condition is satisfied.

$$P = \alpha f_c' bkd + \sum_{i=1}^{n} f_{si} A_{si} + \sum f_f A_f$$

$$M = \alpha f_c' bkd \left(\frac{D}{2} - \gamma kd \right) + \sum_{i=1}^{n} f_{si} A_{si} \left(\frac{D}{2} - d_i \right) + \sum f_f A_f \left(\frac{D}{2} - d_i \right)$$

where

n	Number of reinforcement bars
f_{si}	Stress in the ith bar
A_{si}	Area of ith bar
f_f	Stress in Fiber
A_f	Area of fiber
D	Total depth of section
d	Effective depth of the section
d_i	depth of ith bar from extreme compression fiber

viii. If the force equilibrium condition is satisfied then the assumed value of kd is correct. Else assume a new value of kd and go to step ii.

ix. Calculate the moment of resistance, M and the corresponding curvature,

$$\varphi = \frac{\varepsilon_{cm}}{kd}$$

x. Repeat steps i to ix for a range of ε_{cm}.

xi. Plot M–φ curve.

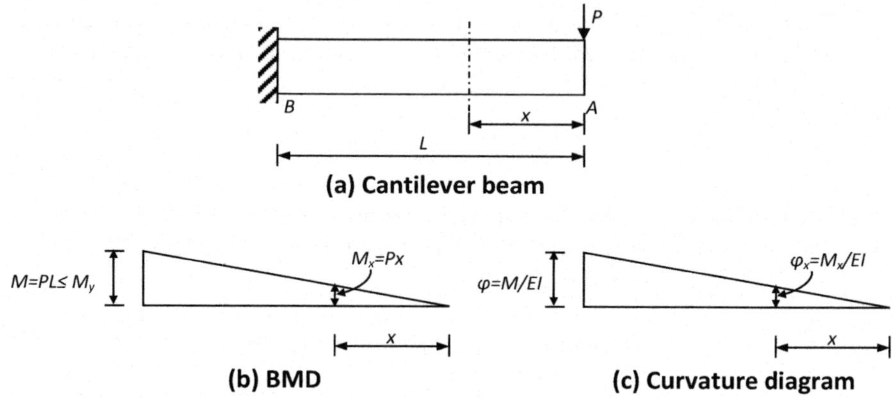

Fig. 11 Cantilever beam and its curvature distribution within elastic range

Determination of Moment-Rotation from Moment–curvature Characteristics

Curvature is defined as rotation per unit length of the member. Therefore, the rotation of a member may be calculated by integrating the curvatures along the length of the member. The rotation between any two points *A* and *B* as shown in Fig. 11 of the member is given by

$$\theta_{AB} = \int_{A}^{B} \varphi dx \tag{14}$$

where *dx* is an element of length of the member.

Since $\varphi = M/EI$,

$$\theta_{AB} = \int_{A}^{B} \frac{M}{EI} dx \tag{15}$$

This is a generalization of first moment-area theorem and it applies to both elastic and plastic curvatures. These equations can be used to calculate rotation of a member by knowing the moment-curvature characteristics and the distribution of bending moment.

Relationship Between Curvature and Rotation

Consider a cantilever beam subjected to a concentrated load at its free end as shown in Fig. 11a. The bending moment diagram and the curvature distribution within elastic range are shown in Fig. 11b, c, respectively.

From Eq. 15,

$$\theta_{AB} = \int_0^L \frac{M_x}{EI} dx = \int_0^L \frac{Px}{EI} dx$$

Since P and EI are constant (within elastic range),

$$\theta_{AB} = \frac{PL^2}{2EI} = \frac{ML}{2EI} = \frac{\varphi L}{2} \tag{16}$$

Equation (16) is valid till the beam is within elastic range, i.e., up to yield point. Therefore the yield rotation is given as;

$$\theta_y = \frac{\varphi_y L}{2} \tag{17}$$

For the reinforced concrete member that has reached the ultimate curvature and bending moment at critical section, the curvature distribution is no more linear, owing to the cracking in the member. Figures 12a, b, c show a cantilever beam, its bending moment diagram and curvature distribution respectively at ultimate moment.

As can be seen, the region of inelastic curvature is spread over a length of the beam as shown in Fig. 12. This region is that, where the bending moment as shown in Fig. 12b exceeds the yield moment of the section. The curvature fluctuates as shown in Fig. 12c because of the increased rigidity of the member between the cracks. The rotation of the member at the ultimate condition can be obtained from the actual curvature distribution using Eq. (14). Since the curvature distribution along the length of the member is complex, it is very difficult to use Eq. (14) to calculate the rotation. Therefore, the curvature diagram is idealized as shown by dotted line in Fig. 12c. The inelastic area at the ultimate stage can be replaced by an equivalent rectangle having height equal to $(\varphi_u - \varphi_y)$ and width equal to the plastic hinge length, l_p. The plastic hinge length, l_p is the equivalent length of the plastic hinge over which the plastic curvature is assumed to be constant. The width l_p is so chosen that the area of the assumed rectangle is equal to that of the actual inelastic curvature distribution. Therefore, the plastic rotation is given as

$$\theta_p = (\varphi_u - \varphi_y) l_p \tag{18}$$

The total rotation at the ultimate moment is given as

$$\theta_u = \theta_y + \theta_p \tag{19}$$

Therefore, for a cantilever beam as shown,

$$\theta_u = \varphi_y \frac{L}{2} + (\varphi_u - \varphi_y) l_p \tag{20}$$

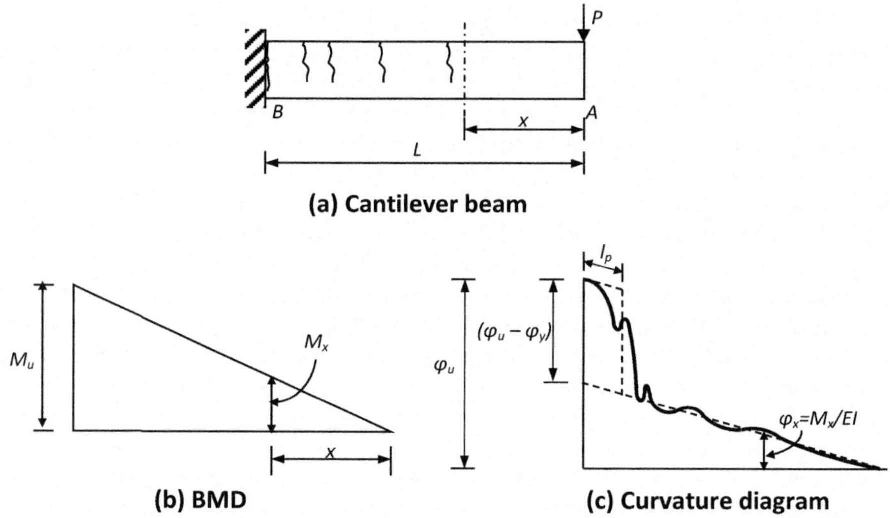

Fig. 12 Cantilever beam and its curvature distribution at ultimate moment

As seen by above equations, if the moment–curvature characteristics and the geometry of the member are known, the only unknown to determine the moment-rotation characteristics is the plastic hinge length, l_p. Pauley and Priestley [10] also recommend using an approximation of $l_p = 0.5d$. In the present problem half of the beam length of the central beam is used as hinge length as a special case due to uniform bending moment. The left and right beam elements are modelled as linear elements since it has confinement steel and subjected to less bending moment compared to central portion. Central portion with constant bending moment as shown in Fig. 13 is modelled with hinges and shear springs. Linear translational shear stiffness as explained in Annexure A1 and shown in Fig. 9 are used to model the concrete central beam portion. The matrices are assembled and numerical analysis is performed using Newton Raphson method as explained in Annexure A2.

For the central beam hinges the yield rotation form the curvature is obtained as follows.

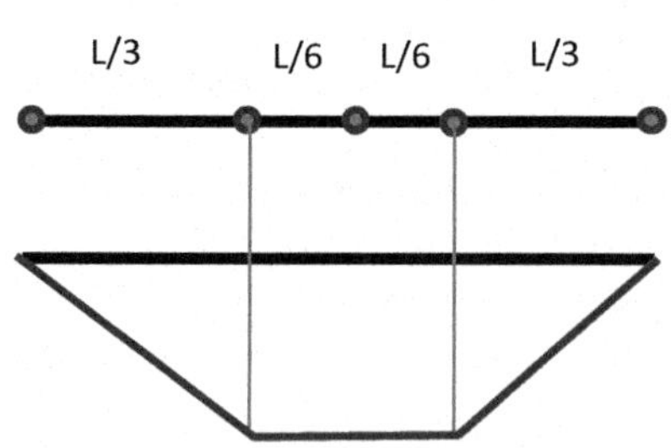

Fig. 13 Bending moment diagram of the loaded beam shown in Fig. 8

Using Eq. 13, rotation of half of the central loaded beam with constant moment is evaluated as follows.

$$\int_0^{L/6} \frac{M}{EI} dx = \frac{ML}{6EI}$$

which results into following yield rotation

$$\theta_y = \frac{\emptyset_y}{6} L \tag{21}$$

Using unconfined properties of concrete as given in Fig. 7, steel reinforcement properties and CFRP properties as shown in Fig. 14 are used to estimate the Moment curvature characteristics using the procedure explained above is evaluated and shown in Fig. 15.

Further M–θ characteristics are obtained for central portion of beam hinges using Eqs. 14–20 and shown in the Fig. 16.

As mentioned before Figs. 15 and 16, respectively shows curvature and rotation characteristics of central half beam for the case of without and with CFRP plate. The moment rotation characteristics are assigned to the hinge as shown in Fig. 9. As seen in Fig. 9, left and right portion of the beam has confinement and hence has strength more than central portion and not expected flexure failures as expected in central part. Due to this reason left and right portion are modelled with linear Euler beam. The matrices as explained in Annexure A1 are assembled and solved on EXCEL plot form using Newton–Raphson iteration procedure as explained in Annexure A3. The results obtained are shown in Fig. 17.

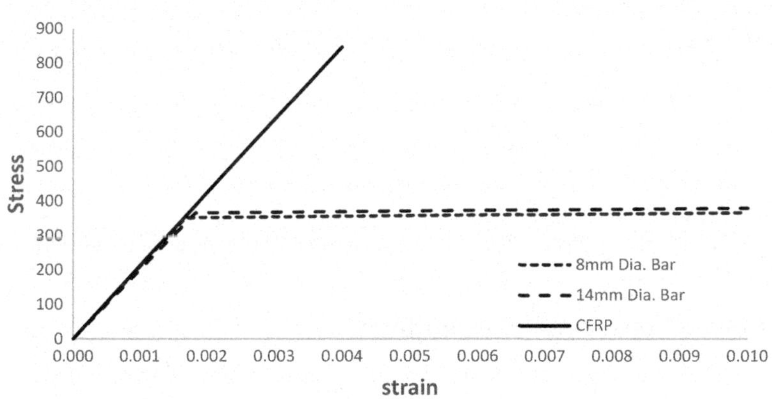

Fig. 14 Stress–strain characteristics of CFRP and steel

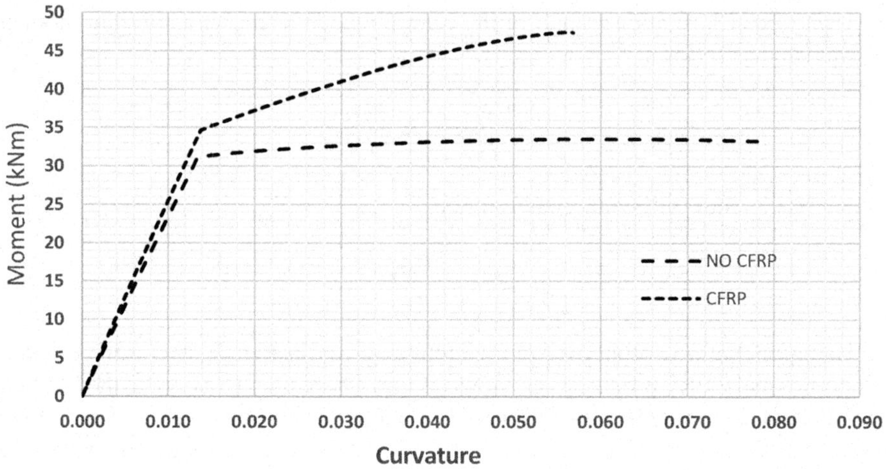

Fig. 15 Moment curvature chracteristics of central beam hinges without and with CFRP

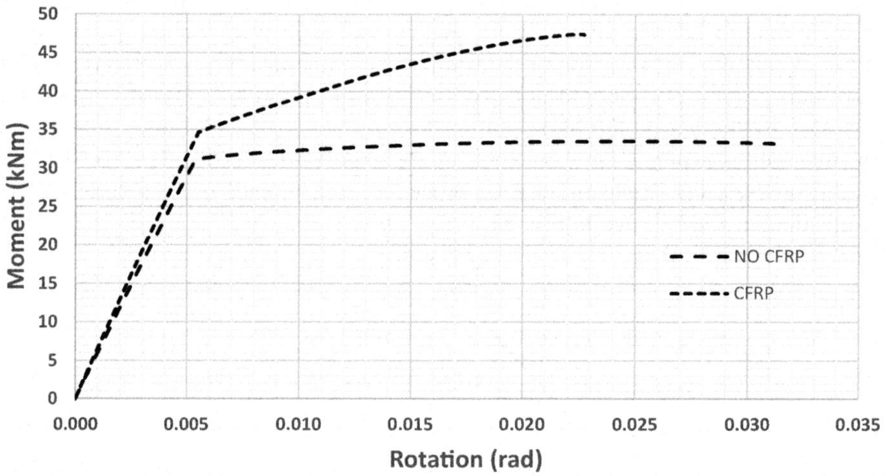

Fig. 16 Moment rotation characteristics of central beam hinges without and with CFRP

RCC Components—Beam-Column Joint

A RCC structural component called beam column joint as shown in Fig. 18a was tested [11] with ductile detailing as per Indian standard IS 13920 [12] shown in Fig. 18b and strengthened with CS and CP as shown in Fig. 18c, d [11, 13]. Table 6 provides the properties of concrete, steel and CFRP . Figure 19 shows the joint model for analysis.

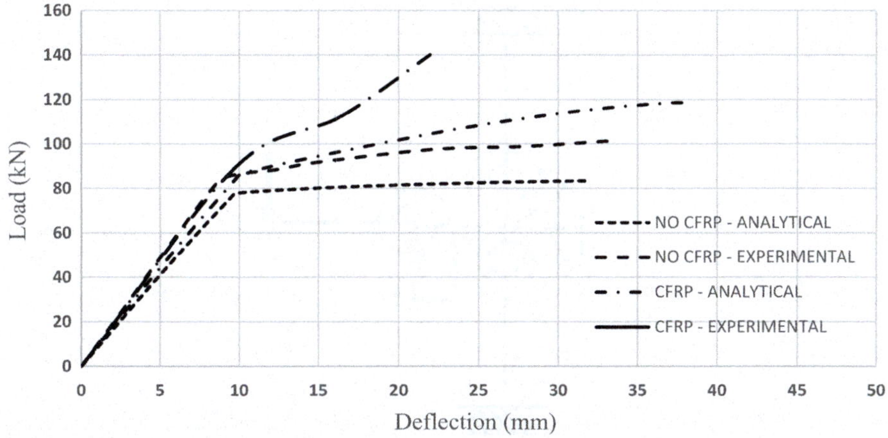

Fig. 17 Variation of deflection with load without/with CFRP plate

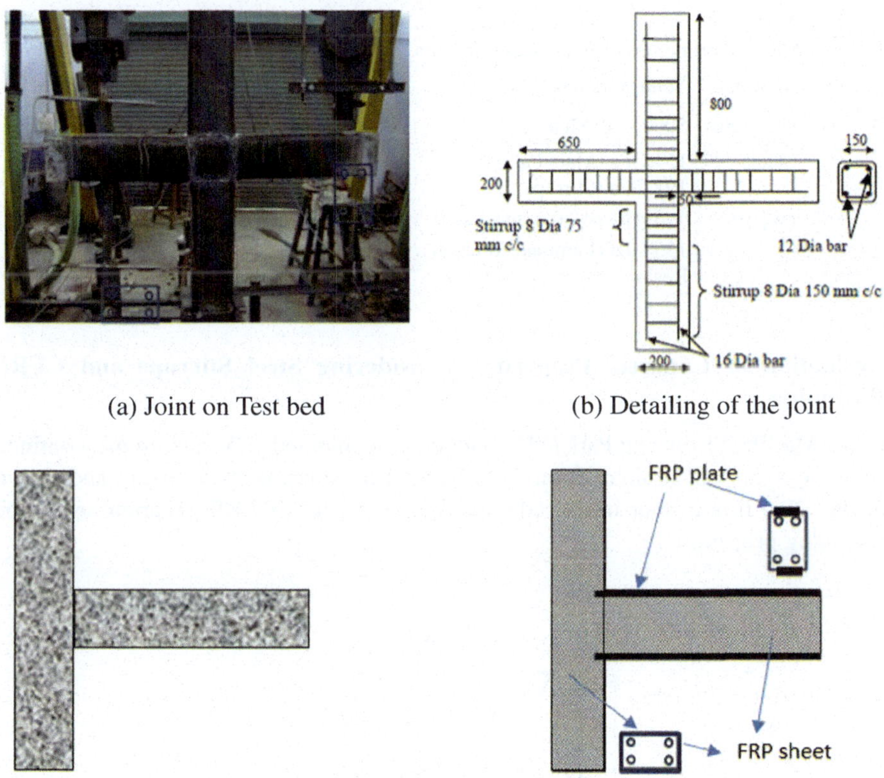

(a) Joint on Test bed

(b) Detailing of the joint

(c) Schematic of effective joint for analysis

(d) Effective Joint with FRP

Fig. 18 Details of Beam-column joint strengthened with FRP sheets and plates (laminates)

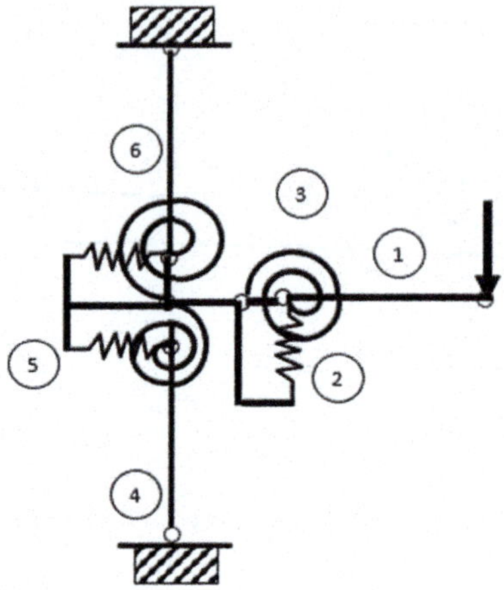

Fig. 19 Joint model with hinges

Table 6 Material properties of Beam Column Joint

Sl. no.	Material	Youngs modulus	Strength	Strain
1	Concrete	$5000\sqrt{30}$ MPa	30 MPa (Cube strength)	–
2	Steel	154 GPa (Main bars) 193 GPa (Stirrups)	0.65 GPa 0.5 GPa	0.062 (Ultimate) 0.043 (Ultimate)
3	CFRP	230 GPa (Carbon Sheets) 155 GPa (Laminates/Plates) Resin	3790 MPa 2790 MPa 21.4 MPa	0.017 0.018 0.05

Evaluation of Concrete Properties Considering Steel Stirrups and CFRP Wrapping

Using Modified Kent and Park (MKP) model as explained in Annexure A2, confined concrete stress and strain relations are obtained considering steel stirrups and shown in Fig. 20a. It is further improved considering effect of CFRP wrapping using the following procedure.

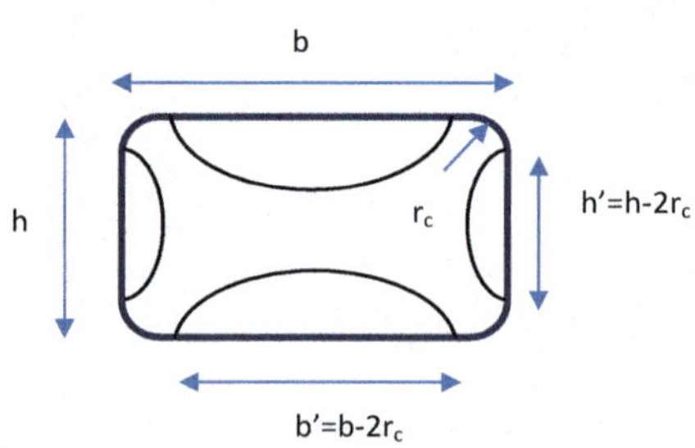

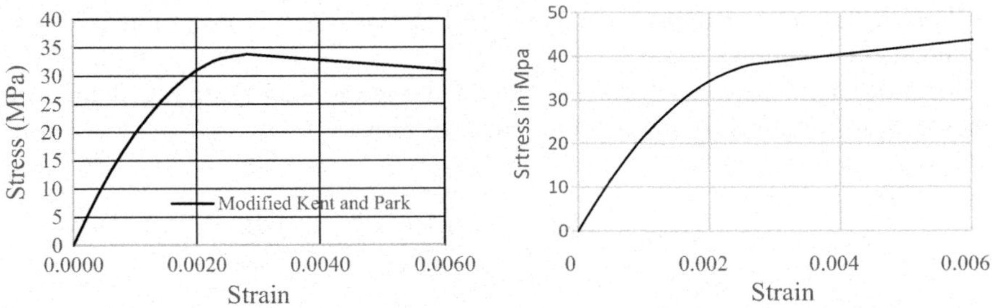

(a) Modified Kent and Park without CS

(b) With CS wrapping

Fig. 20 Material properties without and with CS wrapping

$$k_{eff} = k_H.k_v.k_\alpha$$

$$K_{II} = 1 - \frac{b'^2 + h'^2}{3A_g} = 0.6$$

$$b' = 150 - 40 = 120\,\text{mm}$$

$$h' = 200 - 40 = 160\ \text{mm}$$

$$A_g = \text{Gross Area of the section}$$

$$\rho_f = \frac{2.t_f.(b+h).b_f}{b.h.p_f} = \frac{2 \times 0.23 \times (200 + 150) \times 1.0}{200 \times 150 \times 1.0} = 0.0054$$

For b_f and P_f refer Fig. 5 and for continuous wrapping it is equal to 1.0

$$f_l = \frac{1}{2}.\rho_f.E_f.\varepsilon_{fd,rid} = \frac{1}{2} \times 0.0054 \times 230000 \times 0.004 = 2.484$$

$E_f = 230{,}000$ MPa.
$\epsilon_{fd,rid} = 0.004$.
For continuous wrapping $k_v = 1.0$ and $k_\alpha = 1.0$ with spiral angle zero.
$f_{l,eff} = k_{eff}.\ f_l = 0.6 \times 2.484 = 1.5$

$$\frac{f_{ccd}}{f_{cd}} = 1 + 2.6\left(\frac{f_{l,eff}}{f_{cd}}\right)^{2/3} = 1 + 2.6\left(\frac{1.5}{33..82}\right)^{2/3} = 1.332$$

$$f_{ccd} = 1.332 \times 33.82 = 44.72\,\text{MPa}$$

The stress–strain parameters as explained in Eqs.8 and 9 are evaluated for CS confinement as follows.

Where a and b are evaluated using tangent modulus as shown in Fig. 3 and using unconfined strength and strains of concrete as follows.

a = 1 + γ b = γ-1

$$\gamma = \frac{f_{cd} + E_t.\epsilon_{co}}{f_{cd}} = \frac{33.815 + 1649.86 \times 0.0028}{33.815} = 1.14$$

$$E_t = \frac{f_{ccd} - f_{cd}}{\epsilon_{ccu}} = \frac{44.79 - 33.815}{0.0066} 1649.86\,\text{MPa}$$

where

$$\varepsilon_{ccu} = 0.0035 + 0.015\sqrt{\frac{f_{leff}}{f_{cd}}} = 0.0035 + 0.015\sqrt{\frac{1.49}{33.815}} = 0.0066$$

a = 1 + 1.14 = 2.14

b = 1.14–1 = 0.14.

Using above parameters and using Eqs. 8 and 9, the stress and strain characteristics of concrete beam and column are obtained and shown in Fig. 20b.

Using above material properties, beam hinge and column hinge moment rotation characteristics are obtained using the procedure as explained in Annexure A2 and force balance equations as explained under the section 'Steps for estimating M–φ, M–θ, and P–Δ Diagram of the simply supported beam with and without carbon fiber-reinforced polymer strengthening'. Considering hardening characteristics of concrete with CS wrapping, equivalent stress blocks are obtained with the basic principle as given in Equations A.2.15 and A.2.16. Figure 21a, b respectively show the hinge moment rotation characteristics for beam and column for normal confined concrete sections and Fig. 22 for strengthened section with CS (carbon sheets) and CP (carbon plates).

Evaluation of Capacity of the Joint (Fig. 18)

Considering above properties of material and geometry as shown in Fig. 18, mathematical model of the joint is prepared as shown in Fig. 19. It is assumed that the portion of the beam or column where hinge is not forming is considered as linear beam elements. Only hinge has nonlinear moment-rotation characteristics as shown in Fig. 19. Stiffness of shear and axial springs at hinges are estimated considering axial and shear areas and hinge lengths. Till the concrete reaches the compression strength, full area and shear area of the section is considered and later, area and shear area of main reinforcement is considered for stiffness evaluation. Newton–Raphson iterative method of SAP 200 [18] as explained in Annexure A3 is adopted to estimate the capacity of the joint and the deformation characteristics with load so obtained are shown in Fig. 23.

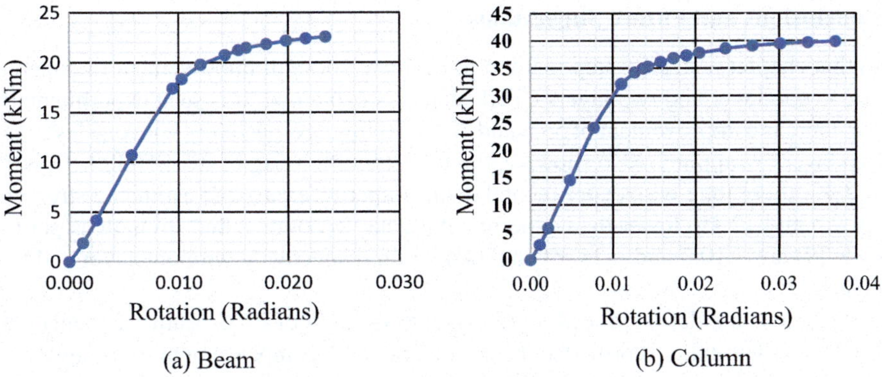

Fig. 21 Moment-rotation characteristics considering stirrups effect in beam and column

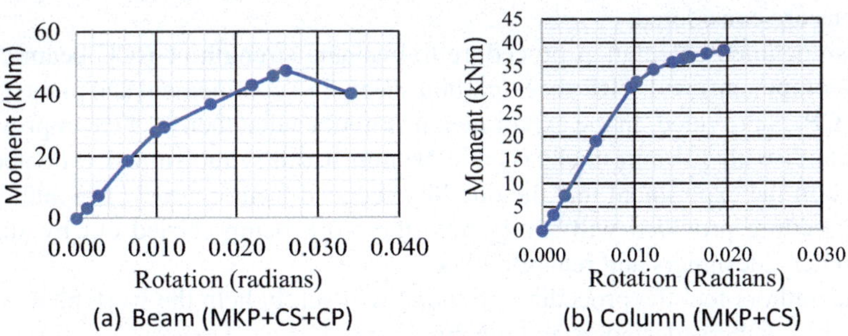

Fig. 22 Moment-rotation characteristics considering stirrups + CS + CP effects in beam and column

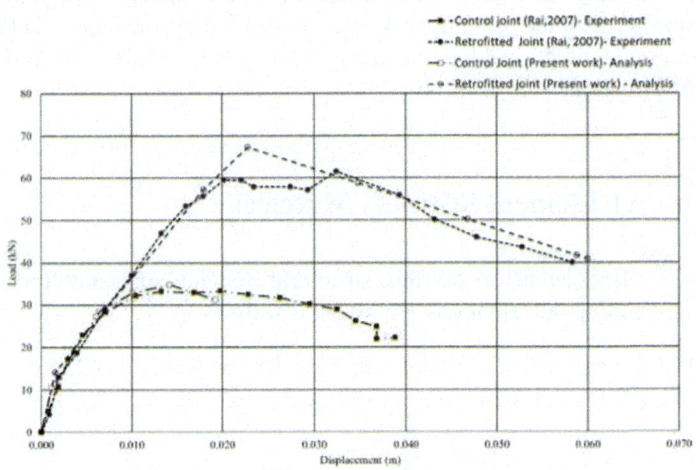

Fig. 23 Capacity of the Joint (Fig. 18) without and with Strengthening

Conclusions and Future Suggestions

In this chapter a step by step method to evaluate strength improvement in concrete due to one layer and two layers CFRP sheets is explained. It is seen experimentally that there is a improvement of 53.8 and 104% strength respectively with one layer and two layers wrapping. Also shown estimated strength using CNR-DT00R1 [9] is 15 and 9% more than with experimental data. As per reference [8] estimated strengths are 9.5 and 12.4% lower than experimental data. Recommended strength as per the code [9] are −10.3 and −27.3% at 0.004 strain respectively with one layer and two layers.

Detailed calculation procedure to evaluate strength of RCC beam (element) with FRP plate (laminate) is provided. Experimentally 11% improvement in strength with 0.15 mm thickness and 150 mm width is shown and compared with numerical analysis. The results may improve if nonlinear effects in left and right side confined beam portions are considered.

Also detailed calculation procedure to evaluate strength of RCC beam-column joint (component) with FRP sheet (carbon sheet, CS) confinement and plate (Carbon plate, CP) is provided. In the experiment it is noticed that there is 87% improvement in strength with 1.4 mm thickness and 50 mm width is shown and validated. It is noticed in the experiment that the ductility reduced from 4.3 to 3. The authors feel the procedure provided will surely help the work being carried out by students, researchers, designers and rehabilitators.

The authors feel the procedure provided will surely help the work being carried out by researchers, designers and rehabilitators.

As a future contribution, it is suggested to give a procedure considering shear effects and bidirectional loading on elements and components in addition to in plane loading considered in the present work. Also suggested to consider a 3D RCC frame tested under three directional loading and provide detailed analysis procedure for the same with validation.

Annexure A1 Element Stiffness Matrices

The element stiffness matrices of frame structural elements such as column and beam considering bending deformations are given as follows.

Euler theory

$$\begin{bmatrix} 12C_i & 6L_iC_i & -12C_i & 6L_iC_i \\ & 4L_i^2C_i & -6L_1C_i & 2L_i^2C_i \\ & & 12C_i & -6L_iC_i \\ \text{Symmetric} & & & 4L_i^2C_i \end{bmatrix}$$

$$\delta_1 \qquad \theta_1 \qquad \delta_2 \qquad \theta_2$$

Where $C_i = \frac{EI_{yi}}{L_i^3}$

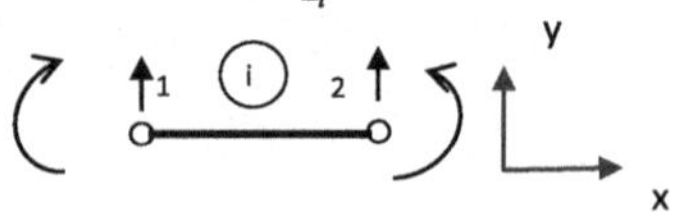

Timoshenko theory

$$\begin{bmatrix} 12\,C_1 & 6\,L_1C_1 & -12\,C_1 & 6L_1C_1 \\ & (4+\phi_{y1})L_1^2C_1 & -6L_1C_1 & (2-\phi_{y1})L_1^2C_1 \\ & & 12\,C_1 & -6\,L_1C_1 \\ \text{Symmetric} & & & (4+\phi_{y1})L_1^2C_1 \end{bmatrix}$$

where $\phi_{yi} = \frac{12\,EI_{yi}}{GA_{syi}L_i^2}$ $\quad C_i = \frac{EI_{yi}}{L_i^3(1+\phi_{yi})}$

where

L_i	Length of the ith beam element
Iyi	Moment of inertia of ith beam element about z-axis
Asyi	Shear area of the ith beam element along y-axis
E	Modulus of elasticity
G	Shear Modulus

In the case to consider axial and torsional deformations in beam or column the following stiffness matrices can be added suitably.

$$\begin{bmatrix} \frac{AE}{L_i} & -\frac{AE}{L_i} \\ -\frac{AE}{L_i} & \frac{AE}{L_i} \end{bmatrix} \text{ or } \begin{bmatrix} k_{axial} & -k_{axial} \\ -k_{axial} & k_{axial} \end{bmatrix}$$

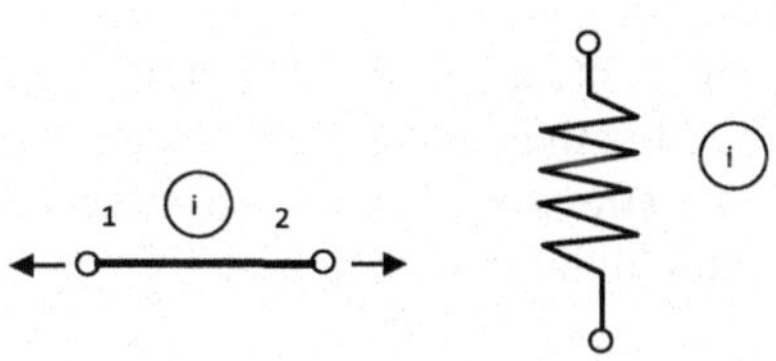

$$\begin{bmatrix} \frac{GJ}{L_i} & -\frac{GJ}{L_i} \\ -\frac{GJ}{L_i} & \frac{GJ}{L_i} \end{bmatrix} \; or \; \begin{bmatrix} k_{rotation} & -k_{rotation} \\ -k_{rotation} & k_{rotation} \end{bmatrix}$$

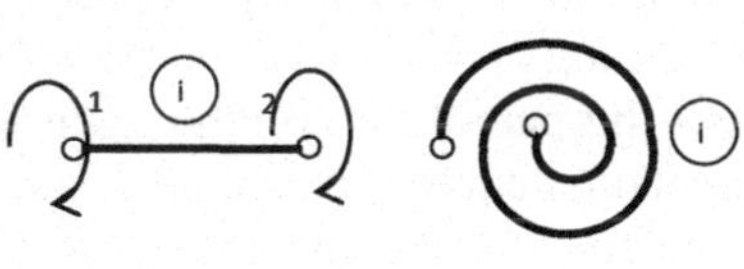

Annexure A2 Concrete Stress Strain Constitutive Relations and Equivalent Compressive Stress Block Parameters

Stress–Strain Models for Confined Concrete

In practice, the concrete in structures is always confined by transverse reinforcement commonly in the form of closely spaced steel spirals or rectangular hoops. In this case, at low levels of stress in concrete, the transverse reinforcement is hardly stressed; hence the concrete is unconfined. The concrete becomes confined when at stresses approaching the uniaxial strength, the transverse strains become very high because of progressive internal cracking and the concrete bears out against the transverse reinforcement, which then applies a confining reaction to the concrete. Thus the transverse reinforcement provides passive confinement [14].

The confinement by transverse reinforcement has little effect on the stress–strain curve until the concrete reaches its maximum stress. The shape of the stress–strain curve at high strains is a function of many variables, the major ones being the following:

(i) The ratio of the volume of transverse steel to the volume of concrete core, because a high transverse steel content will mean a high transverse confining pressure.
(ii) The yield strength of the confining steel, because this gives an upper limit to the confining pressure.
(iii) The ratio of the spacing of the transverse steel to the dimensions of the concrete core, because a smaller spacing leads to more effective confinement.
(iv) The ratio of the diameter of the transverse bars to the unsupported length of the transverse bar, because a large bar diameter leads to more effective confinement. If the flexural stiffness of the hoop bar is small (small diameter compared to unsupported length), the hoops bow outward rather than effectively confining the concrete.
(v) The content and size of longitudinal reinforcement, because this steel will also confine the concrete.
(vi) The strength of the concrete, because low-strength concrete is more ductile than high-strength concrete.

Kent and Park Model

In 1971, Kent and Park [16] proposed a stress–strain curve for concrete confined by rectangular hoops as shown in Fig. A2.1 The suggested relationship combines many of the features of previously proposed curves. A second-degree parabola represents the ascending part of curve and assumes that the confining steel has no effect on the shape of this part of curve or the strain at maximum stress. This essentially means that the ascending curve is exactly the same for both confined and unconfined concrete. It is also assumed that the maximum stress reached by confined concrete is equal to the cylinder strength f_c' that is reached at a strain of 0.002.

The relationship for the ascending parabola is given as,

Region AB, $\varepsilon_c \leq 0.002$

$$f_c = f_c'\left[\frac{2\varepsilon_c}{0.002} - \left(\frac{\varepsilon_c}{0.002}\right)^2\right] \tag{A2.1}$$

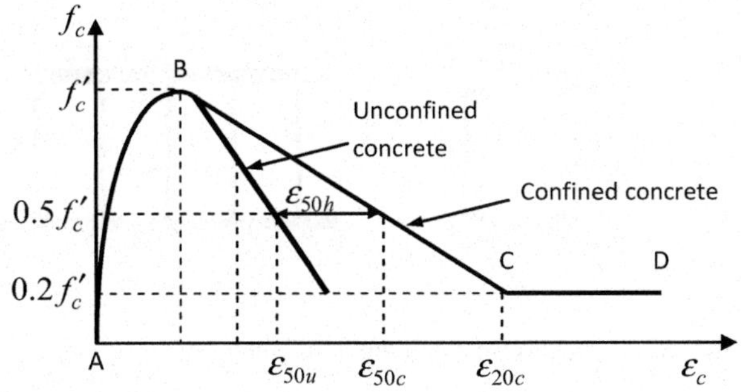

Fig. A2.1 Kent and Park [16] model for Stress–strain curve for concrete confined by rectangular hoops

The descending part of the confined concrete is modelled as per following formulations.

Region BC, $0.002 \leq \varepsilon_c \leq \varepsilon_{20c}$

$$f_c = f'_c[1 - Z(\varepsilon_c - 0.002)] \tag{A2.2}$$

where,

$$Z = \frac{0.5}{\varepsilon_{50u} - \varepsilon_{50h} - 0.002} \tag{A2.3}$$

$$\varepsilon_{50u} = \frac{3 + 0.002 f'_c}{f'_c - 1000} \tag{A2.4}$$

$$\varepsilon_{50h} = \frac{3}{4}\rho_s \sqrt{\frac{b''}{s_h}} \tag{A2.5}$$

f'_c Concrete cylinder strength in psi
ρ_s Ratio of volume of transverse reinforcement to volume of concrete core measured to outside of hoops.

$$\rho_s = \frac{2(b'' + d'')A_s}{b''d''s_h} \tag{A2.6}$$

A_s Cross-sectional area of the stirrup reinforcement
b'' Width of confined core measured to outside of hoops
d'' Depth of confined core measured to outside of hoops
s_h Spacing of hoops

Figure A2.2 shows the various parameters and symbols. The parameter Z defines the slope of the assumed linear falling branch. ε_{50u} is the value of the strain when the stress has fallen to $0.5f'_c$ (50% of the strength is lost) for the case of unconfined

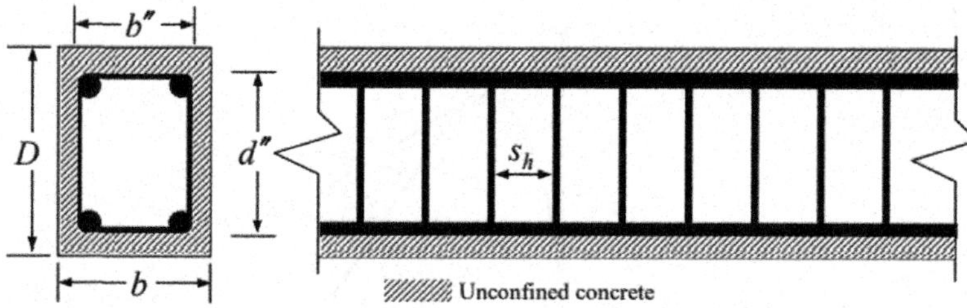

Fig. A2.2 Transverse confining in RC members

concrete. The corresponding value of strain for confined concrete is ε_{50c}. ε_{50h} is the additional ductility due to transverse reinforcement ($\varepsilon_{50c} = \varepsilon_{50u} + \varepsilon_{50h}$). It is assumed that the cover concrete has spalled off by the time the stress had fallen to $0.5 f_c'$.

Region CD, $\varepsilon_c \geq \varepsilon_{20c}$

$$f_c = 0.2 f_c' \tag{A2.7}$$

This equation accounts for the ability of concrete to sustain some stresses at very large strains.

Putting, $f_c = 0.2 f_c'$ and $\varepsilon_c = \varepsilon_{20c}$ in Eq. (A2.2), we get

$$\varepsilon_{20c} = \frac{0.8}{Z} + 0.002 \tag{A2.8}$$

This concludes the Kent and park stress–strain model.

Modified Kent and Park Model.

The modified form of Kent and Park model was proposed by Park et al. [15]. This model makes an allowance for the enhancement in the concrete strength due to confinement. Figure A2.3 shows the modified Kent and Park model.

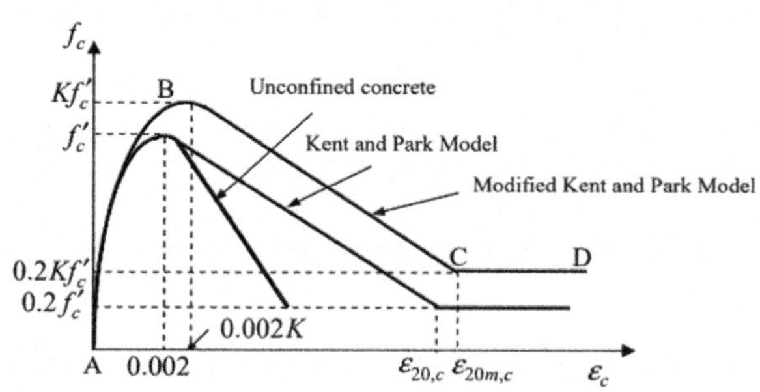

Fig. A2.3 Modified Kent and Park model for concrete confined by rectangular hoops

The maximum stress reached (at point B) is assumed to be Kf_c' at a strain of $\varepsilon_0 = 0.002$ K, in which,

$$K = 1 + \frac{\rho_s f_{yh}}{f_c'} \tag{A2.9}$$

f_{yh} yield strength of steel hoops.

The modified Kent and Park stress–strain curve can be defined as,
Region AB, $\varepsilon_c \leq 0.002$ K.

$$f_c = Kf_c'\left[\frac{2\varepsilon_c}{0.002K} - \left(\frac{\varepsilon_c}{0.002K}\right)^2\right] \tag{A2.10}$$

Region BC, $0.002K \leq \varepsilon_c \leq \varepsilon_{20m,c}$ $0.002K \leq \varepsilon_c \leq \varepsilon_{20m,c}$

$$f_c = Kf_c'[1 - Z_m(\varepsilon_c - 0.002K)] \geq 0.2Kf_c' \tag{A2.11}$$

where,

$$Z = \frac{0.5}{\frac{3+0.29f_c'}{145f_c'-1000} + \frac{3}{4}\rho_s\sqrt{\frac{b''}{s_h}} - 0.002\,K} \tag{A2.12}$$

f_c' Concrete cylinder strength in mega Pascal (N/mm^2)

Region CD,

$$f_c = 0.2Kf_c' \tag{A2.13}$$

This equation accounts for the ability of concrete to sustain some stresses at very large strains.

Putting $f_c = 0.2Kf_c'$ and $\varepsilon_c = \varepsilon_{20m,c}$ in Eq. (A2.2), we get

$$\varepsilon_{20m,c} = \frac{0.8}{Z_m} + 0.002\,K \tag{A2.14}$$

This concludes the Modified Kent and park stress–strain model.

The equivalent stress block parameters are calculated using Eqs. A2.15 and A2.16 for different values of ε_{cm} depending on whether ε_{cm} lies in zone AB, BC or CD of Kent and Park model or modified Kent and Park model.

1. Stress block parameters for **Kent and Park** model

Region 'AB': $\varepsilon_{\text{cm}} \leq 0.002$.

The mean stress factor, α and the centroid factor, γ for any strain $\varepsilon_{\mathrm{cm}}$ at the extreme compression fiber can be determined for rectangular sections from the stress–strain relationship as follows

$$\text{Area under stress} - \text{strain curve} = \int_0^{\varepsilon_{cm}} f_c d\varepsilon_c = \alpha f_c' \varepsilon_{cm}$$

$$\therefore \alpha = \frac{\int_0^{\varepsilon_{cm}} f_c d\varepsilon_c}{f_c' \varepsilon_{cm}} \tag{A2.15}$$

First moment of area about origin of area under stress–strain curve

$$= \int_0^{\varepsilon_{cm}} f_c \varepsilon_c d\varepsilon_c = (1-\gamma)\varepsilon_{cm} \int_0^{\varepsilon_{cm}} f_c d\varepsilon_c$$

$$\therefore \gamma = 1 - \frac{\int_0^{\varepsilon_{cm}} \varepsilon_c f_c d\varepsilon_c}{\varepsilon_{cm} \int_0^{\varepsilon_{cm}} f_c d\varepsilon_c} \tag{A2.16}$$

Using Eq. (A2.1) in Eqs. (A2.15) and (A2.16) we get

$$f_c = f_c'\left[\frac{2\varepsilon_c}{0.002} - \left(\frac{\varepsilon_c}{0.002}\right)^2\right]$$

$$\alpha = \frac{\varepsilon_{cm}}{0.002}\left[1 - \frac{\varepsilon_{cm}}{0.006}\right] \tag{A2.17}$$

$$\gamma = 1 - \frac{\left[\frac{2}{3} - \left(\frac{\varepsilon_{cm}}{0.008}\right)\right]}{\left[1 - \left(\frac{\varepsilon_{cm}}{0.006}\right)\right]} \tag{A2.18}$$

Region 'BC': $0.002 \le \varepsilon_{cm} \le \varepsilon_{20,c}$
Using Eqs. (A2.10) and (A2.11) in Eq. (A2.15) and (A2.16) we get

$$\alpha = \frac{1}{\varepsilon_{cm}}\left[\frac{0.004}{3} + (\varepsilon_{cm} - 0.002) - \frac{Z}{2}(\varepsilon_{cm} - 0.002)^2\right] \tag{A2.19}$$

$$\gamma = 1 - \frac{1}{\varepsilon_{cm}}\left[\frac{\left[\left(\frac{\varepsilon_{cm}^2}{2} - \frac{(0.002)^2}{12}\right) - Z\left(\frac{\varepsilon_{cm}^3}{3} - 0.001\varepsilon_{cm}^2 + \frac{(0.002)^3}{6}\right)\right]}{\left[\left(\varepsilon_{cm} - \frac{0.002}{3}\right) - Z\left(\frac{\varepsilon_{cm}^2}{2} - 0.002\varepsilon_{cm} + \frac{(0.002)^2}{2}\right)\right]}\right] \tag{A2.20}$$

Region 'CD': $\varepsilon_{\mathrm{cm}} > \varepsilon_{20,\mathrm{c}}$
Using Eqs. (A2.10), (A2.11) and (A2.13) in Eqs. (A2.15) and (A2.16) we get

$$\alpha = \frac{1}{\varepsilon_{cm}}\left[\frac{0.004}{3} + \frac{0.32}{Z} + 0.2\varepsilon_{cm} - 0.0004\right] \tag{A2.21}$$

$$\gamma = 1 - \frac{1}{\varepsilon_{cm}}\left[\frac{1.2667\times 10^{-6} + \frac{0.00064}{Z} + \frac{0.8^3}{6Z^2} + 0.1\varepsilon_{cm}^2}{\frac{0.004}{3} - \frac{0.32}{Z} + 0.2\varepsilon_{cm} - 0.0004}\right] \tag{A2.22}$$

2. Stress block parameters for **Modified Kent and Park** model

Region 'AB': $\varepsilon_{cm} \le 0.002$ K.
Using Eq. (A2.10) in Eqs. (A2.16) and (A2.17) we get

$$\alpha = \frac{\varepsilon_{cm}}{0.002K}\left[1 - \frac{\varepsilon_{cm}}{0.006K}\right] \tag{A2.23}$$

$$\gamma = 1 - \frac{\left[\frac{2}{3} - \left(\frac{\varepsilon_{cm}}{0.008K}\right)\right]}{\left[1 - \left(\frac{\varepsilon_{cm}}{0.006K}\right)\right]} \tag{A2.24}$$

Region 'BC': $0.002\ \mathrm{K} \le \varepsilon_{cm} \le \varepsilon_{20m,c}$
Using equations (A2.10) and (A2.11) in Eqs. (A2.15) and (A2.16) we get

$$\alpha = \frac{1}{\varepsilon_{cm}}\left[\frac{0.004K}{3} + (\varepsilon_{cm} - 0.002K) - \frac{Z_m}{2}(\varepsilon_{cm} - 0.002K)^2\right] \tag{A2.25}$$

$$\gamma = 1 - \frac{1}{\varepsilon_{cm}}\left[\frac{\left[\left(\frac{\varepsilon_{cm}^2}{2} - \frac{(0.002K)^2}{12}\right) - Z_m\left(\frac{\varepsilon_{cm}^3}{3} - 0.001K\varepsilon_{cm}^2 + \frac{(0.002K)^3}{6}\right)\right]}{\left[\left(\varepsilon_{cm} - \frac{0.002K}{3}\right) - Z_m\left(\frac{\varepsilon_{cm}^2}{2} - 0.002K\varepsilon_{cm} + \frac{(0.002K)^2}{2}\right)\right]}\right] \tag{A2.26}$$

Region 'CD': $\varepsilon_{cm} > \varepsilon_{20m,c}$
Using Eqs. (A2.10), (A2.11) and (A2.15) in Eqs. (A2.16) we get

$$\alpha = \frac{1}{\varepsilon_{cm}}\left[\frac{0.004K}{3} + \frac{0.32}{Z_m} + 0.2K\varepsilon_{cm} - 0.0004K\right] \tag{A2.27}$$

$$\gamma = 1 - \frac{1}{\varepsilon_{cm}}\left[\frac{1.2667\times 10^{-6}K^2 + \frac{0.00064K}{Z_m} + \frac{0.8^3}{6Z_m^2} + 0.1\varepsilon_{cm}^2}{\frac{0.004K}{3} - \frac{0.32}{Z_m} + 0.2K\varepsilon_{cm} - 0.0004K}\right]$$

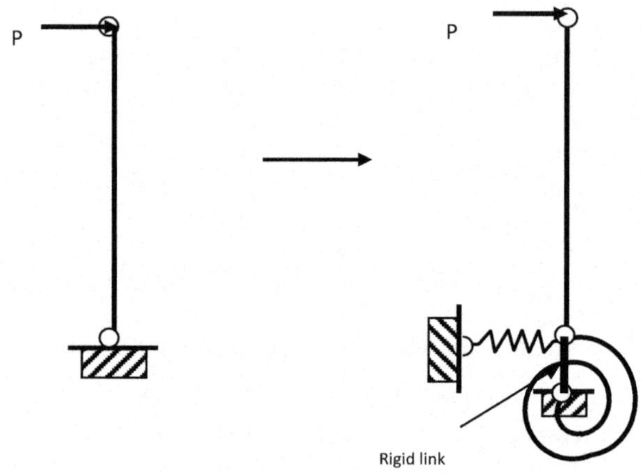

Fig. A3.1 Cantilever and its inelastic model for nonlinear static analysis

Annexure A3 N–Raphson Technique

Consider a cantilever is subjected to monotonic loading with increase in amplitude. Maximum moment will be at the fixed end and when the member reinforcement starts yielding hinge forms and mathematical model is modified as shown in Fig.A3.1. The hinge is replaced with translational and rotational springs and to evaluate rotation induced translation within the hinge length a rigid link is introduced.

Equilibrium equation is written as:

$$[K]\{\delta\} = \{P\}$$

Stiffness matrix will have single element stiffness till yielding and will get modified with hinge springs. The stiffness matrices are given in Appendix 1. If hinge moment rotation characteristics are known, at every load step, the stiffness will get modified with the iterations [17] as explained in Fig. A3.2.

$$\theta_{i+1}^{(0)} = \theta_i \quad \mathrm{M_S^{(0)}} = (M)_i \quad \Delta\mathrm{R}^{(1)} = \Delta M_e \quad \mathrm{K_T} = K_i \quad j = 1, 2, 2$$

1. $K_T \times \Delta\theta^{(1)} = \Delta R^{(1)}$
2. $\theta_{i+1}^{(j)} = \theta_{i+1}^{(j-1)} + \Delta\theta^{(1)}$
3. $\Delta M_p^j = \mathrm{M_s^{(j)}} - \mathrm{M_s^{(j-1)}}$
4. $\Delta R^{(j+1)} = \Delta R^{(j)} - \Delta M_p^j.$

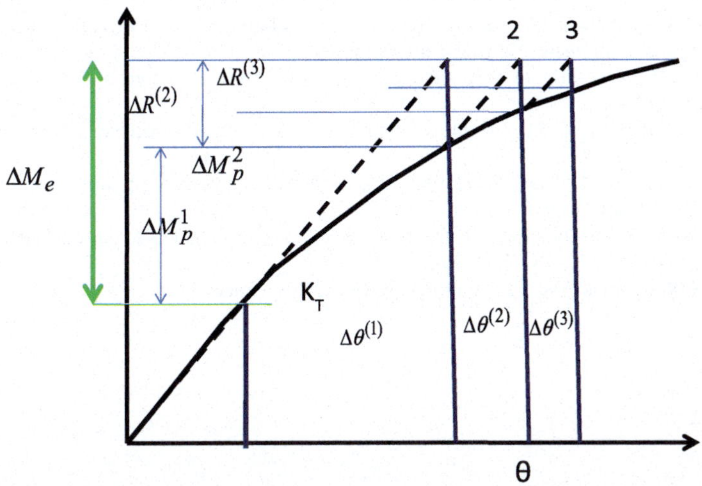

Fig. A3.2 Inelastic moment-rotation characteristics of beam/column hinge

References

1. Önder Çetin K, Ilgaç M, Can G, Çakır E (2023) Preliminary reconnaissance report on February 6, 2023, Pazarcık Mw = 7.7 and Elbistan Mw = 7.6, Kahramanmaraş-Türkiye Earthquakes, February 20, 2023, Report No: METU/EERC 2023-01
2. Earthquake Engineering Research Institute, LFE Program GEER Association Report 082. https://doi.org/10.18118/G6PM34
3. Reddy GR (1998) Advance approaches for the seismic analysis of nuclear power plant structures, equipment and piping systems, PhD Thesis, Tokyo Metropolitan, University, Tokyo
4. Reddy GR (2023) Rehabilitation and Retrofitting of reinforced concrete structures using fibre reinforced polymers-experiments. In: Singh SB et al (eds) Fiber reinforced polymeric materials and sustainable structures, composites science and technology
5. Guide for the Design and Construction of Externally Bonded FRP Systems for Strengthening Existing Structures, Materials, RC and PC structures, masonry structures, CNR-DT 200 R1/ 2013, 2013
6. Lin H-J, Chen C-T 92001) Strength of concrete cylinder confined by composite materials. J Reinf Plast Compos 20(18)
7. Moodi Y, Mousavi SR, Sohrabi MR (2019) New models for estimating compressive strength of concrete confined with FRP sheets in circular sections. J Reinf Plast Compos 38(21–22):1014–1028
8. Wenwei W, Guo L (2006) Experimental study and analysis of RC beams strengthened with CFRP laminates under sustaining load. Int J Solids Struct 43:1372–1387
9. Sharma A, Reddy GR, Vaze KK, Ghosh AK, Kushwaha HS (2009) Experimental Investigations and evaluation of strength and deflections of reinforced concrete beam-column joints using nonlinear static analysis. Bhabha Atomic Research Centre, report no. BARC/2009/E/012
10. Sharma A, Reddy GR, Vaze KK (2013) Pushover experiment and analysis of four storey full scale reinforced concrete structure before and after retrofitting. Bhabha Atomic Research Centre
11. Rai G (2007) Short-term and long-term performance of externally prestressed RC Beams and Joints. PhD Thesis, Department of civil engineering Indian Institute of Technology, Bombay
12. IS 13920:2016 (2016) Ductile design and detailing of reinforced concrete structures subjected to seismic forces-code of practice. Bureau of Indian Standards

13. Reddy GR, Hari Prasad M, Varma AK (2019) A text book of seismic design, structures, piping systems and components. Springer International
14. Pauley T, Priestley MJN (1992) Seismic design of reinforced concrete and masonry buildings. Wiley, New York
15. Kent DC, Park R (1971) Flexural mechanics with confined concrete. J Struct Div ASCE 97(ST7):1969–1990
16. Park R, Priestley MJN, Gill WD (1982) Ductility of square-confined concrete columns. J Struct Eng ASCE 108(ST4):929–950
17. Chopra AK (2001) Dynamics of structures, theory and applications to earthquake engineering. Prentice Hall of India Pvt. Ltd., New Delhi
18. SAP 2000 Ultimate V-21, Education and Research Licence, 2021

The manufacturer's authorised representative in the EU is Springer Nature Customer Service Centre GmbH, Europaplatz 3, 69115 Heidelberg, Germany. If you have any concerns regarding our products, please contact ProductSafety@springernature.com

Printed and bound by CPI Group (UK) Ltd, Croydon, CR0 4YY
01/10/2026
02225870-0004